Digital Entrepreneurship and Global Innovation

Ioan Hosu
Babeş-Bolyai University, Romania

Ioana Iancu
Babeş-Bolyai University, Romania

A volume in the Advances in Business Strategy and Competitive Advantage (ABSCA) Book Series

IGI GLOBAL
DISSEMINATOR OF KNOWLEDGE
www.igi-global.com

Published in the United States of America by
IGI Global
Business Science Reference (an imprint of IGI Global)
701 E. Chocolate Avenue
Hershey PA, USA 17033
Tel: 717-533-8845
Fax: 717-533-8661
E-mail: cust@igi-global.com
Web site: http://www.igi-global.com

Library of Congress Cataloging-in-Publication Data

Names: Hosu, Ioan, 1971- editor. | Iancu, Ioana, 1983- editor.
Title: Digital entrepreneurship and global innovation / Ioan Hosu and Ioana Iancu, editors.
Description: Hershey, PA : Business Science Reference, [2017] | Series: Advances in business strategy and competitive advantage | Includes bibliographical references and index.
Identifiers: LCCN 2016032835| ISBN 9781522509530 (hardcover) | ISBN 9781522509547 (ebook)
Subjects: LCSH: Technological innovations--Economic aspects. | Entrepreneurship. | Strategic planning. | Information technology--Management.
Classification: LCC HC79.T4 D53169 2017 | DDC 658.4/21--dc23 LC record available at https://lccn.loc.gov/2016032835

This book is published in the IGI Global book series Advances in Business Strategy and Competitive Advantage (ABSCA) (ISSN: 2327-3429; eISSN: 2327-3437)

British Cataloguing in Publication Data
A Cataloguing in Publication record for this book is available from the British Library.

All work contributed to this book is new, previously-unpublished material. The views expressed in this book are those of the authors, but not necessarily of the publisher.

For electronic access to this publication, please contact: eresources@igi-global.com.

Advances in Business Strategy and Competitive Advantage (ABSCA) Book Series

Patricia Ordóñez de Pablos
Universidad de Oviedo, Spain

ISSN: 2327-3429
EISSN: 2327-3437

Mission

Business entities are constantly seeking new ways through which to gain advantage over their competitors and strengthen their position within the business environment. With competition at an all-time high due to technological advancements allowing for competition on a global scale, firms continue to seek new ways through which to improve and strengthen their business processes, procedures, and profitability.

The **Advances in Business Strategy and Competitive Advantage (ABSCA) Book Series** is a timely series responding to the high demand for state-of-the-art research on how business strategies are created, implemented and re-designed to meet the demands of globalized competitive markets. With a focus on local and global challenges, business opportunities and the needs of society, the **ABSCA** encourages scientific discourse on doing business and managing information technologies for the creation of sustainable competitive advantage.

Coverage

- Cost Leadership Strategy
- Value Chain
- Entrepreneurship & Innovation
- Tacit Knowledge
- Globalization
- International Business Strategy
- Strategic management
- Strategy Performance Management
- Business Models
- Innovation Strategy

IGI Global is currently accepting manuscripts for publication within this series. To submit a proposal for a volume in this series, please contact our Acquisition Editors at Acquisitions@igi-global.com or visit: http://www.igi-global.com/publish/.

The Advances in Business Strategy and Competitive Advantage (ABSCA) Book Series (ISSN 2327-3429) is published by IGI Global, 701 E. Chocolate Avenue, Hershey, PA 17033-1240, USA, www.igi-global.com. This series is composed of titles available for purchase individually; each title is edited to be contextually exclusive from any other title within the series. For pricing and ordering information please visit http://www.igi-global.com/book-series/advances-business-strategy-competitive-advantage/73672. Postmaster: Send all address changes to above address.

Titles in this Series

For a list of additional titles in this series, please visit: www.igi-global.com

Case Studies as a Teaching Tool in Management Education
Dominika Latusek (Kozminski University, Poland)
Business Science Reference • copyright 2017 • 326pp • H/C (ISBN: 9781522507703) • US $190.00 (our price)

Analyzing the Relationship between Corporate Social Responsibility and Foreign Direct Investment
Marianne Ojo (North West University, South Africa)
Business Science Reference • copyright 2016 • 322pp • H/C (ISBN: 9781522503057) • US $185.00 (our price)

Comparative Perspectives on Global Corporate Social Responsibility
Dima Jamali (American University of Beirut, Lebanon)
Business Science Reference • copyright 2017 • 367pp • H/C (ISBN: 9781522507208) • US $215.00 (our price)

Crowdfunding for Sustainable Entrepreneurship and Innovation
Walter Vassallo (MC Shareable, Monaco)
Business Science Reference • copyright 2017 • 336pp • H/C (ISBN: 9781522505686) • US $205.00 (our price)

Handbook of Research on Driving Competitive Advantage through Sustainable, Lean, and Disruptive Innovation
Latif Al-Hakim (University of Southern Queensland, Australia) Xiaobo Wu (Zhejiang University, China) Andy Koronios (University of South Australia, Australia) and Yongyi Shou (Zhejiang University, China)
Business Science Reference • copyright 2016 • 743pp • H/C (ISBN: 9781522501350) • US $365.00 (our price)

Ethical and Social Perspectives on Global Business Interaction in Emerging Markets
Minwir Al-Shammari (University of Bahrain, Bahrain) and Hatem Masri (University of Bahrain, Bahrain)
Business Science Reference • copyright 2016 • 372pp • H/C (ISBN: 9781466698642) • US $185.00 (our price)

Strategic Approaches to Successful Crowdfunding
Djamchid Assadi (Burgundy School of Business, France)
Business Science Reference • copyright 2016 • 378pp • H/C (ISBN: 9781466696044) • US $205.00 (our price)

Handbook of Research on Entrepreneurship in the Contemporary Knowledge-Based Global Economy
Neeta Baporikar (University of Pune, India)
Business Science Reference • copyright 2016 • 607pp • H/C (ISBN: 9781466687981) • US $345.00 (our price)

Competitive Strategies for Academic Entrepreneurship Commercialization of Research-Based Products
Anna Szopa (Jagiellonian University, Poland) Waldemar Karwowski (University of Central Florida, USA) and David Barbe (University of Maryland, College Park, USA)
Business Science Reference • copyright 2015 • 317pp • H/C (ISBN: 9781466684874) • US $195.00 (our price)

IGI GLOBAL
DISSEMINATOR OF KNOWLEDGE
www.igi-global.com

701 E. Chocolate Ave., Hershey, PA 17033
Order online at www.igi-global.com or call 717-533-8845 x100
To place a standing order for titles released in this series, contact: cust@igi-global.com
Mon-Fri 8:00 am - 5:00 pm (est) or fax 24 hours a day 717-533-8661

List of Reviewers

Constanţa-Nicoleta Bodea, *Bucharest University of Economic Studies, Romania*
Camelia Cmeciu, *Bucharest University, Romania*
Anca Constantinescu, *Technical University of Cluj-Napoca, Romania*
Lorina Culic, *Babeş-Bolyai University, Romania*
Jorge Colvin Díez, *Universidad Antonio de Nebrija, Spain & INVESTEAM Asesores Registrados, Spain*
Bogdan Iancu, *Technical University of Cluj-Napoca, Romania*
Oguzhan Irengun, *İstanbul University Faculty of Economics, Turkey*
Radu-Mihai Meza, *Babeş-Bolyai University, Romania*
Alina Neamtu, *Bucharest University of Economic Studies, Romania*
Anisoara Pavelea, *Babeş-Bolyai University, Romania*
Adrian Peculea, *Technical University of Cluj-Napoca, Romania*
Oana Madalina Popescu, *Bucharest University of Economic Studies, Romania*
Rauno Rusko, *University of Lapland, Finland*
José Manuel Saiz-Alvarez, *Tecnologico de Monterrey, Mexico & Nebrija University, Spain*
Farhan Shaikh, *Capgemini Consulting, India*
Stelian Stancu, *Bucharest University of Economic Studies, Romania*
Seyedeh Khadijeh Taghizadeh, *Universiti Sains Malaysia, Malaysia*
Ramayah Thurasamy, *Universiti Sains Malaysia, Malaysia*

Table of Contents

Detailed Table of Contents

In a context characterized by an inflation of marketing messages, it is imperious to understand how consumers succeed in making the buying decision. Starting by briefly describing the structure and the role of the brain and the differences between consciousness and unconsciousness, the paper aims to investigate the way neuromarketing can help in comprehending the feelings of the consumers, the way products or services match the consumers' needs, and the way companies can discover the insights of decision-making process. This paper can be perceived either as a guide for the companies that aim to find more on the way people manage information and make decisions or as a comprehensive description on human being marketing behavior that can serve both business, academic environments and consumers.

Microcredit has been studied from many perspectives. In this chapter, we analyze KIVA, the most important Person-to-Person microfinance organization from the perspective of social change, and we study how it has impacted on the nascent of a new wave of entrepreneurs known as digital entrepreneurial charity. Applied to KIVA, we analyze the impact of the digital space and its Internet-based Peer-to-Peer Lending to create social change in the poor, while alleviating the poverty thanks to solidarity and charity. This work concludes affirming that banking the poor and education, with the intensive use of Internet-based devices, is the best way to alleviate poverty in our digital and globalized economic world.

Universities play a fundamental role in promoting entrepreneurship and directly contributing to the economic development of the country. This chapter presents the case of EAN University, an institution that from its foundation has focused on entrepreneurship as its mission. Over time the university has taken concrete actions to develop its three mission areas: training, research and social outreach. At EAN University, the entrepreneur training model is mandatory for all programs, making it part of the student's core education. In addition, the university promotes and creates a culture of entrepreneurship in its community through permanent business incubation and acceleration services, which are available to all students and graduates. EAN's accompaniment model begins with a diagnostic which determines the stage the company is in: preincubation, incubation or acceleration. The university's effort is focus on directing and accompanying companies in the preincubation and incubation stages.

The chapter proposes an e-Learning solution for the entrepreneurship education, based on several simulation modules integrated into a classical learning management system. The originality of the approach is that the solution is domain independent and applied advanced IT technologies, such as agent-based simulations and extended graphical support. Using this e-learning solution, the students can learn how to choose characteristics/aspects for particular type of business and how important is each of them according specific criteria; how to set realistic values for different characteristics/aspects of the business, how a business scenario can be changed in order to fit better to the business context (business reality), modeled through by the scenario pattern and how to assess/evaluate business scenarios.

The second most used page on a newly found website from which a client might buy an item is the ABOUT US page. Using real images of real employees has an overall positive effect on the outcome of that visit. However, it is not always as simple as that. Subtle differences in the content of the pictures may have more of an impact than previously thought. Online advertising has long been considered as a media where creativity is not needed, with the advertising agencies focusing on a few basic rules for displaying ads: call to action, numbers, colors, human figures or symbols and so on. The results are analyzed only on their conversion rate value, this being one of the main reasons why online advertising is being regarded as the future of advertising worldwide. This study is focusing on the Click Through Rate (CTR) impact of different ads using different Facial Symmetry Indices (FSI) on the human figures used. The hypothesis proposed is that a person with a higher FSI, noticeable only at a subconscious level, will generate more clicks on an online ad.

Preface

In a context characterized by a high number of companies, of products and of messages, the concept of strategic entrepreneurship arises. While on one hand, strategy implies extracting value from existing domains and creating competitive advantage, on the other hand, entrepreneurship enables identification of opportunities or of new niches (Webb, Ketchen, & Ireland, 2010, pp. 67-68). Thus, strategic entrepreneurship is defined as an effort of a company to both exploit the competitive advantage and explore for innovation (Ireland & Webb, 2007, p. 50).

Within the literature, strategic entrepreneurship refers to three main elements: developing an appropriate mindset within a company, finding a balance between exploration and exploitation, and continuous innovation. First, mindset implies cognitive frameworks through which the new knowledge is used and interpreted. Second, while exploration is defined as a process that involves "variation, search, creativity, experimentation, and the integration of diverse knowledge stock," exploitation is a process that involves refinement, speed, precision, and a focus on existing competencies. Third, the concept that perfectly resonates with the content of this volume, continuous innovation, is perceived as a pathway through which strategic entrepreneurship contributes to the capacity of a company to produce prosperity (Webba, Ketchen, & Ireland, 2010, p. 69). Although innovation implies risk taking, it helps increasing the amount of utilized knowledge (Audretsch, Bönte, & Keilbach, 2008, p. 688).

As a new notion within the innovation context, digital economy (also known as e-economy, "information economy" or "the new economy") is perceived as a driver of structural change and economic growth at the international level (Johansson, Karlsson, & Stough, 2006). Confronting with a permanent and rapid technological development and considering the need for enduring innovation in a digital economy, the concept of digital entrepreneurship must be developed. In this new environment, Information Technology, mobile technology, and new media are considered to be competitive advantages. Put it differently, it means having a high capability of accessing, creating, processing and strategically disseminating knowledge (Davidson & Vaast, 2010, p. 1).

While the initial definition of entrepreneurship, back in the 1700s, emphasizes the risk of buying at a certain price and selling at uncertain prices, nowadays digital entrepreneurship underlines the need to pursuit the opportunities based on the digital media and information and communication technologies (Davidson & Vaast, 2010, p. 2).

In this respect, digital entrepreneurs are considered to combine business, knowledge and institutional opportunities. While business entrepreneurship is a venture aiming to generate financial profit (e.g. Amazon.com), knowledge entrepreneurship implies searching for information and knowledge-related opportunities in order to develop, expand and circulate a domain-related knowledge (e.g. consultants, journalists, academics, and pundits). Institutional entrepreneurship refers to the activities of creating new institutions or transforming the existing ones, generating legitimacy (Davidson & Vaast, 2010, p. 3).

Considering this theoretical background, the present volume aims to investigate the role of digitalization and of global innovation on the entrepreneurial field. It represents an interdisciplinary work, binding together research from a wide range of disciplines. Regardless of the different approaches, all of the papers within this volume have, as a central concept, the entrepreneurship one. They all converge around this notion in the attempt to offer a comprehensive overview on entrepreneurship and on its applicability on the market. The main topics of this volume rely on the impact of new digital technologies on business, on collaborative processes and on social change, on the role of neuromarketing on measuring unconsciousness marketing influence, and on the correlation between educational field and entrepreneurship.

The contributors of this volume have different profiles. They have specializations in computer science, artificial intelligence, economics, business administration, strategic product design, entrepreneurship, strategic research, advertising and communication. In spite of these differences, by contributing to the present volume, all the authors have a common ground, namely developing knowledge on the digital entrepreneurship and global innovation.

The target of the volume is comprised of both academic specialists and industry professionals. The academic specialists can use this volume as a teaching guide by presenting students the latest developments within the digital entrepreneurship. At the same time, they can use the knowledge in the volume either to replicate the existing research instruments or to further develop other research ideas. The industry professionals can find a high amount of inspiration that can be useful for their future work or business ideas. Thus, they can find applicability of all the presented ideas in a wide range of domains.

Considering this context, the volume on Research on Digital Entrepreneurship and Global Innovation is composed on twelve chapters that have been reviewed during 2015-2016.

The article written by Farhan Shaikh, *The Benefits of New Online (Digital) Technologies on Business: Understanding the Impact of Digital on Different Aspects of the Business*, analyzes the impact of digital technology on the businesses and the challenges of the industrial leaders and the entrepreneurs. The overall findings are structured under three broad sections – Customer Experience, Process Optimization and Business Innovation. First, Customer Experience section captures the way companies are using the data from digital technologies, enhancing the way they interact with the customers. Second, the section on Process Optimization highlights the significant improvements achieved by using digital technologies for the existing business processes. Finally, Business Innovation captures the transformation power of digital, through new business models, business areas and data monetization.

Maria Ramirez-Salazar and Rafael Pérez-Uribe, in their article, *Open Collaborative Innovation: Booster of Solutions for the Challenges of a Global World – Collaborative Processes and Innovation Networks Create Value*, emphasize the role of open collaborative innovation on dealing with the challenges of a global world. Open innovation is considered to be a way that allows the incorporation of external knowledge into a company. Open collaborative innovations are adding an external point of view from different interinstitutional and transdisciplinary entities. These entities rely on the common goal of provide practical solutions to business challenges that arise in a global world. There are seven components in the Open innovation collaborative model: Innovation challenges, internal-external knowledge, paradigm change, leadership, interinstitutional and transdisciplinary teams, storytelling, and creative solutions. It also contains six principles, thus having thirteen interactive variables: identity, agreements, flexibility, commitment, recognition, and trust. This research has identified, among the previously mentioned variables, that in order to be more successful leading projects (taking into consideration the millennia's generation), it is necessary to strengthen creativity and communication as essential factors.

Aiming to call attention to the concept of competitive intelligence, Ainul Mohsein Abdul-Mohsin develops the paper *The Relationship between Entrepreneurial Competencies, Competitive Intelligence, and Innovative Performance among SMEs from an Emerging Country: Competitive Intelligence in SMEs.* Competitive intelligence is considered to be vital to an organization because it emphasizes on understanding and measuring an organization's competitive environment and the unstructured information that influence an organization's strategic decision making. In fact, it also assists in creation of ideas and innovation. Yet, research on competitive intelligence practice among entrepreneurs is quite scanty. There is a need to investigate if entrepreneurial competencies and innovative performance is influenced by competitive intelligence. Thus, this study aims to analyze the relationship between entrepreneurial competencies, innovative performance and if competitive intelligence is the missing link among the entrepreneurs within the Malaysian SME context.

A technical contribution to the volume is the chapter *Intercloud: Delivering Innovative Cloud Services*, written by Alexandru Aurel Costan, Bogdan Iancu, Petru Cosmin Rasa, Alexandru Radu, Adrian Peculea, and Vasile Dadarlat. The article begins from the idea that cloud computing is considered the next step in the evolution of technological applications, services and products development, in the Future Internet and Internet of Things context. The permanent connection of users to the Internet has become widespread, so almost all available resources (data, software) can be placed on the Internet, shared between users - sometimes completely independently of others, and brought from the Internet to the user's computer on demand. The chapter highlights the benefits of using hybrid technology with a focus towards entrepreneurs and business innovators. Intercloud systems are built of heterogeneous services, offered by different providers of cloud computing. To emphasize the business opportunities offered by innovative Intercloud approaches, an architecture for the interconnection of cloud services offered by multiple cloud providers, with applications to e-health, is presented. The main concepts and business benefits of using Intercloud systems and an architecture for interconnecting cloud services from multiple cloud providers is described.

Stelian Stancu, Constanta-Nicoleta Bodea, Laura Elly Naghi (Novac), Oana Madalina Popescu (Predescu), Alina Neamtu (Idorasi) have develop the chapter *Use of New Innovative Technologies in Business by All Age Groups.* The paper is covering the topic of the innovative technologies and their use at business level in an entrepreneurial environment, by all age groups. A number of areas are discussed, such as the demographic evolution and the age pyramid, the ageing process and a number of aspects regarding the flexi-security phenomenon in Romania compared to its evolution at European level. The chapter provides an overview about the demographic evolution, the age pyramid and the ageing process. The labor flexi-security at European level is also discussed covering the main flexi-security related indicators and few related econometric models are proposed.

By analyzing the case of Mexico, Verónica Baños-Monroy, Edgar Ramírez-Solís, and Lucia Rodríguez-Aceves are developing the chapter *Entrepreneurship and Innovation in Latin American Family Firms: The Case of México.* It examines the role of entrepreneurship and innovation in the development of family businesses in Latin America. The socio-economic dynamic of such region is an interesting field for research due to following reasons: it is a major manufacturing hub with growing importance in retail; it experienced a huge economical shift in the last ten years causing the growth of the middle class (an increase of 50 million people); family firms in the region are the main driver of growth and employment. The authors focus the analysis on Mexico due to its singular characteristics, making it a representative example of the region. Based on secondary sources, a characterization of the entrepreneurship and innovation in family owned and managed firms in the region is reviewed. Moreover, based on the Global

Entrepreneurship Monitor database, the authors present some recent data related with entrepreneurship levels in the region and its impact in terms of innovation. Finally, public policy implications are provided in order to encourage entrepreneurship and innovation in family businesses.

With a strong accent on the concept of *pitching*, the chapter *Pitching and the Other International Practices of Innovation Competitions: Channel for Youth Entrepreneurship* has been written by Rauno Rusko, Katja Härkönen, and Sofia Petäjäniemi. This chapter focuses on the international practices of innovation competitions. Pitching is a truly global phenomenon in the education and youth entrepreneurship, which has similar structures and contents all over the world. Thus, this international practice follows the ideas of Ohmae about the borderless world. In other words, the micro-activities and the practices of innovation events are following the joint global macro-level frameworks and practices. However, there seems to be slight differences of these events basing on how these events are named or marketed in the Web. Some of these events are web-based while others involve face-to-face happenings. All these forms of pitching events provide important channel for young entrepreneurship.

By emphasizing the increasing role of neuromarketing in the nowadays market, Ioana Iancu is developing the chapter *A Neuromarketing Perspective on Measuring Marketing Influence at the Unconsciousness Level*. In a context characterized by an inflation of marketing messages, it is imperious to understand how consumers succeed in making the buying decision. Starting by briefly describing the structure and the role of the brain and the differences between consciousness and unconsciousness, the paper aims to investigate the way neuromarketing can help in comprehending the feelings of the consumers, the way products or services match the consumers' needs, and the way companies can discover the insights of decision-making process. This paper can be perceived either as a guide for the companies that aim to find more on the way people manage information and make decisions or as a comprehensive description on human being marketing behavior that can serve both business, academic environments and consumers.

José Manuel Saiz-Alvarez and Jorge Colvin-Díez bring into discussion the correlation between digital entrepreneurship charity and social change. They are the contributors of the chapter *Digital Entrepreneurial Charity and Solidarity for Social Change: KIVA and the "Social-on-Linezation" of Poverty*. Microcredit has been studied from many perspectives. In this chapter, the authors analyze KIVA, the most important Person-to-Person microfinance organization from the perspective of social change. Moreover, the authors study how it has influenced the nascent of a new wave of entrepreneurs known as digital entrepreneurial charity. Applied to KIVA, they analyze the impact of the digital space and its Internet-based Peer-to-Peer Lending to create social change in the poor, while alleviating the poverty thanks to solidarity and charity. This work concludes affirming that banking the poor and education, with the intensive use of Internet-based devices, is the best way to alleviate poverty in our digital and globalized economic world.

The role of education in entrepreneurship is developed by Omar Patiño Castro, Catalina Ruiz Arias, Jose Jimenez Ibañez, and Francisco Matiz Bulla in their chapter *Universities Fostering Business Development: The Role of Education in Entrepreneurship*. They underline that Universities play a fundamental role in promoting entrepreneurship and directly contributing to the economic development of the country. This chapter presents the case of EAN University, an institution that from its foundation has focused on entrepreneurship as its mission. Over time, the university has taken concrete actions to develop its three mission areas: training, research and social outreach. At EAN University, the entrepreneur training model is an obligatory part of all programs, making it part of the student's core education. In addition, the university promotes and creates a culture of entrepreneurship in its community through permanent business incubation and acceleration services, which are available to all students and graduates. EAN's accompaniment model begins with a diagnostic, which determines the stage the company is in: preincubation, incubation or acceleration. The university's efforts then focus on directing and accompanying companies in the preincubation and incubation stages.

Constanţa-Nicoleta Bodea, Stancu Stelian, and Radu-Ioan Mogos are discussion the importance of e-learning solutions in their chapter *E-Learning Solution for Enhancing Entrepreneurship Competencies in the Service Sector*. The chapter proposes an e-Learning solution for the entrepreneurship education, based on several simulation modules integrated into a classical learning management system. The originality of the approach is that the solution is domain independent and applied advanced IT technologies, such as agent-based simulations and extended graphical support. Using this e-learning solution, the students can learn how to choose characteristics/aspects for particular type of business and how important is each of them according specific criteria; how to set realistic values for different characteristics/aspects of the business, how a business scenario can be changed in order to fit better to the business context (business reality), modeled through by the scenario pattern and how to assess/evaluate business scenarios.

Dorin Spoaller, in his paper *Online Facial Symmetry and Achieving Success in a Digital Entrepreneurship World*, is conducting a comprehensive research on the importance of facial symmetry in advertising. The second most used page on a newly found website from which a client might buy an item is the ABOUT US page. Using real images of real employees has an overall positive effect on the outcome of that visit. However, it is not always as simple as that. Subtle differences in the content of the pictures may have more of an impact than previously thought. Online advertising has long been considered as a media where creativity is not needed, with the advertising agencies focusing on a few basic rules for displaying ads: call to action, numbers, colors, human figures or symbols and so on. The results are analyzed only on their conversion rate value, this being one of the main reasons why online advertising is being regarded as the future of advertising worldwide. This study is focusing on the Click Through Rate (CTR) impact of different ads using different Facial Symmetry Indexes (FSI) on the human figures used. The hypothesis proposed is that a person with a higher FSI, noticeable only at a subconscious level, will generate more clicks on an online ad.

Editing this volume was equally a complex and an amazing challenge. We have electronically met people all over the world, people eager to contribute to our knowledge experience and people eager to help us in pursuing the goal of developing a valuable book. Overall, it was a great experience, with many lessons to learn. We wish all the readers to enjoy the research, to get inspiration from them, to replicate the ideas and to further write their own works.

Ioan Hosu
Babeş-Bolyai University, Romania

Ioana Iancu
Babeş-Bolyai University, Romania

REFERENCES

Audretsch, D. B., Bönte, W., & Keilbach, M. (2008). Entrepreneurship capital and its impact on knowledge diffusion and economic performance. *Journal of Business Venturing*, *23*(6), 687–698. doi:10.1016/j.jbusvent.2008.01.006

Davidson, E., & Vaast, E. (2010). Digital entrepreneurship and its sociomaterial enactment. In *Proceedings of the 43rd Hawaii International Conference on System Sciences*. IEEE Computing Society.

Ireland, R. D., & Webb, J. W. (2007). Strategic entrepreneurship: Creating competitive advantage through streams of innovation. *Business Horizons*, *50*(1), 49–59. doi:10.1016/j.bushor.2006.06.002

Johansson, B., Karlsson, C., & Stough, R. (Eds.). (2006). *The emerging digital economy: Entrepreneurship, clusters, and policy*. Springer-Verlag Berlin Heidelberg. doi:10.1007/3-540-34488-8

Webb, J. W., Ketchen, D. J. Jr, & Ireland, R. D. (2010). Strategic entrepreneurship within family-controlled firms: Opportunities and challenges. *Journal of Family Business Strategy*, *1*(2), 67–77. doi:10.1016/j.jfbs.2010.04.002

Acknowledgment

The editors of the volume would like to acknowledge the valuable help of all the people involved in the process of developing the volume of *Digital Entrepreneurship and Global Innovation*. We appreciate the contribution of the authors and of the reviewers. First, we would like to thank all the authors for their investment of time, effort, expertise, and knowledge. We do value every single input and we believe that all the delivered knowledge is equally important. Second, we would like to show appreciation to all the reviewers. They have significantly contributed to the improvement of the book by adding quality and coherence.

Ioan Hosu
Babeş-Bolyai University, Romania

Ioana Iancu
Babeş-Bolyai University, Romania

Chapter 1

The Benefits of New Online (Digital) Technologies on Business:

Understanding the Impact of Digital on Different Aspects of the Business

Farhan Shaikh
IIT Bombay, India

ABSTRACT

This chapter explores how digital technologies are impacting the businesses – challenging the industry leaders while enabling entrepreneurs to do so. The overall findings have been bucketed under three broad sections – Customer Experience, Process Optimization, and Business Innovation. The Customer Experience section captures the way companies are using the data from digital technologies to enhance the way they interact with the customers. The section on Process Optimization highlights the significant improvements achieved by using digital technologies for existing business processes. Business Innovation captures the transformation power of digital, through new business models, business areas, and data monetization.

DOI: 10.4018/978-1-5225-0953-0.ch001

INTRODUCTION

Digital technologies have deeply impacted the way we live, deal, shop, travel, socialize, interact, choose, purchase and make other important decisions. They have enabled a large number of entrepreneurs to solve real world problems and in most of the cases – in real time. A majority of startups today have a digital gene giving them the flexibility of having an agile business model, which can be tweaked as per the requirements and needs of the market. For example, Uber operates as an aggregator in some regions but provides ride sharing facilities in others, to avoid legal hassles. It also provides services such as cargo delivery along with pick up and drop facility in some cities. In parallel, existing and established companies are investing in digital, either to optimize the existing business models or to create new ones by lapping up the opportunities provided by the digital citizens of the world. To give an example from the travel industry, established car manufacturers like Ford are trying to establish themselves in peer to peer car sharing service industry, in some regions to battle stiff competition from purely digital startups like Uber and Lyft. Other than the new business models, digital also impacts the way companies reach out to their customers, understand what their customers are saying and what their needs are. Using digital platforms, companies can empower their employees to collaborate in real time, reach out to colleagues in different parts of the world and break the silos, which prevent innovation. Digitalized processes also mean increased efficiency, new data points for optimized calibration and most importantly – increased profitability. As Gartner predicts, by 2018, digital business will require 50% fewer business process workers and 500% more key digital business jobs, compared with traditional models.

Also, once thriving industry leaders such as Blockbuster, Kodak and Borders that couldn't adapt to the digital shift by providing any alternatives to the market, no longer exist.

Every business speaks about being digital or going digital; however, very few convey what they actually mean by it. Is it about the digital marketing and online presence, is it just the use of social media to reach out to customers or is it about using cloud or about applying the science of analytics to understand their customers better – the definition is wide open. That exactly explains how it positively impacts multiple aspects of business, internally as well as externally.

Digital is less of a 'thing' and more 'a way of doing things'. It is the strategy that drives digital, not the technology. Digital is orchestrated through technologies like social media, analytics, cloud and big data. It is demonstrated in the ease of usage, connectedness of customer and business resources, multi-channel experience, end-to-end integration, facilitation of storage for dissemination of data and information; and policies such as BYOD (Bring Your Own Device) which have led to consumerization of IT. Gartner defines consumerization as, "the specific impact that consumer-originated technologies can have on enterprises. It reflects how enterprises will be affected by, and can take advantage of, new technologies and models that originate and develop in the consumer space, rather than in the enterprise IT sector." Digital technologies are reaching the maturity stage with introduction of wearables, virtual reality, drones, video analytics and connected devices.

Digital helps companies understand their customer in ways not known earlier, in proper context of information and location, to provide a delightful experience, at different touch-points on every single occasion. It enables companies to engage their employees and empower them with the right tools to connect on a platform where they can provide quicker solutions by going breaking the silos. Digital also allows the company to keep their business models agile, have a faster go to market strategy, reach out to larger customer base and at the same time, have real time feedback. It helps companies hear what their customers are saying across multiple platforms about their existing brands, react and take measures to keep a check on their sentiments, or to nurture a new brand with which customers can relate to. It helps companies to generate revenues from new streams, supported by digital technologies

Need for Digital

An average employee uses much more advanced digital tools and applications at home than in office. Population in technologically mature markets like UK and US have lead the digital adoption with mobile penetration as high as 138% and 112% respectively, while Germany leads globally with 141% share. Similarly, mobile penetration in emerging markets has grown 321% compared to 46% in developed countries in 2002. Globally, the number of smartphones is expected to increase from 2 billion in 2014 to almost 5 billion in 2019. In today's time, a smartphone has more computational power than what a computer had just a few years back. This has increased the consumption on mobile devices. Out of 890 million daily active users, 745 million access Facebook from mobile. Similarly, Twitter just crossed 302 million Average Monthly Active Users (MAUs) for the first quarter of 2015, up 18% year-over-year and compared to 288 million in the previous quarter and approximately 80% of total users logged in through mobile. Google searches on mobile devices now outnumber those on personal computers in 10 countries, including America and Japan.

The need for digital is further expected to increase as Millennials have started joining the work force. Millennials, as demography are those who were born between 1980 and 2000. They are also known as digital natives, given their ease at leveraging digital technologies for personal as well as professional use. Millennials represent the largest demographic group in history and by 2025; they are expected to comprise almost 75% of the workforce that is expected to have a tremendous impact on the entire business landscape. Their affinity for technology means they prefer to interact and work for those companies, which offer maximum convenience through the use of digital technologies. They want the information to be available to them with complete ease, as and when required by them.

This has made companies change the way they function and they have started adopting digital technologies such as cloud, social media and internet of things. According to forecasts by Forrester, cloud computing will increase from approximately $41 billion this year to more than $240 billion in 2020. Gartner reports that there would be 25 billion things connected over internet, by 2020, clocking in close to $2 trillion of economic benefit globally. All of these will have a great impact on the economy by transforming many enterprises into digital businesses and facilitating new business models, improving efficiency and new sources of revenue.

And digital also means profit. An in depth research undertaken by Capgemini Consulting along with MIT Center for Digital Business in 2012, surveying over 400 major global companies worldwide, reveals that digital leaders outperform industry competitors on multiple financial metrics. They generate, on average, 9% more revenues through their existing assets; outperform their peers by 26% in terms of profitability; and achieve significantly higher (12%) market valuations.

Impact of Digital

By and large, majority of practitioners equate social with digital. While social is an important subset of digital, digital goes much beyond that. Most of the large scale companies were already working with standard technologies as per their business requirements such as ERP, CRM, PLM and CAD systems.

Figure 1. Impact of digital on business

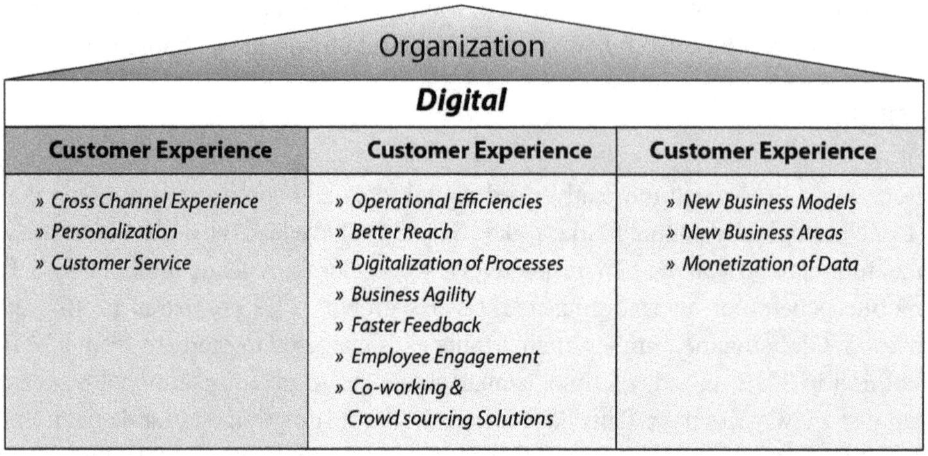

However, with the introduction of digital tools, the way companies were working has changed drastically as it has connected machines, employees and customers together through big data, analytics, social media, mobility and cloud. While initially these technologies were being used to support the existing systems within the company, the digital tools are now driving business forward, creating new opportunities and revenue models. Briefly, there are three aspects to how digital impacts businesses – Customer Experience, Process Optimization and Business Innovation.

Companies which are digitally mature are focused on integrating their different digital initiatives to transform their business, while the less mature ones are focusing on solving discrete business problems using individual technologies.

Customer Experience

Digital has changed the way companies deal with customers. From the demand side, the customer expectations have witnessed a sea change. Customers now expect personalized experience at all the touch points – offline as well as online. They expect companies to be aware of their likes, dislikes and preferences. With rise in emerging technologies customers now expect services to be available on demand, at the pull of the customer, while being more relevant and responsive. Similarly, customers are now active users of social media and mobile penetration has been huge. Buying decisions are made on the move and the customer journey has evolved to become complex as customers keep use a mix of mobile, social media and store before completing a purchase. There has been a 'mobile mind shift' in customers and they now expect to get what they want in their immediate context and moments of need.

In most of the industries, the competition has been so intense, that it has reduced the scope for traditional product and service advantages. With rapid technological evolution and innovation, the products across segments have moved towards commoditization, making customer experience the new battlefield. In the digital age, innovation in customer experience thus becomes the most strategically important aspect for value creation and is proving to be the only truly durable competitive advantage. According to Gartner, by 2016, 89% of companies expect to compete mostly on the basis of customer experience and by 2017, 50% of consumer product investments will be redirected to customer experience innovations.

Cross Channel Experience

Digital has lead to complex customer journeys with customers using multiple channels to connect with the companies. Engagement on digital channels like social media, web and mobile means companies need to be forever listening and be available for the customers. Digital technologies have empowered customers to compare prices, specifications and features on the go, before making the purchase decision. This has resulted into customers using mobile devices within the store to make the final decision and often moving out of the funnel before reaching the purchase stage. Initially when there were store-only retailers and pure ecommerce players, many retailers suffered from the concept of 'showrooming' – where customers would research about the product online, walk in to the stores for checking it out physically and then end the purchase journey by ordering it online at a discounted price. As ecommerce players adopted the mobile first strategy, the 'showrooming' actually increased significantly as customers would check the product in the store and directly order it online from ecommerce players, even while being at the store itself.

Burberry uses multiple digital channels to ensure a perfectly orchestrated cross channel experience for all its customers. Using digital channels, Burberry brings its brands to life within the store premises. Sales personnel within the stores are equipped with iPads, which provides customers access to the entire global collection, irrespective of their availability in the store. On the other hand, 60% of its customers complete the purchase journey online but prefer collecting their products offline, from the stores.

Burberry also uses digital technologies to enhance the multimedia experience within the stores. Whenever a customer approaches a clothing item, the RFID system gets activated which shows the details about the clothing item on the display. Customers can also place their products on RFID platforms which project the same products on the large display. Mirrors double as audio-visual displays that show footage from run-way shows while the fitting rooms have screens which show specific product information such as a bag's stitching detail or how a skirt was worn on the catwalk. Thus the overall experience, across offline and online channels, remains seamless, creating a memorable experience for customers while increasing satisfaction and minimizing annoyances.

Similarly, pure ecommerce players are looking at ways to provide a platform where customers can physically experience the products to counter the growing phenomenon of 'reverse showrooming'. 'Reverse showrooming', also known as 'webrooming' is when customers browse, compare and finally choose the product online, but complete the purchase offline, in a store. The reasons for completing the purchase journey offline can be varied – from better customer service, to immediate ownership of the product, assurance of receiving a genuine product or even the level of trust in the local shopkeeper. Also, for most of the customers, there's no substitute for being able to see, touch and test a product in person before buying it.

However, many companies, especially retailers, have invested in uniform cross channel experience. As customers have started using multiple channels for shopping, the more channels they find a particular brand, the more they develop a familiarity with the product or service associated with the company and, ultimately, a loyalty towards the brand. It has positively contributed towards the revenues by ensuring brand loyalty and has helped companies to leverage 'showrooming' and 'webrooming' in their favor. Companies use data from customer's web presence and online behavior to personalize the experience offline, in the store. Similarly, companies are leveraging customer preferences disclosed and mapped in the stores to send relevant mailers, offers and newsletters to the customers – online, on social platforms as well on the mobile.

Digitalization is also enabling manufacturers to establish a more direct link with the consumers and manufacturers have already started leveraging it. They have also started using online marketing and advertising, in compared to their earlier traditional ways of reaching the customers. Significant numbers of manufacturers have started selling online, especially higher-margin products in high-growth markets. This has helped them with increased profitability (by recovering some of the margin given up to retailers) and has also empowered them to get valuable new insights as they engage in more direct relationships with their customers than they have ever had before.

Personalization

Personalization is a process that creates a relevant, individualized interaction between two parties designed to enhance the experience of the recipient. It uses insight based on the recipient's personal data, as well as behavioral data about the actions of similar individuals, to deliver an experience that meets specific needs and preferences. – Gartner (n. d.)

Industrialization enabled companies to manufacture at scale, thereby reducing costs, while allowing them to maintain high standards in quality. Henry Ford took the industry by storm using mass production for his Fords Motor Company using assembly line method for producing just one car – in one single color, black. Over a period of time, they got displaced in the market by General Motors which understood and celebrated the existence of different customer segments which had different expectations from the same product – car. More number of companies started realizing the importance of segmentation and they tried categorizing customers using different parameters. With evolution of computers and databases, companies were able to reach out to customers directly and store structured information about them. Similarly, more processing power empowered companies to generate better insights using data mining, artificial intelligence and clustering analysis.

With the advent of digital technologies, the customer touch points increased significantly, giving companies an opportunity to have a more holistic view of the customer. Mobile and digital channels not only provide details such as real time location, in app behavior, on site preferences, but also empower companies to understand the relevance, context and customer sentiment at given point in time along with their purchase intention and aspirations. The segmentation has thus evolved from the simple geographic segmentation to the more advanced and technology dependent behavior-based segmentation (Oliviera-Brochado & Martins, 2008). Integration of structured data from enterprise systems and CRM along with the unstructured data from digital channels is helping companies create better personas of their potential customers.

Caesars Entertainment, one of the world's largest gaming and entertainment companies, has personalized the entire experience – right from booking till the actual stay in the hotel. Company starts with collating customer details from the vast amounts of transactional, demographic and gameplay data through its loyalty program for creating a detailed profile for each of the customers. It also provides the customers with mobile check in facilities and uses their mobile data for sending location based offers. The mobile app of the company helps customers to track their loyalty points and to check where they can use those points. The app is also the first point of contact for any concierge and room services, which helps the company in understanding individual behavior to personalize the experience accordingly. Even the wait staff uses mobile app for ordering drinks to reduce the time for reaching the counter. As company uses multiple data points, it is well aware about customer's experience even at the table and it tries to personalize the experience accordingly – so a player with big losses gets freebies such as food and extra night accommodation—that can ease the pain of the losing, without affecting the gaming for anyone else.

All these can lead to more sales per customer, higher retention and greater profitability. It has also helped companies to identify where the next marketing investment should be made and while ensuring higher return on investment by implementing behavior-based segmentation and personalization (Tsai & Chiu, 2004). According to Gartner (n. d.), by 2018, organizations that have fully invested in all types of personalization will outsell companies that have not by more than 30%. Companies are leveraging this data to envision new products and services, and ways to bundle and price them in different markets and categories. Almost 20% of durable goods "e-tailers" are expected to use 3D printing for creating personalized product offerings by the year 2017.

Large organizations which value the growth of their employees are exploring data generated from digital training channels to personalize the learning through programs which are paced according to individual's capabilities to grasp and use the knowledge imparted through such programs. With digital technology being available for instructional data and tools for analyzing, such data points that help understand immediate and larger context are increasing the adoption of personalized and adaptive learning.

Customer Service

Customer service involves engaging with customers for a variety of reasons – it could be before purchase to answer any queries and concerns, it can also be for handling post purchase issues and eventually for up sell based on the existing purchase. For any of which reasons, customer service is generally expected to be pain-free, proactive at minimum and preemptive at best. Company needs to value the customer's time spent during the engagement and needs to ensure maximum productivity. For customer focused companies, customer service has been one of the key pillars on which the brand has been built over a period of time through exceptional service record.

BMW launched ConnectedDrive technology that serves as an in-car navigation system with mobile components. It also offers a wide range of intelligent services and apps that provide customers with all the relevant information and entertainment during their journey. BMW car come with SIM card and basic telematic features along with facilities such as BMW Teleservices, BMW Breakdown Call, TeleService Battery Guard and Intelligent Emergency Call (Motoring, 2014)

BMW Teleservices keeps the driver and the preferred dealer informed of the car's condition – and notifies them if there's any need for a service. Driver can use BMW Breakdown Call using the iDrive menu and relay the vehicle's functional state and location to a technician for roadside assistance. The system will also provide vehicle's driver with further information to connect with technician. TeleService Battery Guard automatically updates the vehicle's owner of unexpected deterioration of the battery, sending the information by text or email, according to the preference selected in the iDrive menu. Intelligent Emergency Call contacts a call centre in the event of a crash and provides details of the severity and circumstances of the crash. Customers can also book other optional services and applications such as Concierge Services, Internet or Real Time Traffic Information. The car is connected to Google and can be used to locate best restaurants in town along with recommendations from BMW Online (Motoring, 2014).

The digital technologies have redefined the times to be called the age of the customer. With everything expected to be at the pull of the customer, companies don't decide how customer centric they are, but the customers do. The social, mobile and big data have changed the customer expectations from companies they have to deal with. Also, providing good customer service is a win-win situation for both - the customers as well as the company. Great customer service makes customer more loyal and loyal customers are less likely to switch companies as they will prefer considering purchase from the same company. As the company's efforts to establish loyalty increases, customers are also more likely to recommend the company to friends and colleagues, thus establishing advocacy for the brand. In digital age, customers expect their questions to be answered in real time, with painless and minimum service interactions, irrespective of their position in the purchase journey and across different channels. They get frustrated if they have to struggle for accessing customer services and they even drop off the purchase journey if the experience is unsatisfactory at any stage.

By 2020, the number of connected devices is expected to cross 50 billion. The kind of customer information companies will have will be unmatched and so will be the expectations of the customers. Companies will be in a position to be in forever listening mode of the connected devices. This will help them notify customers in advance, preemptively diagnose and fix small issues with minimal human intervention. The emerging digital technologies will thus change the face of customer service resulting into faster resolution at lower costs and extremely personalized engagements. It will also help companies to plan accordingly and anticipate future service requirements of their customers.

Process Optimization

While digital is largely associated with digital transformation and disruption, digital technologies are also being used to optimize the existing business processes. By leveraging digital technologies, companies are able to benchmark the repetitive processes against the standards or industry leaders and map the deviation, if any. It has even empowered companies to rectify the processes on the go with real time feedback. This has increased the overall operation efficiencies of the long established processes.

Some companies entirely transforming their existing business process to rethink it in terms of final output. This allows companies to break away from the business strategies that have been deemed redundant in the digital age. However, not all companies have the necessary mandate or will to change from the scratch. A lot many companies are using digital to enhance and optimize the prevailing business models by replacing some modules with digital technologies. Breaking larger processes in to small independent ones, which can be digitized without disrupting the overall system, is helping companies to increase efficiency. It has enabled companies to have faster access to information, which on processing in correct context, helps them to access accurate insights, almost in real time. It the companies are leveraging digital technologies like cloud; it also empowers them to have agile business models in place.

Operational Efficiencies

The kind of data provided by digital technologies is humungous and depending on the capabilities companies can use it to improve operational efficiencies of different business units. While some companies would like to focus on the supply chain aspect, others would be more interesting in looking how their marketing campaign is being received by different segments of the customers. For example, the financial organizations use data to reduce risk, improve fraud detection and to ensure complete compliance. Companies working with employees operating out of different locations, in different time zones are successfully leveraging digital technologies to run the processes in a more efficient manner, with a direct impact on the business outcomes – making it more profitable and customer friendly.

ABB Process Automation Service (USA) used different spreadsheets to manage approximately 300 field service engineers (FEs) managed through regional coordinators. The spreadsheets were used to aggregate data to streamline and improve processes which led to inefficient resource utilization. Therefore, the company invested in a new cloud based technology platform to run the service application. A service dispatch board was created to provide a consolidated view of all regional service requests. The software allows the company to assign FEs based on skill set, availability, and geographic proximity. It also allows the customers choose FEs from a more distant region by paying an additional fee, depending on urgency. As a result of this, these services have become more responsive and have delivered improved uptime. The productivity applied by field engineers increased from 70-80% to a whopping 120%, post the deployment of cloud based application. The digital upgrade has also resulted in an improvement of customer satisfaction ratings from 79% to 84%. (Capgemini Consulting, 2014)

Better Reach

As digital captures every single interaction with customers, it helps companies to understand their customers better. Given the rate of adoption of digital technologies, customers are present at different channels, round the clock and are engaging with other customers and companies. As the number of digital technologies being used by customer increases, it gives companies new avenues to reach out to customers. The data generated from these touch points also helps companies to understand their target customers and tailor the communication accordingly, which helps in creating engaging conversation instead of blink-and-you-miss traditional channels.

Nike introduced Nike+ product range in 2006, consisting of a sports watch, a wrist band and a mobile phone running app. The products track wearer's location and their activity level throughout the day. The data is collected on a central platform that the athletes use to make training schedules, track sports achievements and share them with friends across multiple digital platforms. This service enables Nike to achieve strong customer engagement with all its existing customers. Over and above this, the digital platform also helps Nike to reach the customers in a direct manner for advertising its products while gathering valuable data about how customers use their products.

Companies are also using the digital reach for different business reasons. Industrial companies use the digital channels to convey the technological advancements and innovation to the end customers as well to inspire potential employees. Hotels and restaurants use different digital mediums such as Instagram and Vine to share pictures of new děcor and dishes. Those employed in entertainment industry leverage these channels to share behind the stage scenes to create a special bonding with the users and these days it is also used to interact in real time through platforms like Google Hangout, Twitter and Facebook. Digital also makes it faster. Cloud enables companies to have a faster time-to-market at a pace, which could not be achieved using traditional data center, empowering companies to deliver results in a short span of time.

Digitalization of Processes

Digitalization of processes is one of the initial steps when companies set out for digital transformation. In such initiatives, companies start with introducing digital technologies to their existing business processes and workflows to replace manual and physical efforts, such as introducing mobile application instead of paper-pen entry work to capture data in digital form. This tremendously increases the velocity of business process and enables quicker decision making. While digitalization being about huge benefits to the organization, it is not transformational in nature as it simply improves the existing process, by and large following the same structure of implementation and governance.

The aviation industry commonly uses milling process to manufacture complex aviation parts. But milling produces large amounts of waste in the form of swarf. This makes prototyping complex, consuming more time for retailing and produces large amount of waste. Leveraging the new digital technologies GE adopted rapid prototyping approach using 3D printing for these parts. Instead of machining parts from solid billets of metal, only the material needed to shape that specific part is used. (Capgemini Consulting, 2014)

Digitalizing processes not only replaced paper and manual processes with software and digital technologies, but also allows businesses to automatically collect accurate data that can be analyzed to better understand process performance, cost drivers and causes of risk, if any. Further, it also enables companies to deliver real-time reports and dashboards on digital-process performance, empowering managers to address problems before they become critical in nature. According to a report released by McKinsey & Company in 2014, digitalizing information-intensive processes can help save costs by up to 90% and improve turnaround times by several orders of magnitude. It also helps companies in improving their bottom lines while delighting customers with faster, accurate and predictive processes.

Business Agility

Legacy systems, which earlier helped the companies to evolve as they became huge enterprises, have often posed a challenge to agile business strategy. Established process often followed top down approach with decisions taking forever to percolate to the ground where the real action takes place. With the expansion of businesses, companies got department oriented which created organizations working in multiple silos, further slowing down the strategic decision making process. While such a setup ensures cost benefits and business efficiency, demands of the customers have changed and they now expect agility and real time capabilities from companies they deal with.

Uber being a purely digital company is one of the most agile companies in the market. Its ecosystem consists of the back end system, which leverages big data and analytics, a driver willing to work as a partner using mobile application and the end customer that pays for these services. The technology helps Uber to modify its business model on the go – so in some countries it uses the ride sharing model while in others it offers carpooling services. This model helps Uber to provide other services such as UberCARGO for moving cargo in the city and UberRUSH, a document delivery service. Uber also uses analytics to understand the demand in a particular region and increases the price through its surge pricing program, where it offers higher fees to drivers to attract them in these particular areas, all in real time.

Digital technologies when integrated with the existing enterprise systems provide companies with a much needed agility to handle the new age connected customers. Deep social listening with advanced analytics and cloud-based infrastructure enable fast paced deployment of new products and services. They also offer the ability to quickly learn from and fix mistakes. The explosion in smartphones, tablets, wearable devices with internet-connected objects and mobile apps means that a company should have the ability to interact personally, directly and in real time with consumers and business partners. Companies need to move with great speed to meet evolving customer needs, adapt to market and deal with dynamic regulatory changes. Companies are investing in processes which are unstable in nature and can be adapted dynamically with changing customer demands. They are moving away from stable processes which are based on typical legacy systems having limited information about the customer to unstable and ever evolving ones which capture customer information on the go and adapt.

Faster Feedback

Before the advent of internet, companies would launch product and services, which would have a long curve of adoption. The cycle to gain some ground level feedback about the offering would therefore be much longer. In case of factory based production, companies would have to complete one cycle of quality check before flagging off mass production. With digital technologies, companies have redefined their feedback processes – in terms of manufacturing as well as customer feedback.

The digital processes create continuous cycle of improvement fed by continuous feedback during the manufacturing process. The entire process cycle fetches data from multitudes of sensors that monitor every move and feed those observations directly into predefined and tested models. As the feedback is in real time, the model itself can rectify the glitch or there is scope for manual intervention to make real time adjustments in the manufacturing process. Such digital processes also enable mass personalization of products as the interventions can be made in real time to customize products on the go.

BMW's prototype cars produce an average of 15,000 data points from the engine and transmission down to the suspension and brakes. Using big data analytics, BMW Group can detect and even fix vulnerabilities that show up during the manufacturing process. In absence of digital technologies, it would have taken months to analyze reach the actual root cause of the defect, if any. However, with digital, the entire process of error detection to correction can now happen in just days. Based on the data collected during the test drives and other processes, BMW can find faults and fix them before new cars go into full production. BMW Group is thus able to enhance customer satisfaction by producing higher quality cars through digital technologies as the feedback is quick and data can be mapped to the source easily (Motoring, 2014).

Digital process can also be leveraged to sense the feedback from the customers and make adjustments in the current manufacturing cycle without much effort. If the products are enabled with internet or if the customers are speaking about the product on digital channels, companies can leverage such data to do real time listening, analyze the cause and try to fix up in the next production batch itself. This helps companies to fail fast and reiterate at much faster rate. Some retailers have actually replaced the price marking with digital boards as it helps them fluctuate the pricing of the products in real time – based on the insights from the sales generated in the store at that very point of time. In a traditional ERP or CRM based system it would be difficult to leverage such real time feedback unless the systems are able to speak with the new digital technologies (Motoring, 2014).

Employee Engagement

Digital technologies have changed the way employees work in office. The concept of on duty hours and off duty hours has been impacted as digital allows employees do some of their personal work online during office hours and vice versa. Employees end up carrying doing thing – work or entertainment while being in office and often complete the task assigned to them from home. Depending on the maturity level of bring your own device (BYOD) policy in the office, employees have been provided with enough flexibilities to choose a device as per their own liking and consideration. While some companies limit this option to phones, some even allow employees to choose their work stations (computer and laptop). Social media has also helped to establish an employee-to-employee relationship, which often goes beyond the work.

Companies are now leveraging all these aspects to engage employees, which studies have proved, have a direct impact on productivity, efficiency and attrition rate of the company. The digital technologies have now empowered companies with data to design, asses and redesign their engagement activities in continuum. Some companies are using off the shelf digital tools such as Yammer and Chatter for providing a platform for employees to connect and share knowledge with each other, while some companies are investing in bespoke mobile applications which are hardwired according to the needs and functioning of the company.

The Cheesecake Factory, a well-known national chain of restaurants, gamified its operational training, to change training from a typical "operational training" process to a Disney-like gaming portal. Instead of notes and presentations, employees had to post videos of their experiences in solving common restaurant problems in dedicated digital platform called as Video Café. Other employees accessing this content had options to comment and rate it along with sharing their videos or blogs. The contributors with highest ratings are celebrated on a "leaderboard" and are presented with "slices" (for example, of cheesecake) for their contributions to the team (Engaming, 2012).

It also uses a mobile application to ensure that the employees know all the ingredients of their burgers. The Cheesecake factory launched a mobile app where employees actually "build the burger" on their iPhones, instead of practicing it offline. They have to increase the speed as they play further and after that a new burger recipe is introduced. Playing the game for quite a few times ensures that the employees remember all the important aspect about creating a perfect burger for their customers (Engaming, 2012).

Using gamification and analytics, companies are mapping the performance of individual employees against fixed parameters, while benchmarking them against colleagues to choose the leading players. These players are then awarded with points or rewards based on the company policy, which inspires other colleagues to join in as the engagement continues further. Companies are even using engagement activities to improve the health of their employees by tracking their calories and activity details using fitness bands which keep sending the data to the company. Such companies are utilizing the data to make informed decisions which help them save on insurance.

Co-Working, Collaboration, and Crowdsourcing Solutions

Digital tools have not only enabled companies to connect with their customers better, they also strengthened the intra company networks of employees like never before. Large of companies have invested in such digital tools which enables employees to connect with each other beyond the email. Cloud based collaboration tools have provided the connectivity on the go. They have eased communication and thus given a huge impetus to collaboration. This has helped companies to increase resource efficiency even for widely distributed teams.

Citi launched Citi Mobile Challenge to foster digital and mobile innovation in banking. Citi Mobile Challenge is a next-generation accelerator that combines a virtual hackathon with an incubator, a worldwide network of FinTech experts and Citi's unparalleled global sponsors and clients to discover solutions across more than 100 markets. It invites developers from around the world to help build innovative solutions based on the Citi digital platform. Citi opens up a variety of APIs from the bank and other leading financial and technology companies to help developers create real-world innovations that could function with existing Citi technology. This helps Citi to reach beyond its own employees, crowd sourcing innovation from the global community of developers on its digital platforms. (PwC, 2015)

In 2014, Citi brought 60 finalists to Silicon Valley, New York City and Miami to present their ideas to a panel of judges who could decide whether these were innovations that could benefit its clients. The finalists competed for an opportunity to take their idea to production, a share of $100,000 and other services to help get some of their ideas off the ground. Citi is currently working on several top innovations from this competition to launch pilot solutions for consumer and business clients in various markets. (PwC, 2015)

Companywide collaboration has sped up innovation and development processes as the digital tools are being used to support idea generation. Bringing multiple stake holders from different geographies on one single platform, virtually together, has further reduced the turnaround time of an idea to get implemented in the organization.

Business Innovation

As more number of companies are embracing digital technologies and deepening the engagement with the customers, they are generating piles of data which is often left unused. However, few companies understand the importance of this data and are investing in technologies which will help them innovate and to create new sources of revenues. While some companies are using data to create new business models, which are premium because of the nature of information they carry, others are trying to monetize this trove of data by either selling it in the raw format or by processing it to deliver in on demand – as a service. Entrepreneurs are leveraging digital technological advancements to create platforms where users can share and offer underutilized resources in creative ways. Most of such companies are growing rapidly across the globe, entering new markets, without having to invest in any resources other than technology. Being asset-light makes these companies agile and they are able to transform the business strategy as per the needs and demands of the market.

New Business Models

Digital technologies have helped connect people in ways which were not thought of earlier. Companies have created dedicated platforms where users having a common need, can directly get in touch with other users and can share, barter, buy or loan products and services, without having to deal with any middlemen in the entire process. Such platforms have paved the way for the new sharing economy, also known as peer to peer economy, which encourages collaborative consumption. The sharing economy is an emerging economic-technological phenomenon that is fueled by developments in digital technology, growing consumer awareness, proliferation of collaborative web communities as well as social commerce/sharing (Botsman & Rogers, 2010; Kaplan & Haenlein, 2010; Wang & Zhang, 2012). The potential sustainability benefits associated with such sharing economies are thus interesting from an organizational and environmental perspective, particularly in the context of the increasing urbanization many countries experience today.

According to MetroMile, 70% of American car owners overpay for insurance because they drive much less than rest who actually drive more. The insurance is then designed in a manner that those who drive less often end up paying the same amount as those who drive much more, creating a steady balance for the insurance companies. With digital, companies can now access data directly from the vehicles – be it the distance covered in the car or the way it was driven, using details about the fuel consumed. MetroMile has thus started distributing a Metronome device that plugs onto the dashboard of the car and capture such detailed information in real time. These details are also available to the customers on their mobile through a dedicated app which information such as miles per gallon, time spent behind the wheel and other weekly statistics. MetroMile uses this data to provide a personalized deductible and liability options along with the pay by mile alternative (Constantine, 2014).

The kind of data being available to companies is helping them to define new business models as well. With internet of things, companies are able to analyze the performance of anything and everything, which has internet enabled on it. From cars to patients, from tennis racket to sports car, all objects which transmit data on regular basis can be optimized to perform better or prevented from failure by timely intervention. These scenarios have encouraged existing companies and entrepreneurs alike, to establish business models which would have not been possible without digital technologies. As per the predictions from Gartner, by 2017, a significant and disruptive digital business will be launched that was conceived by a computer algorithm.

New Business Areas

Companies which adopted digital technologies early on, started accumulating data in large quantities, which when processed right, provides insights about costs, profits, operations, supply chain and even the end customer behavior. The early adopters are leveraging data to disrupt the markets and are threatening the traditional value propositions of their respective industries. Similarly, such companies are challenging the competitors in adjacent industries as well, while they are being themselves getting threatened by the digitally agile startups.

GE resolved to expand its business into the software and analytics domain in 2011 and since then it has developed and introduced nearly 40 software products under its "Predictivity" brand. It started embedding sensors to all the machines for capturing their performance data. The data is then analyzed for providing real time information which helps in improving machine efficiency. It also helps clients to prevent downtime and enables them to effectively schedule preventive maintenance. The company has also introduced a wide range of big data products which predictive software products, a Hadoop-based big data software for ingesting and managing industrial data. GE has partnered with Amazon Web Services to share industrial data in public cloud.

GE expects the revenues from 'Predictivity' solutions to reach $5 billion by 2017.

Companies are redefining their value propositions and unique selling points, based on the new data. For example, an automobile company which records detailed data of the connected car for providing efficient customer service will know the driving habit and risk attached with the driver more than anyone else. Such an automobile company is thus, in a much better position to offer personalized and robust insurance offerings than traditional insurance companies. Similarly, social media companies, with magnitude of data about the customers, is much better equipped to provide feedback about the newly launched product or service than a dedicated market research company. Digital companies which not only capture such diversified data, but also have the processing power, are challenging the traditional players in horizontal industries.

Monetization of Data

For data rich digital companies, data is the new product, which they either directly sell to other organizations or process it and sell it as a service. As data reflects the customer preferences, patterns and behaviors, a company from one industry can benefit from using data of another company from a different industry altogether. This kind of data actually allows companies to understand customers from a more holistic point of view. For example, data from telecom companies is helpful for retailers to deliver real time location based marketing. Similarly, data collected from automobile companies can help insurance companies bundle their packages in a more personalized manner where every customer pays according to his driving habits.

Vodafone Netherland wanted to monetize the mobile data it had of its customers. However, Dutch Telecom law restricts telecom companies from using customer-generated personal data for non-revenue assurance purposes, since the personal data is also stored in non-anonymized form within the same company. So Vodafone partnered with Mezuro and started supplying crowd analytics to the public sector. Mobile phone data is sold by Vodafone in anonymized and aggregated form to Mezuro. Mezuro further accesses data from companies specializing in camera-based person counting systems (ViNotion) in order to provide high-level estimates of crowd density, mobility and footfall. With the combined network and video-count data, Mezuro is able answer questions related to city marketing, event safety and traffic density. Because Vodafone provides network data around the clock, Mezuro can offer its customers trend analyses, such as the effects of weather or changing shopping hours on footfall in a city district. Thus, mobile network data is used for crowd analytics without endangering the privacy of Vodafone's own customers, while it earns a share of Mezuro's revenues billed by processing the data further for the end clients.

FUTURE TRENDS

Digital technologies have the capability to revolutionize the way industries have been operating, but companies are finding it challenging to leverage it completely. While some companies have successfully implemented digital for driving down costs, improve efficiency, very few have used it to create new revenue streams and business models. Companies will need a deep shift, more than technological; it has to be strategically and culturally driven for being digital in truest sense. Since 2000, 52% of the companies in the Fortune 500 have either gone bankrupt, been acquired, ceased to exist, or dropped out of the Fortune 500. With data being the new oil, or gold, as some thought leaders call it to be, companies with such data will be the new leaders across different industries. Companies which are aware and accept the new data based economy will adapt and continue to flourish.

According to Forbes, by 2018, 35% of IT resources will be spent to support the creation of new digital revenue streams and by 2020 almost 50% of IT budgets will be tied to digital transformation initiatives. Digital will also transform the way companies chose their IT partners to implement strategies in action. Bank of America, for example, is working to leverage Facebook's Open Compute project, which applies the principles of open source software to hardware, and is expected to have a major tech overhaul by 2018. Standard IT organizations are expected to spend 30% of its budget on risk, security and compliance by 2017 while allocating 10% of IT staff to these functions.

Companies are working on innovation centres where they can attempt and fail faster to in order to understand the impact of emerging technologies on different aspects of the business. These centres, working as startups within the organization, are working closely with entrepreneurs and venture capitalists to bank upon the next wave of digital technologies. Another example can be of Bitcoin, a type of digital currency in which encryption techniques are used to regulate the generation of units of currency and verify the transfer of funds, operating independently of a central bank, which is now being accepted for transactions by larger corporations such as Target and Amazon. Companies, such as Tesla, are also a sneak view of how digital transformation could be manifested in hardware, on the roads. Not only the cars from Tesla have been truly digital in nature, they have also upped the customer experience expectations, and with its frequent software upgrades, it has revolutionized the way cars would be treated in the future.

Artificial intelligence is also expected to drive the wave of digital transformation. After Apple's Siri and Microsoft's Cortana, even Facebook is leveraging artificial intelligence and has launched virtual assistant, M. This wave of digital transformation initiatives will shift focus from gathering and mining data to creating new models and algorithms that augment existing work activities and support consumers when they shop, trade, and make decisions in real time. Similarly, government agencies will also be more open to leverage such technologies either by adoption or through strategic partnerships with startups to improve the efficiency of governance. Altamonte Springs, just outside of Orlando, has adopted Uber as its public transport and partially subsidized fares, paying 20% of the total fare, starting or ending within the city. It also provides a twenty 25% waiver for rides to and from the local railway station.

However, according to research by Arthur D. Little, eight out of ten companies are lagging behind when it comes to digitalization. Most of the leading companies are being reactive, rather than proactive, with lack of knowledge and sense of urgency, being the biggest obstacles. While every industry is trying to respond to the digital disruption, automotive industry seems to be leading the pack followed by telecom and media industry. The energy and utilities industry comes a close third, but the travel & transport along with EPC & manufacturing industry appear to be the slowest in chasing up with digitalization. Overall, Accenture estimates that digital technologies in the world's top 10 economies can add a combined $1.36 trillion to their GDP in 2020 – more and more companies are digitalizing their operations.

As more companies embrace digital transformation, across different industries, the future looks exciting!

REFERENCES

Buvat, J. & Bisht, A. (2015, May 28). *Going Digital: General Electric and its Digital Transformation.* Retrieved from https://www.capgemini-consulting.com/general-electric-and-its-digital-transformation

Constine, J. (2013, November 19). *MetroMile Launches Per Mile Car Insurance And Free Driving Analytics Device*. Retrieved from https://techcrunch.com/2014/07/16/per-mile-car-insurance/

Consultancy.uk. (2015). *80% of firms face lagging digital transformation maturity*. Retrieved from http://www.consultancy.uk/news/3040/80-percent-of-firms-face-lagging-digital-transformation-maturity

Engaming. (2012). *"Gamifying" Training* [With The Cheesecake Factory]. Retrieved from https://engaming.wordpress.com/2012/06/15/gamifying-training-with-the-cheesecake-factory/

HBR. (2015). *The digital transformation of business*. Retrieved from https://hbr.org/resources/pdfs/comm/microsoft/the_digital_transformation_of_business.pdf

Hempel, J. (2015). *Facebook Launches M, Its Bold Answer to Siri and Cortana*. Retrieved from http://www.wired.com/2015/08/facebook-launches-m-new-kind-virtual-assistant/

Manintveld, B., & Schalekamp, J. (2014, November 3). *Make money by buying and selling company data*. Retrieved from http://www2.deloitte.com/nl/nl/pages/data-analytics/articles/make-money-buying-selling-company-data.html

Markovitch, S., & Willmott, P. (2014). *Accelerating the digitization of business processes*. Mckinsey.com. Retrieved from http://www.mckinsey.com/insights/business_technology/accelerating_the_digitization_of_business_processes

Motoring. (2014). *BMW rolls out ConnectedDrive Services and Apps*. Retrieved from http://www.motoring.com.au/bmw-rolls-out-connecteddrive-services-and-apps-43096/

Npr.org. (2016). *NPR*. Retrieved from http://www.npr.org/sections/alltechconsidered/2016/04/09/473422686/the-newest-public-transportation-in-town-uber

Press, G. (2015). *6 Predictions About The Future Of Digital Transformation*. Retrieved from http://www.forbes.com/sites/gilpress/2015/12/06/6-predictions-about-the-future-of-digital-transformation/2/#e64562d750c

PwC. (2015). *The Sharing Economy. Consumer Intelligence Series*. Retrieved from https://www.pwc.com/us/en/industry/entertainment-media/publications/consumer-intelligence-series/assets/pwc-cis-sharing-economy.pdf

Secure-retail.com. (2015). *Secure Retail*. Retrieved from http://www.secure-retail.com/blog/reverse_showrooming_why_the_web_is_actually_driving_traffic_into_the_store

Shastri, R., Waddiparthi, K., Nagarajan, S., Narayanan, A., & Ahamed, A. (2014, December 1). *Building a Code Halo Economy for Insurance*. Retrieved from https://www.cognizant.com/InsightsWhitepapers/building-a-code-halo-economy-for-insurance-codex1072.pdf

TCS Digital Enterprise. (2014). *Digital Reimagination ™ - The Art of Reimagining Business*. Retrieved from http://www.tcs.com/SiteCollectionDocuments/White%20Papers/Art-reimagining-business-0714-1.pdf

Thomas, R., Kass, A., & Davarzani, L. (2015). *How digital technologies are changing the way we work | Accenture Outlook*. Accenture.com. Retrieved from https://www.accenture.com/us-en/insight-outlook-how-digital-technologies-are-changing-the-way-we-work.aspx

Chapter 2

Open Collaborative Innovation:
Booster of Solutions for the Challenges of a Global World – Collaborative Processes and Innovation Networks Create Value

Maria del Pilar Ramirez Salazar
EAN University, Colombia

Rafael Ignacio Pérez-Uribe
EAN University, Colombia

ABSTRACT

Open innovation is a way that allows the incorporation of external knowledge into a company (Chesbrough, 2003). Open collaborative innovation allows to add an external point of view from different interinstitutional and transdisciplinary entities which look the common goal of provide practical solutions to business challenges that arise in a globalized world (Ramirez, 2016). There are seven components in the Open innovation collaborative model: 1) Innovation challenges, 2) Internal-external knowledge, 3) Paradigm change, 4) Leadership, 5) Interinstitutional and transdisciplinary teams, 6) Storytelling, 7) Creative solutions. It also contains six principles which become thirteen interactive variables: 1) Identity, 2) Agreements, 3) Flexibility, 4) Commitment, 5) Recognition, 6) Trust. This research has identified, among the previously mentioned variables, that in order to be more successful leading projects (taking into consideration the "millennials" generation), it is necessary to strength creativity and communication as essential factors.

DOI: 10.4018/978-1-5225-0953-0.ch002

INTRODUCTION

In order to provide entrepreneurs with strategies for the better performance in a globalized and digitalized word, this article presents an Open innovation collaborative model in which collaborative processes and innovation networks create value. This study is based on an analysis of students from the Open Collaborative Innovation Program from Colombia.

First, there is an exploration and characterization of current generations, aimed to inform businessmen and leaders about attitudes and behaviors of young people; this information may help businessmen and leader to create strategies to motivate, retain and involve young people in their entrepreneurial endeavors, getting the best from their talents. Next, the Open innovation collaborative model is presented; this methodology allows businessman and entrepreneurs to design projects to obtain practical solutions to business challenges. It is advised that entrepreneurs join universities, technological development centers, and development banking, in order to be involved with regional innovation systems. The model was statistically validated using multiple regression; the results show that communication and creativity are the main variables when it comes to leading successfully an open innovation collaborative project. By the end of the chapter, specific recommendations are presented, so leaders can propose clear and effective action plans to improve communicative abilities and to stimulate worker´s creativity.

The authors are university professors in Colombia, recognized for their work in the MMOM (model of modernization for organizational management); such model has been tested and validated in over 300 companies, and for the Open innovation collaborative model, currently being applied in a number of Colombian universities, supported by the Colombian development banking sector.

The objectives of this chapter are:

1. To demonstrate that Open Collaborative Innovation is a booster of solutions for global challenges,
2. To present a theoretical framework about the importance of Open Collaborative Innovation considering the role of "millennials" in a global world,
3. To consolidate arguments through statistical validation in order to find the most significant variables required for the performance of a good leader, and
4. To present a proposal to improve the competencies of leaders that inspire undertaking Open Collaborative Innovation projects.

THEORETICAL BRIEFING

Entrepreneurship and Innovation in Different Generations

With the economic development of this decade, entrepreneurs and innovators must be alert towards two main aspects:

1. Current technological developments (Leontief, 1983), and
2. New needs that arise in new generations because of such technological developments (McCrindle, 2014).

Since the XVII Century, it is possible to notice that the human race, by its entrepreneurial spirit, looked to create machinery to improve land´s productivity; the only objective was to obtain good harvests to feed the own family and to exchange the surplus for other products. Technology is evolving fast from its function as helper to satisfy human's basic needs to its role in creating what used to be unimaginable. Technology has gone from manufacturing to the development of transportations equipment, to building using different materials, to mega structures to facilitate transportations such as suspension bridges and subways, etc. The development of high tech for defense, security and communication starts after WWII. Then, computerized equipment, later, automatized, robotized and digitalized that are becoming the norm in the XXI Century (Leontief, 1983).

Technological developments are showing the change in the needs of different generations. In Latin America, there are 157 million millennials –around 26% of total population- born between 1979 and 1999, interacting with builders –born between 1925 and 1945-, baby boomers –born from 1946 to 1964-, X Generation –born from 1965 to 1979-, Y Generation –born from 1980 to 1995- Z Generation –born from 1995 to 2009- and soon, the Alfa Generation –born from 2010 to 2025 (Valles, 2014, page 43).

People belonging to each generation has lived different events that has shaped the way they see the world:

1. **Builders (1925-1945):** They went through world wars and therefore they created air, land and sea transportation means, their needs were based on their countries' survival
2. **Baby Boomers (1946-1964):** They saw man on the moon, therefore, they started the search for high technology, the way to find other planets, they focused on investing on scientific research in fields like health and public security.
3. **X Generation (1965-1979):** They lived the fall of the Berlin Wall, and they started changes opening economies; their needs arose from the competition based on products quality in order to position them in new markets.
4. **Y Generation (1980-1994):** They lived 9/11, finding themselves in a world of radicalism and fundamentalism; their needs focus on international mobility and enjoyment of video games.
5. **Z Generation (1995-2009):** They lived the Iraq War and the digitalization of technology; they are global interactors and their needs are focused on consumerism of short life-cycle products.
6. **Alfa Generation (2010-2025):** They live in the time of virtual social networks and wireless tablets; their needs will be focused on finding virtual solutions to their problems; environmental-friendly way of live becomes a must, they are people who wish to live in different parts of the world, so therefore, they are multilingual and multicultural.

The 21st Century started when people of Y Generation (1980-1995) became adults, then, they are beginning to be known as "*Millennials*", nickname that is still used for the Z and Alpha generations (Sterbenz. 2015), young people and kids today use technology intensively in their daily lives. According to Deloitte (2014), in 2025 this *generation* will account for 75% of the world´s labor force. That is why people who belong to Y, Z and Alpha Generations must be a target market of future entrepreneurs.

Characteristics of "Millennials"

Innovative Attitude

They look for challenging experiences at their work places; 78% of them say when choosing a place to work they rather work at a Company that invest in innovation (Ramírez, 2016b). To stimulate entrepreneurship becomes a must in order to motivate this generation; companies must adapt themselves or design innovation platforms that allow collaborators to work on their ideas in order to propose solutions to companies' challenges; becoming part of an innovation system is stimulating for all (Zahra & Nambisan, 2011). To retain young people from this generation, companies must pay good salaries and keep constant processes of innovation; it is dangerous when a company loses focus and fails to motivate employees (stops paying good salaries and does not provides challenging activities), because other companies can hire those unmotivated employees (Linden, et al, 2009).

Collaborative Work

They enjoy collaborative work; a motivation for them is the impact their work brings to the companies for which they work. They consider their bosses as their equals, therefore their communication is open and transversal; they like flexible working hours. Their sense of belonging to a company depends on their enjoyment and personal interest towards the company. A difference between *"Boomers"* and *"millennials"* is that the formers become *"workaholics"* and pretend to work at their organizations until retirement, while the latter just pretend to work for short time in each organization (Myers, & Sadaghiani, 2010).

Digital "Appdicts"

Being connected is the base of their daily lifes; 78% of Latin American *"Millennials"* own a cell phone (10% more than 2013), 37% own a tablet, 70% own a laptop and 57% own a desktop. Almost 50% of them say they cannot spend a day without their Smartphone" (Telefónica, 2014). They rather using Internet at any device instead of watching T.V., therefore, marketing and advertisement companies must develop new ways to call their attention.

People, Interactive Networks

Social networks are part of their lives, with 88% of Latin American *"Millennials"* having profiles on social networks. They are visible through the use of these platforms; they used them to plan their parties, events, to share their opinions about their experiences, etc.

Before going to a new restaurant or place they search for information in social networks, paying attention to other people's comments. To contact a company, they prefer to use social media than *"contact center"*. (Telefónica, 2014b).

Experientials

To make decisions, they prefer experiences rather than personal service or a written manual; 20% of companies say that customer's experience will be the great opportunity, "businesses are not only goods and services anymore, they are also relations, and purchase experiences are almost as relevant as products themselves" (Econsultancy. *2014*)[1].

Multitask Behavior

They enjoy performing multiple activities at the same time, so, they use many media, channels and devices when working. They are connected about seven (7) hours per day. Therefore, this characteristic must be taken into consideration in order to provide them with activities and challenges that keep them active, proactive and working on fulfilling the companies´ objectives. New communication channels must be searched in order to motivate them and commit them as much as possible with the company. (Millward Brown, 2014)[2].

Nowadays, entrepreneurs must know that, in order to have committed workers, they need to understand the generations they belong to, and motivate them accordingly. Workers in the past (Builders, Boomers, and X Generation) valued ethics, banking account, focus on tasks, independence, formal structures, autonomy and loyalty. Now, Y, Z and Alpha Generations prefer a balance between life and work, team participation, enjoyment, support, flexibility, creativity, innovation, global, multitask, variety, and life-long learning (Mac-Crindle,2014b).

According to Barry Salzberg, CEO at Deloitte Global, (2015) in the Millennials Survey "The message is clear: when facing their career goals, Millennials place the same importance on how companies develop their human resources and contribute to the society as well as on how they deal with their products and their profitability. Otherwise, there is the risk of losing them and stay behind." This is a warning for all entrepreneurs: they must understand millennials or there is the risk of losing them.

More and more, new generations look for the internationalization and globalization of their companies. Entrepreneurs want to be everywhere regardless of race, religion or culture; they feel they are part of a planet not just of a single place. That is why it is important, at their workplaces, to let them explore links and relations, which with no doubt will bring good results for the company (Knight, 1996).

Table 1. Characteristics of generations

Item	Boomers, 1946-1964	X Generation, 1965-1979	Millennials, Y, Z and Alfa Generations, 1980-2025
Motivation to undertake projects	Money	Life plan	Creativity
Training style	Formal- Monologue	Programmed- Dialogue	Interactive. Multimedia
Leadership style	Control- think	Coordination-do	Consensus- feel
Influence- values	Local- long term needs	Regional. Mid term needs	Global. Short term wants
Technology types	Color TV- audio casetes	Walkman-VCR- IBM PC	Internet, Xbox, Plasma, Wireless technology, mp3, tablets

Source: Authors, taken from Mc-Crindle, 2014.

OPEN INNOVATION COLLABORATIVE MODEL AND ITS RELATION WITH NEW GENERATIONS OF DIGITAL ENTREPRENEURS

The Open innovation collaborative model is a methodology developed to contribute, through collaboration processes and innovation networks, to the creation of value in companies (Ramírez, 2016c).

The model pretends to show the way for companies to present their innovation challenges; it also pretends that the components and principles derived from it will help companies to undertake programs to boost competitiveness, differentiation, increase in market share and therefore the sustainability of companies.

"Millennials" are the fittest to participate in such collaborative projects, because they: like to participate in innovation programs, are creative, think they are helping the planet with their knowledge, perform multiple tasks at the same time, use social networks to communicate, use mobile devices, among others.

New "millennials" and the Open innovation collaborative model fit well one each other, because of the flexibility and the values that add experience and affective feelings to those who participate in an Open innovation collaborative project. The model is described below.

Description of the Model

The model is composed by three dimensions, seven components and six principles. The central figure of the model is a flask, like a metaphor in a business lab; the open triangles frame the philosophy of the value triad; there are two virtue circles. One of the circles shows the steps for the development of a project, the other circle corresponds to the spiritual part, the values that must exist when projects of open collective innovation are undertaken.

Model Dimensions

The model has three dimensions:

1. Collaborative processes,
2. Innovation networks, and
3. Value creation.

Collaborative processes are all the actions that participants of an open innovation project manage, and/or the knowledge with which they contribute to achieve a common project. The activities performed by the participants are complementary, their objective being the creation of value.

The development of different actions is required for each specific project; the following are generic actions that must be taken when participating in Open innovation collaborative projects:

1. To specify times and spaces for meetings;
2. Agreements to regulate arrangements among participants
3. Commitment towards the objectives of the project;
4. Gives innovation challenges;
5. To clarity benefits and recognitions for participants;
6. Planning, organization, implementation and evaluation of the project;
7. Project follow up, adjustment and feedback.

Innovation networks. It refers to the interactions manage internally in the Open innovation collaborative projects, which are visible by means of digital tools.

The actions that must be taken to formalize networks are those inherent to the relation actions that are set by:

1. Delivery of resume and summary of each participant,
2. Presentation to the group,
3. Creation of a guide for communication in networks.
4. Choose a leader that consolidates ideas.
5. Creation of guidelines for content elaboration,
6. Creation of participant´s database.
7. Meetings (face to face and online) for doubt clarification.
8. Choosing of the multimedia interaction channel.

Value creation refers to the added value that participant entities generate from collaborative participation and innovation networks.

The actions that arise from this dimension are:

1. Search and keep intellectual capital that participates in the project,
2. Evaluation of solutions proposed, in terms of pertinence, relevance and viability.
3. Recognition to participants and
4. Exposure of results in media of private business and Government interest.

Figure 1. Open innovation collaborative model
Source. Authors, taken from Ramirez, 2016.

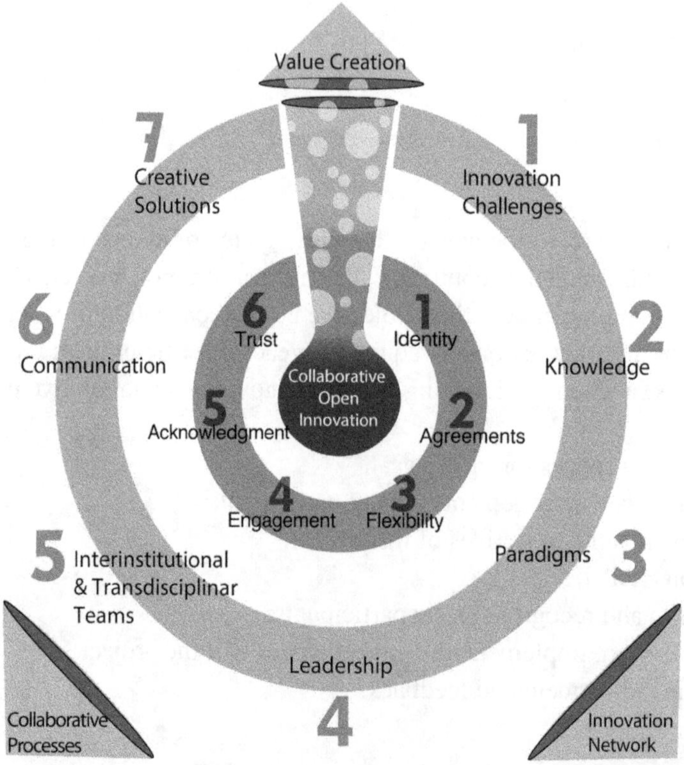

Model Components

The components of the model correspond to the moments structured in an open collaborative project; they always start with an innovation challenge and finish by finding a solution to a need, problem or opportunity posed. For this model, seven steps have been structured. Next, they are described:

- **Innovation Challenges:** An innovation challenge is a need, problem or opportunity posed at a given moment in a company (Euchner, 2014). The organization that proposes the Open innovation collaborative project is the one that stablishes the challenge or challenges for entrepreneurs and innovators.
- **Internal and External Knowledge:** Knowledge from people or entities outside the company. Different view mixed with the company´s internal knowledge (Chesbrough, 2009). People who become part of an innovation project bring their talent and knowledge, which are added to those already owned by the company. This way, there is a bidirectional knowledge transfer that adds value to both parts. It is then when expert´s networks and interactions arise.
- **Paradigms Change:** Different ways to act (Senge, Smith, Kruschwitz, Laur & Schley. 2011). Each person knows how to do things, when to work in teams; some ways of acting to standardize actions or simplify tasks must be changed. Learn to unlearn is a skill easy for millennials.
- **Leadership***:* Who directs a project stimulating people´s creativity (Zarate, 2015). To mobilize people is an art from which those who lead must learn strategies in order to have more followers. "Millennials" know well tools such as Twitter, Instagram, Facebook, etc., that allow their ideas to spread and motivate others to evaluate trends. A leader must know and use these new technologies that provide interesting spaces for motivation.
- **Interinstitutional and Transdisciplinary Teams***:* Work groups between different organizations and various fields of knowledege (Ezkowitz and Leydersdof, 2006) (Sábato and Botana, 1968). What is interesting about a project of collaboration is that it is possible to learn from others, and everybody´s actions are complemented. To know diverse cultures and to participate from the same challenge makes possible to have many new ideas for challenges solving.
- **Communication:** Written, non-verbal and verbal ways that allow the narrative expressions with other people (Euchner, 2014b). It is necessary to have clear communication practices; text messages, WhatsApp, emails, online meetings, and Dropbox work well and they are basic tools for projects in which many people work.
- **Creative Solutions:** Answer chosen by a group of experts to a company´s need or problem (Prahalad and Ramaswany, 2004a). The most interesting part of an Open innovation collaborative project is when solutions are received. Being able to find a concrete solution to a problem adds a special value to the company. It is important to have a good technical team that evaluates the appropriateness and the viability of the solution for its future implementation.

MODEL FOUNDATIONS

A model must be founded on values, beliefs and principles that makes it different from others in order to have an own identity. Values strength relationships. For this model, principles that unite working teams and consolidate long term ties in Open innovation collaborative projects were chosen.

Model's principles are the values that must be interiorized by participants so the Open innovation collaborative project can achieve its general objectives. Such values are fundamental for collaborative processes and innovation networks that arise inside the project (Ramírez, 2016d). The six principles suggested are described below:

- **Identity:** Set of own social and interinstitutional features from a person or organization (Cooke, 1996). Each person is unique and, as previously described, "millennials" are different from one each other, and some belong to the "Z Generation" others to the "Alpha Generation Alpha". Therefore, in order to take the maximum advantage from their talent, their characteristics must be well-known.
- **Trust:** Security environment (Cheikhrouhou, Pouly and Madinabeitia, 2013) (Saiz, 2009). It is the sense of calm felt in a specific environment. When there is trust, the person contributes with his/her knowledge without thinking somebody else is going to plagiarize him/her. An environment of trust is achieved by building long term relationships, and it results in people's recognition and reputation (Pérez-Uribe, 2012:115).
- **Agreements:** Consensus among parties (Euchner, 2014c). To achieve consensus among parties is an art learnt when people work in teams; this brings maturity to people and companies, consolidating professional and relational ties.
- **Flexibility:** Easiness to participate and interact in the development of the innovation project. Millennials enjoy situations without much requisites neither norms; flexibility is the speed and easiness with which they can move from one place to another, with the least number of obstacles.
- **Commitment:** Being responsible of tasks (Pérez-Uribe, 2012b:276). For a project to be successful, assigned tasks must be fulfilled according to schedule; this way, the required knowledge to design the solution is built.
- **Recognition:** To value people's work (Largacha et al, 2014). When delivering the final product, whether is the best or not, it is always necessary to stimulate the work with which each person contributes; this way, people will always be ready to keep participating in Open innovation collaborative projects, contributing generously with their knowledge.

The Open innovation collaborative model proposed has been though carefully, considering what authors have written about innovation. Even though the components are similar to the steps proposed by Davila, Epstein and Shelton (2008); by VanGundy (2007) and by Chesbough (2003b), and what other authors propose, what is new about this model is the inclusion of six principles and three dimensions that form the philosophical axis.

METHODOLOGY AND STATISTICAL ANALYSIS

Leaders, entrepreneurs and businessmen must follow all components and principles suggested in the model in order to successfully motivate and develop Open innovation collaborative programs. However, it is necessary to focus the leader strategy on the more important elements to assure success. That is why, it is decided to run a statistical validation presenting the following hypothesis:

In order to manage an Open innovation collaborative project it is essential that the Leadership will be supported by the variables of innovation challenges, external knowledge, paradigm change, leadership, interinstitutional and transdisciplinary teams, creative solutions, identity, communication, trust, agreements, flexibility and recognition. (Ramirez.2016e)

Statgraphics Centurión[3], a software that by running multiple regression allows finding the variables that are most relevant and have the highest relation with the chosen dependent variable (leadership in this case), was used for statistical validation.

Multiple regression is used with a single dependent variable, which is supposed to be related to one or more metric variables (Hair, Anderson, Tatham & Black. 2007). The method "crossed impact matrix - - Multiplying Applied to a Classification" (MICMAC) (Pérez-Uribe & Vargas. 2015) (Pérez-Uribe, Nieto, Vargas & Diaz. 2015) was used for data collection. The objective of the Structural Analysis MICMAC is to identify he main variables, influential and dependent.

Sample

The Open innovation collaborative model was applied in four Colombian universities that participated in an Open innovation collaborative program. There was a known population of 167 students, corresponding to students enrolled in the Open innovation collaborative program. Since the total enrolled population was known, the size of the sample had to be calculated for a finite population, using the following formula (Spiegel & Stephens, 2009; Pérez-Uribe, 2012c:49).

$$n = \frac{NP(1-P)}{(N-1)\left(\frac{E}{Z}\right)^2 + P(1-P)},$$

where

P = Calculated percentage of those surveyed that could fulfill the stated hypotheses (80%).
E = Margin of error stablished to make the respective inference.
Z = Standard deviation of the normal distribution with a confidence of 95%.
n = Ideal size of the sample.
N = Total population.

In this instance,

$P = 80\%$ (0,8),
$E = 5\%$ (0,05),
$Z = 1{,}96$,
$N = 167$.

$$n = \left(167\left(0,8\right)\left(0,2\right)\right) / \left(166\left(0,05/1,96\right)\right)^2 + \left(0,8\left(0,2\right)\right) = 99,737$$

When sampling is without replacement from a finite population, the correlation for finite population is required" "The finite correction factor was applied to the size of the sample, based on the universe. (Pérez-Uribe, 2012d:49)

Finite factor: n = no / 1+ f

where:

n = Sample corrected,
N = Population (167),
no = Initial sample (99,737),
f = Fraction of sampling, calculated from: no/N,

n= 99,737 / (1 + (99,737/167) =62, 44 students.

Therefore, the sample needed to validate the hypothesis is **62 students**. There were 110 students, so the number was above the statistic sample, obtaining higher confidence.

Statistical Analysis

The system provided 11 analyses, each determining the convenience of eliminating each variable. Paradigm change, innovation challenges, recognition, external knowledge, identity, commitment, agreements, flexibility, trust, and interinstitutional and transdisciplinary teams were eliminated in that order.

The variables *creative solutions and communication* had a P value lower than 0, 05, obtaining a confidence level of 95, 0%. *Therefore, it is validated that such variables are those most related to leadership.*

Ten information analysis were made. The last of such analysis, that tests the statistical validation without the independent variables (paradigm change, innovation challenges, recognition, external knowledge, identity, commitment, agreements, flexibility, trust, and interinstitutional and transdisciplinary teams) is shown in Tables 2 and 3.

Table 2. Multiple regression- leadership

Dependent Variable: Leadership
Independent Variables:
Creative Solutions
Communication
Number of observations: 110

Source: Authors using STATGRAPHICS Centurión XVI version 16.1.11.

Table 3. Variance analysis

Source	Sum of Squares	Gl	Mean Squared	F-Rate	P-Value
Model	2577,3	2	1288,65	65,55	0,0000
Residue	2103,62	107	19,66		
Total (Corr.)	4680,92	109			

Source: Authors using STATGRAPHICS Centuri6n XVI version 16.1.11.
R^2 = 55, 0596%
R^2 (adjusted for g.l.) = 54, 2196%
Standard Error of est. = 4, 43396
Mean Absolute Error = 3, 27287
Durbin-Watson Statistic = 1, 64089 (P=0, 0297)
Autocorrelation of residuals in r 1 = 0,176793

StatAdvisor

The output shows the results of adjusting a model of multiple linear regression to describe the relation between Leadership and 2 independent variables. The equation of the adjusted model is:

Leadership = 6, 06498 + 0, 365267*Creative Solutions + 0,528609*Communication

Since the P value in the ANOVA table is lower than 0, 05, there is a significant statistical relation among the variables with a confidence level of 95, 0%. The R^2 coefficient indicates that the adjusted model explains 55, 0596% of the variation in Leadership. The adjusted R^2 coefficient, which is more appropriate to compare models with different numbers of independent variables, is 54, 2196%. The standard error of the estimate shows that the standard deviation of the residuals is 4, 43396. The mean absolute error (MAE), 3, 27287 is the average value of the residuals.

The Durbin-Watson (DW) examines the residuals to determine if there is any significant correlation based on the order in which they are shown in the data files. Since the P value is lower than 0, 05 there is indication of a possible serial correlation with a confidence level of 95,0%. To determine if the model can be simplified, notice that the highest P value of the independent variables is 0, 0004 for Creative solutions. Since the P value is lower than 0, 05, this term is statistically significant with a confidence level of 95, 0%. *Therefore, probably would not want to eliminate any variable from the model.*

It is inferred from these analyses that:

Leadership = 6, 06498 + 0, 365267*Creative Solutions + 0,528609*Communication

Fifty-four percent (54%)[4] of Leadership is explained by the variables Creative solutions, and Communication. This means that, by making strengthening activities in these two variables, it will be possible to maintain in 54% an adequate level of leadership to develop collaborative processes.

SOLUTIONS AND RECOMMENDATIONS

Contributions and Proposal to Improve the Areas of Communication and Creative Solutions so the Leader Becomes Successful Directing Open Innovation Collaborative Projects for Global Challenges

Since Creative Solutions, and Communication were the two most relevant variables in the statistical analysis, it is inferred that a Leader must have communicative skills and develop strategies that boost the creativity of *"Millennials"*.

Suggestions of specific activities for leaders in areas of Creative Solutions, and Communications are shown below.

Communication of Ideas About Open Collaborative Innovation to "Millennials"

Before communicating, a leader must learn to listen; with the information that is provided to him/her, he/she will be able to make improvements where he/she is informed that things are going wrong. Active listening will offer him/her a better perspective of people, their feelings and interests. He/she will also be able to help the interlocutor to say something that was difficult to be said before. The most important matter is to promote positive relations with his/her collaborators (Puchol, Martín & Núñez. 2006).

An adequate methodology to communicate can be learnt from Glaser, (2015), who proposes the TRUST Methodology, which comes from:

- **Transparency:** Always demonstrates information based on clear data.
- **Relationships:** Learning to get long term relationships, given the faithfulness of communication and the harmony among people.
- **Understanding:** Make sure that the message sent or received in meetings, conversations or reports is understood clearly.
- **Shared Success:** To congratulate everybody for the results achieved, and finally
- **Testing Assumptions/Telling the Truth:** Always tell the truth, regardless of how difficult it is.

Leaders must move from personal power towards the development of relations, and from the uncertainty to comprehension. They must renounce to the need to be right and move towards shaping the shared success through group unity and the creation of partnerships. (Glaser, 2015)

A new challenge of communication nowadays is the social networks; they connect people and become virtual communities, useful for many objectives; leaders must promote their use in favor of collaborative processes and innovation. Social leaders must know how to make information and communication work in favor of all stakeholders of the company (Guglielmo, & Palsule, 2015).

A leader must possess good communication skills to inspire all participants to create different alternatives to the specific needs of the companies. When a leader has all the knowledge of what he/she must do, he/she starts a dialogue between he/she and all internal and external collaborators; such dialogue must be motivating, like a tale well told about the purpose and challenge of innovation that must inform to everybody, so they take active part in the process the company is going to undertake (Taylor & Novelli, 1991).

Story Telling, is definitely one of the best techniques; a leader that motivates the audience with the story told will have many followers, but telling stories requires various steps. First, the sense of belonging; second, sense of relation and then the place in the planet, for example, a mother must definitely have kids, must relate to the school, the bakery, etc., and being in a place in the world. For the story to be well told, the CASTLE Methodology is suggested. It proposes steps for the leader to follow, so he/she by using this methodology can create his/her own stories:

1. **Characterization:** To contextualize the situation.
2. **Actions:** The events that happen in the story.
3. **Structure:** Each story needs a start, a development, an end and a conclusion.
4. **Texture:** Those sensorial descriptions that make the audience feel the story with emotions.
5. **Language:** Must be appropriate as if it were happening at that moment; that makes the audience feel they are living that situation.

The leader must introduce it with an example, look for the audience's attention by staring at it, making sounds and using the appropriate voice tone, in order to guarantee the success of the story telling. He/she will notice that people are listening by the way everybody looks at him/her, smile, or show interest. A story told with a clear and simple language will always be well listened (Ramirez, 2016e).

Referring to the development of creativity to generate creative solutions to managerial challenges in Open innovation collaborative projects, a leader must know that stimulating people in innovation projects is based on the improvement of skills, knowledge visibility and recognition of the contribution delivered. A creativity process can be seen from the acronym CREATE:

1. Check goal, refers to creating the challenge to be socialized among participants.
2. Research facts, related to the research about the context of the problem, and focus on the topic to be solved.
3. Examine issues, follows the observation of all factors related with a problem
4. Acquire ideas, propose, by using different brainstorming techniques, alternatives that bring innovative actions to the challenge posed.
5. Test solutions, means test the idea with participants involved in the problem, in order to observe their reception.
6. Enable action, review the viability of implementation of the proposal (Taylor, 2015).

Boost entrepreneurs to find new solutions is an inspiring action that requires practical examples, as explained in the Blue and Purple Oceans, that is, looking for new market segments, in which products are designed tailored to publics not attended before (Kim & Mauborgne, 2005; Saiz - Álvarez & Colvin, 2015b)

For the innovation to be successful in the company, it is necessary to integrate it to the development plan and be stated as corporate strategy, specifying clear objectives for all departments of the company, so by means of the action plan and clear goals, collaborators can involve strategies that promote processes' innovation in their roles. Innovation must be in the minds of people as a principle to be followed, a shared vision. There will be resilient people, which is normal, but it is necessary to neutralize them using communication and motivation. A system to measure innovation and a clear policy of recognition are a must.

Innovation is not just creativity; it is a corporate strategy that leads to the creation of value, visibility, local, regional, and international recognition of the company (Dávila et al. 2008b; Hamel. 2006).

In collaborative processes, the leader must enrich the system by the incorporation of the academy and the industry in common projects, by leading the relation University-Industry-Government which encourages *millennials* to contribute with ideas and creative solutions that improve the surroundings, generating the consolidation of dynamic innovation systems. Interinstitutional innovations deserve a solid argument from the Government, in order to have a motivation to invest in them. A leader that fosters collaboration produces efficiency and multiplier effects (Kanter, 2012).

A good leader is the one that leads his/her followers to the achievement of the proposed objectives. The most indicated for a business person is to assume innovation as his/her own, and as something necessary for the survival and differentiation of his/her business. The leader must have a strategic plan that involves the planning, organization, execution and evaluation of the innovation process.

The leader must be accompanying the whole process, appoint coordinators that help his/her management as long as such coordinators become motivating team leaders of the process. Ethics, and the capacity to transform participants into creative and innovative people who boost their talents, are the most important characteristics of a leader (Zarate, 2015b).

The leader must promote a flexible, extroverted, critic and reflexive culture; the environment created must be sensible to the person, emphasized on trust, and respectful of the cultural identity of each person. (Saiz-Álvarez & Colvin. 2015b). Nowadays, people prefer to work in companies that focus on the person; humanistic management is a scheme and new managerial model that looks for the authenticity of the person, being a model to be considered by innovative cultures (Largacha et al, 2014b).

The leader must promote policies that favor the implementation of innovation processes that show the path to follow, the resources, mechanisms, and protection of intellectual property (Huizingh, 2010).

To guarantee success, the leader must ensure the good use of innovation networks. Therefore, it is possible to talk about networks between "Ideators and selectors", people who present and evaluate trends to guarantee the success of new products in the market. Social Networks, Internet platforms in which is convenient to make presence in order to be visible, using the right social media content manager that helps to spread and make visible the innovation of the company.

The leader must understand that managing is not the same as leading; managing is related to achieving planned goals that are trustable and fulfill the expected requirements. This is done by the operation management of the company. Managing guarantees the daily operation of the company, its stability and the coordination of hierarchies.

The leader must be visionary and inspiring; his/her most important feature is motivation; he/she must make other feel commitment towards achieving the objectives of the company. The leader makes others follow him/her to find new paths. Managing and leadership are essential to manage companies (Kotter, 2015).

FUTURE RESEARCH

Considering the importance of "millennials" in a global world, it is necessary to keep researching their behavior and social trends in order to boost entrepreneurial projects tailored to their interests.

Another research line has to deal with the way in which businessmen must modify their talents retention programs in order to inspire young people to become part of their companies.

Another research line deals with new business models that are arising due to the "millennials" creativity in using digital products.

Finally, keep researching about the techniques and tools that "millennial leadership" (the leader that serves as inspiration to "millennials") must have.

CONCLUSION

It is unquestionable that to succeed in a global world "millennials" must adapt themselves to older generations (Builders, Boomers and X generation) so together they can harmonize their actions and work toward common objectives.

New generations' entrepreneurship is, dynamic, flexible, and risky by nature, however, it must consider the attributes of business formalization of older generations. Experience is not for granted; organizational structure, strategic planning, budget management, regulations, and policies to manage human resources allow businesses to consolidate, to be trusted, and to have social recognition.

It is decisive for businessmen to adapt themselves to the technology that arises from the connectivity of the global world, as well as to join open innovation systems, where the relation between Universities, Businesses and Government, that benefits business growth, arises.

The leader of Open Collaborative Innovation projects must inspire projects, epic dreams to involve "millennial" dreamers. "Millennials" are not easy to retain. However, according to the results of this research, adequate communication of innovation challenges stimulates their participation. Also, recognizing human talent stimulates the creation of solutions.

Unquestionably, Open Collaborative Innovation is an alternative that favors financial leverage for entrepreneurs; it also helps to acquire additional specific knowledge of the economy, and most importantly, to expand social and relational capital.

REFERENCES

Cheikhrouhou, N., Pouly, M., & Madinabeitia, G. (2013). Trust categories and their impacts on information exchange processes in vertical collaborative networked organizations. *International Journal of Computer Integrated Manufacturing*, *26*(1/2), 87–100. doi:10.1080/0951192X.2012.681913

Chesbrough, H. (2009). *Open Innovation, The New Imperative for Creating and Profiting from Technology*. Harvard Business School Press.

Chesbrough, H. W. (2003). *Open Innovation. The New Imperative for Creating and Profiting from Technology*. Harvard Business School Press.

Cooke, P. (1996). The new wave of Regional Innovation Networks: Analysis, Characteristics and Strategy. *Small Business Economics*, *8*(2), 159–171. doi:10.1007/BF00394424

Davila, T., Epstein, M. y Shelton, R. (2008). *Making Innovation Work, how to Manage It, Measure It, and Profit from It*. Pearson Education.

Econsultancy. (2014). *Digital Trends Briefing. Produced by Econsultancy in partnership with Adobe, highlights the key digital trends. challenges and opportunities which marketers need to be aware of.* Retrieved from https://econsultancy.com/reports/quarterly-digital-intelligence-briefing-2014-digital-trends

Etzkowitz, H., & Leydersdorff, L. (1995). The triple helix of university-industry-government relations: A laboratory for knowledge based economic development. *East Review*, *14*(1), 11–19.

Euchner, J. (2014). Innovation Culture and Open Space Meetings. *Research Technology Management, 57*(1), 9-10.

Glaser, J. E. (2015). *Inteligencia conversacional.* Spanish. Inteligencia Conversacional. Bibliomotion Inc.

Guglielmo, F., & Palsule, S. (2015). *El líder social.* Bibliomotion, Inc.

Hair, J., Anderson, R., Tatham, R., & Black, W. (2007). Análisis Multivariante (5th ed.). Pearson Prentice Hall.

Hamel, G. (2006). The why, what, and how of management innovation. *Harvard Business Review*, (Feb): 72–84. PMID:16485806

Huizingh, E. K. R. E. (2010). Open innovation: State of the art and future perspectives. *Technovation.* doi:10.1016/j.technovation.2010.10.002

Kanter, R. M. (2012). Enriching the ecosystem. *Harvard Business Review*, *90*(3), 140–147. PMID:22299513

Kim Chan, W., & Mauborgne, R. (2005). *La Estrategia del Océano Azul: Cómo desarrollar un nuevo mercado donde la competencia no tiene importancia.* Harvard Business School Press y Grupo Editorial Norma.

Knight, G. (1996). *Born global.* Wiley International Encyclopedia of Marketing.

Kotter, J. P. (2015).*Acelerar.* Harvard Business School Publishing Corporation.

Largacha-Martinez, C., von Kimakowitz, E., Spitzeck, H., Pirson, M., Kierksmeier, C., & Amann, W. (2014).*Tendencias Gerenciales, Humanistic Management Network.* Palgrave Macmillan.

Leontief, W. (1983). Technological Advance, Economic Growth, and the Distribution of Income. *Population and Development Review*, *9*(3), 403. doi:10.2307/1973315

Linden, G., Kraemer, K. L., & Dedrick, J. (2009). Who captures value in a global innovation network? The case of Apple's iPod. *Communications of the ACM*, *52*(3), 140–144. doi:10.1145/1467247.1467280

McCrindle, M. (2014). The ABC of XYZ: Understanding the Global Generations (3rd ed.). McPherson's Printing Group.

Millwarbrown. (2014). *Adreaction.* Retrieved from http://www.millwardbrown.com/adreaction/2014/report/Millward-Brown_AdReaction-2014_Global.pdf

Myers, K. K., & Sadaghiani, K. (2010). Millennials in the workplace: A communication perspective on millennials' organizational relationships and performance. *Journal of Business and Psychology*, *25*(2), 225–238. doi:10.1007/s10869-010-9172-7 PMID:20502509

Pérez-Uribe, R., Nieto, M., Vargas, H., & Díaz, H. (2015). *MIC MAC, aplicado a los componentes del PENTEX, una aproximación a la validación de las matrices*. Vicerrectoria de Investigación. Universidad EAN. Retrieved from https://www.researchgate.net/publication/280881459_MIC_MAC_APLICADO_A_LOS_COMPONENTES_DEL_PENTEX_UNA_APROXIMACIN_A_LA_VALIDACIN_DE_LAS_MATRICES

Pérez-Uribe, R., & Vargas, H. (2015). El uso del método MicMac, para la definición de procesos de intervención en las organizaciones. Revista Ciencia y Poder Aéreo. *Escuela de Postgrados de la Fuerza Aérea Colombiana, 10*(1). Retrieved from http://www.publicacionesfac.com/index.php/cienciaypoder-aereo/article/view/156

Pérez-Uribe R. (2012). *El ambiente laboral y su incidencia en el desempeño de las organizaciones: estudio de las mejores empresas para trabajar en Colombia*. (Tesis doctoral). Universidad Nebrija. Retrieved from http://edicionesean.ean.edu.co/index.php/es/productos-de-investigacion1/tesis-doctorales/23-publicaciones/192-el-ambiente-laboral-y-su-incidencia-en-el-desempeno-de-las-organizaciones-estudio-en-las-mejores-empresas-para-trabajar-en-colombia

Prahalad, C. K., & Ramaswamy, V. (2004a). Co-Creating unique value with customers. *Strategy and Leadership*, *32*(3), 4–9. doi:10.1108/10878570410699249

Puchol, L., Martín, M., & Núñez, A. (2006). *El libro de las habilidades directivas*. España: Díaz de Santos.

Ramírez, M. P. (2016). Modelo de Innovación Abierta Colaborativa. Universidad EAN.

Sábato, J., & Botana, N. (1968). La ciencia y la tecnología en el desarrollo futuro de América Latina. *Revista de la Integración*, *1*(3), 15–36.

Saiz-Álvarez, J. M. (2009). La organización de Cooperación de Shangai (ocs) claves para la creación de un futuro líder mundial. *Revista De Economía Mundial*, (23), 307-326. (in Spanish)

Saiz-Álvarez, J. M., & Colvin, J. (2015). *Purple Oceans and the Mergers & Acquisitions Virtuous Circle. New Concepts for the Theory of Mergers & Acquisitions. In Progress in Economic Research* (vol. 29, pp. 45–58). Nova Press.

Senge, P., Smith, B., Kruschwitz, N., Laur, J., & Schley, S. (2011). The Necessary Revolution. *Necessary Revolution - Business Book Summaries, 1*.

Spiegel, M. R., & Stephens, L. (2009). *Estadística. 4ta edición*. D.F, México: Mc Graw-Hill.

Sterbenz, C. (2015). Here's who comes after Generation Z — and they'll be the most transformative age group ever. *Business Insider*. Retrieved from http://www.businessinsider.com/generation-alpha-2014-7-2

Taylor, R. (2015). Creativity at Work. Kogan Page Limited.

Taylor, S., & Novelli, L. Jr. (1991). Telling a Story About Innovation. *Issues & Observations*, *11*(1), 6–9.

Telefonica. (2014). *Telefonica survey of Millennials*. Retrieved from http://es.slideshare.net/TelefonicaEurope/telefonica-2014-gms-global-master-deck-final-100614?related=1

Valles, Y, (2014). De baby boomers a millennials, la necesaria evolución de las estructuras sociales. *Revista Tecnológico de Monterrey "Generaciones HOY"*.

VanGundy, A. (2007). *Getting to Innovation: How Asking the Right Questions Generates Great Ideas Your Company Needs*. Amacom.

Zahra, S. A., & Nambisan, S. (2011). Entrepreneurship in global innovation ecosystems. *AMS Review*, *1*(1), 4-17.

Zarate, R. (2015). *Liderazgo como factor clave en la sostenibilidad empresarial. In Gestión de la Sostenibilidad en el marco de las organizaciones*. Bogotá, Colombia: Editorial EAN.

KEY TERMS AND DEFINITIONS

Collaborative Processes: Actions that participants of an open innovation project manage, the knowledge with which they contribute to achieve a common project. The activities performed by the participants are complementary, their objective being the creation of value.

Communications: Written, non-verbal, and verbal ways that allow the narrative expressions with other people.

Innovation Networks: It refers to the interactions manage internally in the Open innovation collaborative projects, which are visible by means of digital tools.

Leadership: Who directs a project stimulating people's creativity.

Millennials: A Generation who uses technology us part of their lives. People born since 1990.

Open Innovation: A methodology developed to contribute, through collaboration processes and innovation networks, to the creation of value in companies.

ENDNOTES

1 Digital Trends Briefing (2014), produced by Econsultancy in partnership with Adobe, highlights the key digital trends, challenges and opportunities which marketers need to be aware of during 2014, covering topics ranging from customer experience and mobile to cross-channel marketing and email. The report is based on a global survey of more than 2,500 marketers and internet professionals carried out at the end of 2013. https://econsultancy.com/reports/quarterly-digital-intelligence-briefing-2014-digital-trends

2 MillwardBrown, they administered a 15-question survey, via smartphone or tablet, to more than 12,000 multiscreen users, ages 16 to 45, across 30 countries. Multiscreen users were defined as people who own, or have access to, a TV and a smartphone and/or a tablet. http://www.millward-brown.com/adreaction/2014/report/Millward-Brown_AdReaction-2014_Global.pdf

3 Statgraphics Centurión, (Versión 17.1.02 (64-bit) is a data analysis tool that allows to make complex procedures. The tool was used under Universidad EAN permission. The most noticeable advantage of this software is that it provides and interpretation of the results, helping the user to have a detailed idea of the obtained results. http://www.fce.unal.edu.co/uifce/pdf/Statgraphics%20Centurion%20XVI.pdf

4 This value (54, 22%) corresponds to the adjusted R^2 coefficient, which is more appropriate to compare models with different numbers of independent variables.

Chapter 3
The Relationship between Entrepreneurial Competencies, Competitive Intelligence, and Innovative Performance among SMEs from an Emerging Country:
Competitive Intelligence in SMEs

Ainul Abdul-Mohsin
Universiti Sains Malaysia, Malaysia

ABSTRACT

Competitive intelligence is vital to an organisation because it emphasises on understanding and measuring an organisation's competitive environment and the unstructured information that influence an organisation's strategic decision making. In fact, it also assists in creation of ideas and innovation. Yet, research on competitive intelligence practice among entrepreneurs is quite scanty. There is a need to investigate if entrepreneurial competencies and innovative performance is influenced by competitive intelligence. Thus, this study aims to analyse the relationship between entrepreneurial competencies, innovative performance and if competitive intelligence is the missing link among the entrepreneurs within the Malaysian SME context.

DOI: 10.4018/978-1-5225-0953-0.ch003

INTRODUCTION

Over the years, Malaysia has made significant progress to become a fast growing developing nation. Bolstered by a relatively stable political environment, much stride was made in the economic and social spheres. The country now prides itself with robust industrialisation to further propel its economy that was once heavily dependent on mining and agriculture. In 1991, the idea of realising a fully developed nation status, a nation that is entirely competitive, dynamic, robust and resilient by the year 2020, was mooted (Economic Planning Unit, 2001). This vision, dubbed Vision 2020, is predicated on the assumption that the country will invest in, among other things, policies and programmes to unleash the innovative power of the economy's engines of growth. This is so because innovation, and hence the emergence of an innovation-based economy, is crucial to the competitiveness, dynamism, robustness and resilience of the country (Guha, 2011). An Innovation Economy denotes an economy that emulsifies knowledge, technology, entrepreneurship and innovation in order to accelerate productivity, which is the heart of economic growth (Janeway, 2012; Nelson & Winter, 2002; Schumpeter, 1943). In order to compete with other developed countries and achieve developed country status as enshrined in Vision 2020, it is imperative that Malaysia take steps to develop a more knowledge-intensive and innovation-driven economy.

Towards this end, the government of Malaysia of late has embarked on a new economic strategy known as the Economic Transformation Program (ETP) (PEMANDU, 2010). Specifically, the ETP seeks to transform Malaysia into a high-income nation, a prerequisite of a fully developed nation. One of the most significant and vital factors of the ETP is the leadership of the private sector in building the nation's economy. In contrast, the government acts as the catalyst in policy making and high-impact funding. The plan is for the private sector to fork 92% of the nation's economic investment from 2010 to 2020, where approximately 73% would be domestic direct-investment and 27% foreign direct-investment (PEMANDU, 2010).

An interesting fact disclosed by the Department of Statistics Malaysia is that 97.3% or 645,136 of the total established businesses in Malaysia are SMEs and only 2.7% are categorised as large organisations (Jabatan Perangkaan Malaysia, 2012). As a major investing sector, the importance of the SMEs in the transformation of Malaysia has been duly recognised by the government. In the Third Industrial Master Plan 2006-2020 (IMP3), for instance, special focus was given to them by the government with their needs being addressed (MITI, 2006). The SMEs provide 57.5% of the nation's employment (National SME Development Council, 2014) despite having 97.3% of the total number of business establishments in Malaysia. In terms of Gross Domestic Product (GDP), they contribute a meagre 1/3 or 32.5% (Star Business, 2013).

In Malaysia, the definition of SME revolves around two criteria, namely the number of people employed by the entity or the amount of revenue generated. Prior to 1st of January 2014, an entity is considered a SME if its annual sales turnover does not exceed RM25 million (approximately USD6.1 million) or if the number of employees does not exceed 150 people. Due to changes in the world economic landscape, especially triggered by price inflation, structural shifts and changing business plans, the definition of SMEs has been revised accordingly. The new definition has the impact of including even more organisations into its fold. This revision, which was effective 1st January 2014, is described in Table 1.

Table 1. SME definition effective 1 Jan 2014

Classification	Manufacturing	Services and Other Sectors
Micro	Sales turnover of less than RM300,000 (USD74,000) or employees less than 5	Sales turnover less than RM300,000 (USD74,000) or employees less than 5
Small	Sales turnover from RM300,000 (USD74,000) but less than RM 15 mil (USD3.7 mil) or employees from 5 but less than 75	Sales turnover from RM300,000 (USD74,000) but less than RM 3 mil (USD736,000) or employees from 5 but less than 30
Medium	Sales turnover from RM15 mil (USD3.7 mil) but not exceeding RM50 mil (USD12.3 mil) or employees from 75 but not exceeding 200	Sales turnover from RM3 mil (USD736,000) but not exceeding RM20 mil (USD4.9 mil) or employees from 30 but not exceeding 75

Background

In recent years, the number of businesses in the SME sector has escalated in accordance to global development. A census conducted by the Department of Statistics Malaysia in 2011 established that 89.23% (591,883 establishments) of the SMEs are in the services sector, 5.98% in the manufacturing sector, 3.34% in the construction sector, 1.33% in the agricultural sector and only 0.063% in the mining sector (Jabatan Perangkaan Malaysia, 2012; National SME Development Council, 2014). Most of those in the services sector are involved in wholesaling, retailing, hospitality and food and beverages businesses.

In 2013, the performance of Malaysian SMEs was promising as indicated by many positive indicators such as increase in sales, increase in new orders, introduction of new products, business expansions within the country and even expansion to other countries (National SME Development Council, 2014). Despite this and the proliferation of SME establishments, SMEs generally are facing many challenges ranging from high cost of raw materials, shortage of skilled workers, low productivity and rising overhead costs, barrier to global sourcing, financing and credit issues, no accessibility to latest technology tools, high regulatory costs and lack of managerial capabilities (Mohd Aris, 2007; Muhammad -Mohd Zylkifli, Abdul Kamal, Mohd Rushdan, & Zakiah Hassan, 2010; National SME Development Council, 2014; Samad, 2007).

In order to address these challenges and to concurrently elevate SME competitiveness to be at par with those of developed nations, the Malaysian government has embarked on a commitment to support the SMEs via learning and development programmes, funding, matching grants, and outreach programmes. In addition, the government has stepped in to steer the economy away from being labour intensive to a more capital, knowledge and technology based that has the ability to innovate, design and develop new products and processes. This has the impact of making the SME sector more resilient (Tan, et al. 2011).

Nevertheless, despite all these assistances many SMEs still fail to sustain (Ahmad, Abdul Halim, & Mohamed-Zainal, 2010). Perhaps there are other factors that the SMEs need to consider in order to better grasp the reasons for their failure and their incapability to contribute sufficiently to the development of the Malaysian economy.

Transition into Becoming a High Income Status Nation

Like Malaysia, many other nations aspire to be a high-income nation. A study of thirteen countries that sustained a 7% yearly growth since 1947 noted that only six were able to achieve high income status.

This accomplishment was driven by a revamping of the countries' SMEs. In five of these countries, namely Hong Kong, Singapore, Japan, Taiwan and the Republic of Korea, the SMEs contribute significantly to the respective countries' GDPs. In contrast, Malaysia, which started off at the same starting point as Taiwan and the Republic of Korea, has lagged behind these two countries in the past decades. The Republic of Korea, for instance, has shown tremendous progress due to the rapid emergence of a technology, knowledge and innovation led-economy that fosters entrepreneurship and infrastructure enhancement (Shukran, Sultana, & Rahman, 2011).

Another noteworthy factor is that the Republic of Korea invests heavily in innovation capability by leveraging on its competitive intelligence (Kim & Ha, 2010). The concept of competitive intelligence (CI) is not new but has typically been neglected by many organisations, particularly SMEs. By engaging in CI, the Republic of Korea has successfully transformed itself into a high-income advanced economy and the world's 11th largest economy by 1995 (International Monetary Fund, 2014). More pertinently, 99.8% of businesses in the Republic of Korea are SMEs and they provide the population with 87.5% of employment. The Korean SMEs also contribute to more than 50% of the national GDP in 2011(Asian Development Bank, 2014; Guha, 2011).

The experiences of the other five countries have similarly shown a hike in the SME contribution to their GDPs from 2012 to 2014. For example, Hong Kong SME GDP contribution is at 58% (Chen, 2012), Singapore at 50% (Singapore Government, 2013), Japan at 55.3% (Hafeez, Shariff, & Mad Lazim, 2013) and Taiwan at 85% (Ur-Rahman, 2013). The ranking of these five countries on the Global Competitive Index Report for year 2014-2015 are as follows: Singapore was ranked second, Japan sixth, Hong Kong seventh, Taiwan fourteenth, and South Korea twenty-sixth (World Economic Forum, 2014). On the Global Innovative Index 2014, Singapore was ranked seventh, Hong Kong tenth, South Korea sixteenth and Japan twenty-one (Cornell University, INSEAD, & World Intellectual Property Organization, 2014). Malaysia was ranked twentieth on the 2014-2015 Global Competitiveness Index and thirty-third on the 2014 Global Innovative Index. From these rankings it is clear that Malaysia needs to explore new strategies to be at par with others. The rankings are consistent with the findings of Che-Ha and Mohd-Said (2012) on the lack of empirical evidences of innovation activities being carried out by SMEs in Malaysia.

It is evident that Malaysia needs a lot of catching up to do before it is able to compete with others. The SMEs are still lacking in competitiveness and innovativeness. In this era of globalisation, the survival of Malaysian SMEs depends highly on the organisations' innovativeness, creativity, technology, knowledge and entrepreneurship. The study by Bilton and Cummings (2010) indicate that highly innovative SMEs tend to grow quicker than non-innovative ones. Hence, Malaysian entrepreneurs must be bold enough to reinvent themselves and make themselves more relevant to the needs of the times or else they will face the possibility of being ousted from the market. As the experience of the Republic of Korea shows, Malaysian SMEs should prioritise their strategies to become more competitive and innovative. They should also be more vigilant of what the SMEs in developed countries are doing and be prepared to emulate them.

In addition, to the Asian experience, Malaysia can also look to the European Union (EU) for inspiration. In the EU, 98% or 20.7 million of the total business establishments are SMEs. The SMEs provide 87 million jobs and contribute to 67% of the total EU employment today. Their contribution to the GDP is more than two thirds (Ecorys, 2012). For instance, since the 1990's, France is one of the countries in the world that support its SME expansion by providing intervention programmes, such as the Competitive Intelligence (CI) programme (Larivet, 2009). This programme heightens the awareness of the SMEs on the importance of essential skills in developing CI practices and the benefits to be derived from it.

Consequently, France and Belgium made the decision to include SMEs in their economic policies and CI is part of the national agenda. This has made France and Belgium unique as compared to other countries. In contrast, most countries have yet to focus on CI as part of their strategy to enhance innovation. They only implement general plans involving scientific and technological information or knowledge management when planning their economic agenda (Bergeron, 2000).

According to Ahmad, Ramayah, Wilson, and Kummerow (2010), the studies on SMEs have typically been classified into two categories - external and internal. External studies focus on the role of governments in helping SMEs, particularly in creating the necessary environment that is conducive to their success, while internal studies are those that focus on organisational or individual variables like resource and managerial or planning competencies that influence business success. The present study looks at entrepreneurial competencies (EC) and their relationships to competitive intelligence (CI) and innovative performance (IP). It is thus an internal study.

Competitive Intelligence

Competitive Intelligence (CI) has been established in European countries as well as some Asian countries as a mechanism to enhance the innovative performance of SMEs. It is defined as an ongoing and systematic evaluation of the external environment for opportunities, threats and developments that could have an impact on the organisation and influence reactive decision-making (Strauss & du Toit, 2010). It entails the following of specific steps from the acquisition of raw information to the production of intelligence for decision makers (Roitner, 2008). CI has also evolved into a systemic application discipline (Bartes, 2010). As such, it is not only a product but also a process. CI is not just market research and information gathering. It is an ongoing process of translating data and information into intelligence by using psychological techniques and new technology. This ongoing process is to assist the decision makers to have a holistic understanding of their industry, the stakeholders, competitors' behaviour and capabilities, and finding ways to maintain or develop their competitive edge (Fuld, 2010). Various sources indicate that the generally accepted norm for the CI process includes the constructs of planning, collection, analysis, communication, and organisational awareness for decision makers in deciding action (Strauss & du Toit 2010; Saayman, 2008; Bose, 2008; Miller, 2001; Vedder et al.,1999; Gates. 1999; Kahaner, 1996). Support for this interpretation also comes from Muller (2002) observation that individuals who are knowledgeable in CI refer to CI as 'multistage process' which consists of identification of CI needs, data collection, analysis and distribution of information and feedback. Interestingly, the United States Central Intelligence Agency (CIA) CI cycle also consists of five phases which are planning, collection, analysis, dissemination, and feedback (Bernhardt, 1994; Calof & Dishman, 2002). Regardless of size, organisations normally establish systems to safeguard their internal environment by scanning the surrounding. However, they often miss to accurately assess the external environment (Herring, 1992).

The purpose of CI is to reduce environmental uncertainty and improve management decisions. It is a method by which an organisation aims to obtain strategies advantage by gathering actionable information about the factors that affect the business environment and apply the information to the planning process and decision-making in order to improve the organisation's performance (Saayman et al., 2008).

CI can assist the business in many ways, from the creation of new concepts, products, opportunities, markets, and positioning and the launching of new products, processes or services, to the generation of new ideas, the tracking of trends, mergers and acquisitions and the formulation of strategies (Wadie, 2011). A study conducted in the scientific township of Isfahan, Iran on the effect of CI on innovativeness indicates that the use of CI leads to innovation and ensures the survival of the organisation (Hussein, Farzaneh, & Amiri, 2011). This finding is corroborated by the study of small establishments in Canada that also shows a clear relationship between CI usage and innovative performance (Tanev & Bailetti, 2008). In addition to the effect on innovation, scholars of marketing have noted the importance of CI in shaping strategic decisions, the creation of new products and the building of the organisation (Jaworski, Macinnis, & Kohli, 2002).

Bisson (2010), in his article, remarked that French SMEs are increasingly dynamic in applying CI and getting connected to the globalised economy. For this reason, CI has a strong foothold in the EU. It has proven to be an important source of innovativeness and organisations are willing to invest heavily in this area (Priporas, Gatsoris, & Zacharis, 2005). Previous research on Malaysian SMEs has primarily focused on the external factor of governmental assistance and support in developing the growth of the SMEs (Saleh & Ndubisi, 2006). Too few studies were attempted to uncover and understand the internal factors that would influence the undertaking of CI by Malaysian SMEs.

Entrepreneurial Competencies

Entrepreneurial competencies (EC) refers to the "underlying characteristics such as specific knowledge, motives, traits, self-images, social roles, and skills which result in venture birth, survival, and/or growth" (Bird, 1995). For the purpose of this study, EC refers to the total ability package of an entrepreneur to perform the job role successfully, and this includes entrepreneurial, managerial and technical functional roles (Man, Lau, & Chan, 2002). EC is important for business growth, survival and success because it is capable of minimising the negative impact of business environment and is a key driver in SMEs (Ahmad, Ramayah, et al., 2010). It has been shown that successful entrepreneurs possess a high level of entrepreneurial competencies (Li, 2009). Therefore, EC is a crucial determinant in the examination of CI among Malaysian entrepreneurs.

According to Muller (2002), the success of CI is highly dependent on the availability of a set of required skills since CI is a multidisciplinary subject. Above all, Edelman, Brush, and Manolova (2002) stated that in SMEs all important and vital resources are held by the entrepreneurs and the SME performance depends on their skills, knowledge, abilities, experience and education. The entrepreneurs are the core decision makers and have the greatest influence on the organisation's roadmap formulation and the way the organisation is steered forward (Masurel et al., 2003). A preposition can be made that the result of a performance is affected by applying the appropriate knowledge into action, which is actually the competency.

It is important to examine the relationship of entrepreneur EC and CI because the SMEs' decision making process is a reflection of the entrepreneurs' competencies. An entrepreneur may personalise his approach on how the information is collected, analysed and ultimately used to improve innovation performance. This is pertinent in view of the fact that there exist very little findings on the relationship of entrepreneurial competencies and competitive intelligence to innovative performance relating to SMEs in Malaysia.

Innovativeness of Malaysian SMEs

In this era of rapid globalisation, the survival of SMEs depends highly on the organisations' innovativeness, creativity, technology, knowledge and entrepreneurship. Innovation, in turn, is the crucial building block for economic growth, development, and better employment. Rogers (2008) defined innovation as the introduction of new knowledge or the advancement or modification of existing knowledge. This is in tandem with the view of Damanpour (1996), who posited that organisational innovation can include managerial, in addition to process and product innovation. Innovation, then, is multi-dimensional. Many or most authors would agree that having distinguished and superior competencies are very important for the survival of SMEs because such competencies will spur innovation.

As mentioned before, there is a dearth of information on SME innovative activities (Che-Ha & Mohd-Said, 2012). Although the research by Sveiby (1997) suggests that knowledge transfer is critical to organisational learning and success since knowledge grows when being used and depreciates when not used, Malaysian SMEs are incapable of fully embracing the idea. In general, they do not invest enough in IT infrastructure to promote knowledge management and innovativeness. Part of the reason for this is the large financial cost involved, lack of top management commitment, poor organisational culture, and lack of policies on knowledge sharing also contribute to this malaise (Chong, Chong, & Gan, 2011).

In conclusion, as confirmed by Hussein et al. (2011), CI can lead to innovativeness and this will help to ensure the survival of the organisation. Hence, it is vital to examine the relationship between CI and innovativeness in the Malaysian SME context. Malaysian entrepreneurs must not only be vigilant in scanning the environment for information but must also be smart in intelligence generation. Generating intelligence is a stepping stone to the cultivation of innovation culture in any organisation. Thus, the survival of Malaysian SMEs will depend on how well they are able to operate in an ambiguous environment and their resiliency and ability to create their own competitive edge (Mohd-Said, 2012). Even with all these challenges, SMEs are indeed a source of dynamism and innovation in the new global economy.

Problem Statement

Malaysia is progressing into an innovation-driven economy to achieve the status of high-income nation by 2020. As highlighted in an earlier paragraph, the new transformation is piloted by the private sector (PEMANDU, 2010). Yet, in Malaysia, 97.3 percent or 645,136 of the total business establishments are categorised as micro, small and medium businesses (Jabatan Perangkaan Malaysia, 2012) . Despite the various initiatives provided by the government to escalate their productivity, SME's are not performing to the expected level and contribute to only 1/3 or 32.5% of the nation's Gross Domestic Product (GDP) (National SME Development Council, 2014; Star Business, 2013). Furthermore, Malaysian SMEs are not competitive and innovative in conducting their businesses (National SME Development, 2011). Malaysian entrepreneurs are slow in reacting to the environmental changes, robust competition and new innovation. It is self-evident that innovation is the route to economic growth as it creates new organisations. This creation of new innovations is the fundamental source of growth in business and also for their continued existence (Teece, 2010).

It is obvious that one of the ways for SMEs to become innovative is to rely on the entrepreneur's ability and creativity to innovate. Nevertheless, Malaysia's ranking as reported by the Global Innovation Index for the past five years has been stagnant. In fact, it has declined from 28 to 33 in 2014. This is exhibited in Figure 1.

Previous studies of SMEs are either external or internally focused. Of the latter kind, particularly those relating to organisational variables and individual variables, there is certainly a lack of materials. This gap fuels a strong suspicion that the Malaysian SMEs do not have the pre-requisite entrepreneurial competencies (EC) to engage in CI and consequently improve their innovation performance (IP).Thus, it

Figure 1. The Global Innovative Index from 2010 to 2014
Adopted from Cornell University et al. (2014).

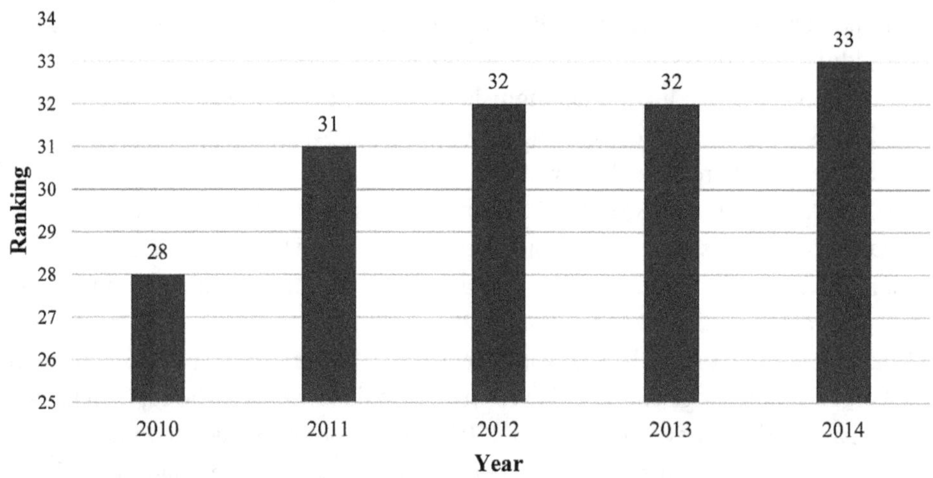

is postulated that greater success can be accomplished if the SMEs were to adopt CI since CI has proven to be effective in engendering business innovativeness and ensuring the survival of the organisation. Based on the above literature review, the following propositions are anticipated. Given the fact that entrepreneurial competencies have similarities with competitive intelligence skills, it can be anticipated that there is a positive relationship between entrepreneurial competencies and competitive intelligence. It can also be anticipated that CI has a positive relationship with innovative performance. This is because CI is also empirically proven to increase innovative performance in SMEs. According to Zaugg and Thom (2003), organisations with a greater pool of competency can convert the skills and knowledge into new or incremental products, services, technology or administrative practice. For an organisation to increase its innovation, it should have the ability to translate information into intelligence.

Hypothesis 1: There is a positive relationship between opportunity competency and competitive intelligence.
Hypothesis 2: There is a positive relationship between relationship competency and competitive intelligence.
Hypothesis 3: There is a positive relationship between conceptual competency and competitive intelligence.
Hypothesis 4: There is a positive relationship between strategic competency and competitive intelligence.
Hypothesis 5: There is a positive relationship between technical competency and competitive intelligence.
Hypothesis 6: There is a positive relationship between competitive intelligence and innovative performance.

METHODOLOGY

This study applied the quantitative methodology because most researches on these topics are conducted by positivists. The population for this study is the SMEs from various industries across Malaysia. This study employed the survey method and data is collected through mail questionnaires which are targeted at the small and medium enterprises (SME) in Malaysia. The respondents are Malaysian entrepreneurs who are also owners of SME businesses that have been in operation for three years or more. Only current active SMEs are considered to reflect the actual present-day SME environment.

In deciding the sample size, this study utilising the statistical G*Power analysis (Faul, Erdfelder, Buchner, & Lang, 2009) and the minimum sample size with a medium effect size of 0.15 is 138. The researcher has decided to collect at least 138 samples because 50 samples as suggested by Hair et al. (2010) is the minimum number that the researcher must have in order to run a multivariate analysis. The measuring tool also plays a very crucial role in empirical research. The items for the questionnaire in this study employed measurements from the literature. Bachman and Schutt (2014) strongly suggest that if there are items already formulated, tested and provided good measurement, the items are recommended to be adopted or adapted. The measure for entrepreneurial competencies are drawn from Man, Lau, and Snape (2008), Ahmad (2007) and Chandler and Jensen (1992). Competitive intelligence is measured through the questions proposed by Saayman et al. (2008). This study also applies the measurement by Johannessen, Dolva, and Kolvereid (1997) to measure innovative performance.

Prior to conducting the quantitative data collection, a preliminary interview was carried out among 10 SMEs to obtain their perception on entrepreneurial competencies, competitive intelligence and innovation performance. This adhered to Morse and Richards (2002) who argued that employing a qualitative approach is appropriate if the purpose is to learn from the participants in a setting or process the way they experience it, the meaning they put on it, and how they interpret what they experience. Data obtained from these preliminary interviews are used to refine the survey instrument based on the local context.

A pretesting is also conducted to test the content validity and relevance of the survey instrument. The survey instrument was in dual languages as many Malaysian SMEs owners are more comfortable in answering surveys in Bahasa Malaysia. The translation in this study used the extended parallel translation procedure known as collaborative translation. This method assisted in achieving equivalence in meaning and captures the intended sense of statement (Douglas & Craig, 2007; Limpanitgul & Robson, 2009; Saunders, Lewis, & Thornhill, 2007).

Another major concern on measuring different constructs with the same method is the effect of Common Methods Biases. For the past fifty years, scholars have been debating about the biases effects that the measurement methods may have on the validity of the measurements (Podsakoff, MacKenzie, & Podsakoff, 2012). Common Method Biases (CMB) is defined as variance that is caused by the measurement procedure rather than the constructs of the measures (Podsakoff, MacKenzie, Lee, & Podsakoff, 2003). As such, this study has taken pre and post remedies suggested by Podsakoff et al. (2012) to minimise CMB. The remedies taken are procedural and statistical techniques.

FINDINGS AND DISCUSSION

The model hypothesis testing was tested by means of the Structural Equation Modelling (SEM) procedure using the SmartPLS. SmartPLS was preferred over other approaches because it is able to handle constructs measured with single and multi-item measures, able to incorporate reflective and formative measurement models and most importantly it can measure complex models with many structural model relations (Hair, Hult, Ringle, & Sarstedt, 2014).

One thousand questionnaires are mailed out to the respective respondents. Only 167 replies meet the specifications of the study, giving a response rate of 16.7%. According to Dennis Jr (2003), a reply rate of 10% is common with SMEs. This finding is consistent with the arguments made by Abid, Rahim, and Scheepers (2011), Bakar and Ahmad (2010), Hilmi, Ramayah, Mustapha, and Pawanchik (2011) and Quek and Sha'ri Mohd Yusof (2003) who have conducted empirical research on Malaysian SMEs and have received a response rate between 5% to 25%. A Harman's one factor single test was conducted to test the presence of CMB. The result showed that 7 factors explained 74.19% of the variance with the first factor explaining for only 41.18% of the variance from the data collected. An assumption was made that the data was not being corrupted by CMB.

Prior to testing the model, the goodness of measure for all constructs is examined in order to test for their validity and reliability. The recommended items' loadings for reflective constructs cut-off value is 0.708 but 0.7 is considered acceptable (Hair et al., 2014). Nevertheless Hair et al. (2010) also cautions social sciences researchers to consider indicators with outer loadings between 0.40 and 0.70 instead of automatically removing them. The rule of thumb is, if deleting the items do not have any effects on the composite reliability and the construct's content validity, then the items can be retained. To measure the convergent validity, the composite reliability (CR) and average variance extracted (AVE) must be approximately 0.70 and 0.50 respectively because these are the acceptable value in many literature (Fornell & Larcker, 1987). Furthermore, Götz, Liehr-Gobbers, and Krafft (2010) argue that CR is a closer approximation of internal consistency compared to the Cronbach's alpha. Following this procedure, any loadings below the requirement are deleted giving the final AVE and CR value above the scale of 0.5 and 0.7. Table 2 demonstrates that the CR values range from 0.871 to 0.959 of which all exceeded the recommended scale. The AVE values for all the constructs range from 0.576 to 0.689 which are also higher than the recommended cut off scale. Thus, a conclusion is made that the constructs explain more than half of the variance of its indicators. Table 2 also summarise the results for the measurement model which clearly suggest that each construct shares greater variance with its own measurement items when compared with other constructs.

Table 2. Reliability of reflective constructs

Constructs	Items	Loadings	AVE	CR	Item(s) Deleted Due to Low Loadings
EC Strategic	ECS1	0.753	0.652	0.944	
	ECS2	0.736			
	ECS3	0.848			
	ECS4	0.785			
	ECS5	0.869			
	ECS6	0.788			
	ECS7	0.853			
	ECS8	0.854			
	ECS9	0.770			
EC Conceptual	ECC10	0.816	0.576	0.871	ECC 12
	ECC11	0.692			ECC13
	ECC14	0.787			
	ECC15	0.753			
	ECC16	0.741			

continued on following page

Table 2. Continued

Constructs	Items	Loadings	AVE	CR	Item(s) Deleted Due to Low Loadings
EC Opportunity	ECO17	0.840	0.680	0.895	
	ECO18	0.793			
	ECO19	0.854			
	ECO20	0.810			
EC Relationship	ECR21	0.744	0.608	0.903	
	ECR22	0.787			
	ECR23	0.826			
	ECR24	0.754			
	ECR25	0.766			
	ECR26	0.797			
EC Technical	ECT27	0.792	0.689	0.898	
	ECT28	0.813			
	ECT29	0.832			
	ECT30	0.880			
CI	CI1	0.678	0.594	0.959	CI15
	CI2	0.765			
	CI3	0.750			
	CI4	0.766			
	CI5	0.584			
	CI6	0.760			
	CI7	0.840			
	CI8	0.840			
	CI9	0.743			
	CI10	0.790			
	CI11	0.844			
	CI12	0.844			
	CI13	0.776			
	CI14	0.840			
	CI16	0.733			
	CI17	0.733			
IP	IP1	0.802	0.588	0.895	
	IP2	0.801			
	IP3	0.794			
	IP4	0.650			
	IP5	0.763			
	IP6	0.777			

Note: AVE - Average Variance Extracted, CR - Composite Reliability.

The Fornell-Larcker criterion is exercised to test the measurement model discriminant validity. Discriminant validity is the extent to which a construct is truly distinct from other constructs by empirical standard (Hair et al., 2014). It is important to establish the discriminant validity because it implies the construct is exclusive and describes observation not represented by other constructs in the model. As shown in Table 3 the square root of the AVE for each construct is higher than the correlation for each construct which indicates adequate discriminant validity for the constructs proposed in this study.

Table 3. Discriminant validity of constructs

	CI	ECC	ECO	ECR	ECS	ECT	IP
CI	**0.771**						
ECC	0.627	**0.759**					
ECO	0.560	0.762	**0.825**				
ECR	0.570	0.722	0.657	**0.780**			
ECS	0.588	0.673	0.668	0.671	**0.808**		
ECT	0.615	0.691	0.570	0.660	0.663	**0.830**	
IP	0.647	0.608	0.578	0.514	0.576	0.508	**0.766**

Note: Values in the diagonal (bolded) represent the square root of the AVE while the off diagonals are correlations.

Figure 2 and Table 4 present the structural model and the path coefficients results which indicate the causal relations among the constructs in this study model. The analysis for the path coefficients and the hypotheses are performed using the bootstrapping method. Four hypotheses are supported with t-values larger than 1.645 and significant $p < 0.01$ and $p< 0.05$. From Figure 2 the R^2 value for innovative performance is 0.418 which suggests that 41.8% of the variance in the dependent variable is explained by competitive intelligence. The competitive intelligence R^2 value is 0.482 which means 48.2% of the variance in the competitive intelligence is explained by the opportunity, relationship, conceptual, strategic and technical competencies.

From Table 4, tests involving opportunity (H1) and relationship (H2) competencies reveal non-significant relationships with competitive intelligence. This perplexing finding is in contrast to the findings by Baron and Markman (2000), Kayakutlu and Büyüközkan (2010). According to these authors, opportunity and relationship competencies are important to the performance of competitive intelligence. Hence, care needs to be exercised in making sense of the result. The first point that needs to be noted is that research on competitive intelligence has mainly been done in western countries where the focus has been mainly on large organisations rather than SMEs. Unlike the SMEs where budget is a constraint, large organisations can afford entertainment and business expense to facilitate client engagement and business networking. Secondly, due to their financial stability and track record, large organisations are always given more attention and opportunities by the government and the business community in Malaysia compared to SMEs. Moreover, large organisations have established their reputation due to their branding and marketing strategies. Thirdly, cultural differences could also be a factor. According to the Hofstede's cultural dimensions theory, a society's culture is affected by the society's values and behaviour (Hofstede, 1984). Malaysia's culture from the Hofstede's lens of the 6D Modal identifies Malaysia as high in power distance, a collectivistic society and a normative culture. This means that businesses in Malaysia are not adverse to a hierarchical set-up, employer/employee relationships are perceived in moral terms and that there is a great respect for traditions. Due to cultural influence and budget constraint, SMEs in Malaysia may less likely invest heavily in relationship building with the business community and government agencies. Furthermore, in a hierarchically ordered society, the CEOs of the business entities are expected to be the contact point for all the higher level appointments and this can create time constraint and limitation on relationship building. Another conceivable reason for the non-significant relationship maybe due to the basic concept of *malu* (shame) in the culture of the Malays in Malaysia (Abdullah, 1993). Malay respondents constituted 79.04% in this study. To the Malays *malu* is an ele-

Figure 2. Structural model

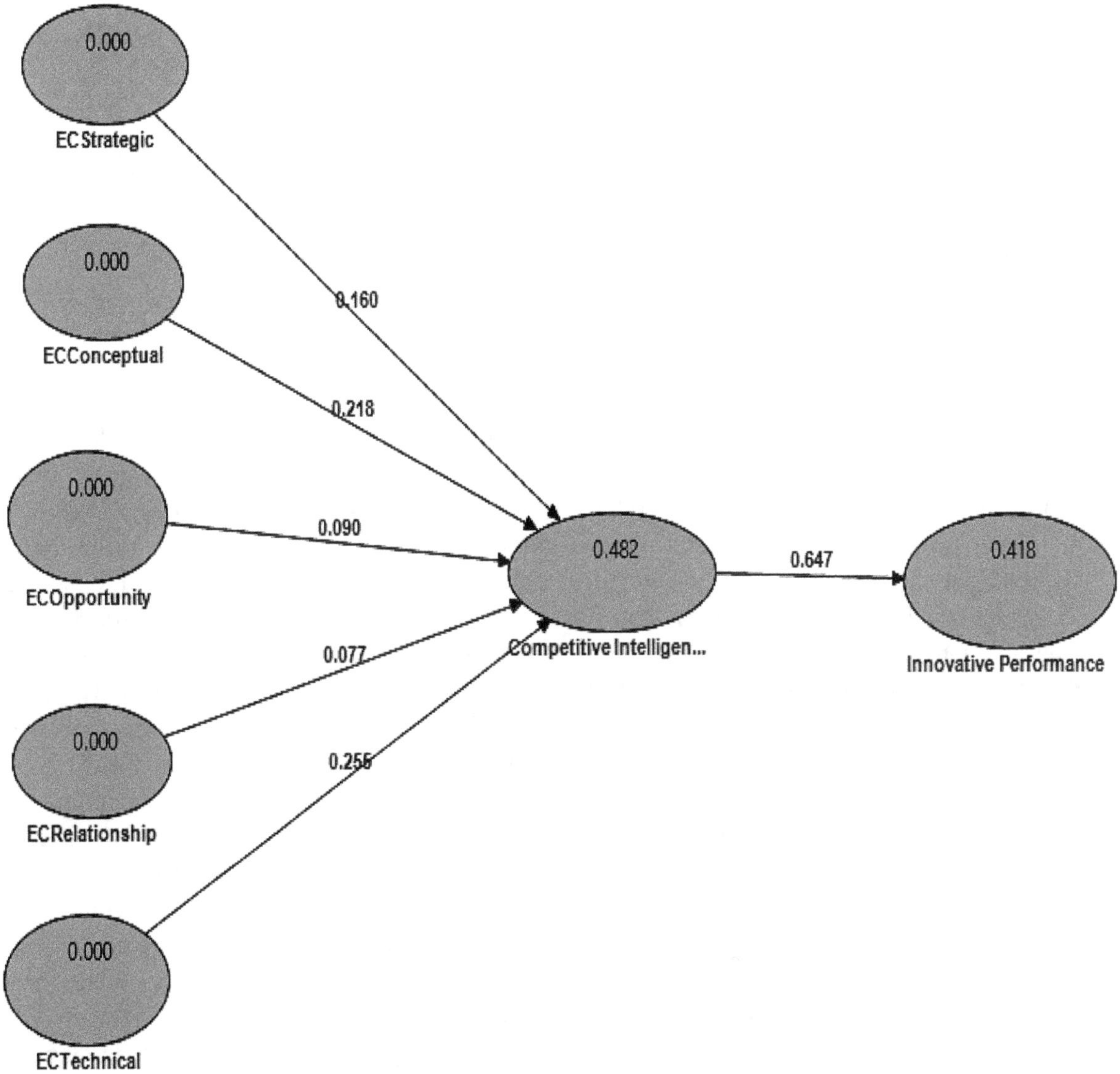

Table 4. Hypothesis testing

Hypothesis	Relationship	Std. Beta	Std. Error	t-Value	Decision
H1	ECOpportunity -> Competitive Intelligence	0.090	0.107	0.845	Not Supported
H2	ECRelationship -> Competitive Intelligence	0.077	0.092	0.839	Not Supported
H3	ECConceptual -> Competitive Intelligence	0.218	0.118	1.846**	Supported
H4	ECStrategic -> Competitive Intelligence	0.160	0.089	1.800**	Supported
H5	ECTechnical -> Competitive Intelligence	0.255	0.086	2.976***	Supported
H6	Competitive Intelligence -> Innovative Performance	0.647	0.041	15.737***	Supported

***p<0.01(2.33), **p<0.05 (1.645), *p<0.10 (1.28).

ment of basic goodness and virtue in society and in one study, the Malays were reported to score highly on self-consciousness than other ethnic groups (Mastor, Jin, & Cooper, 2000; McCrae & Terracciano, 2005). The Malays are found to be *malu* to ask for favours and opportunities. Therefore, it is likely that opportunity and relationship competencies are less developed among the Malay SMEs due to the reasons given. However, mindful of the above discussion, it is plausible to conclude that entrepreneurs with superior conceptual, strategic and technical competencies are more likely to engage in competitive intelligence which may help them in enhancing their innovative performance.

Table 4 result confirms that conceptual competency has a positive effect on competitive intelligence with ß value at 0.218, t-value equals to 1.846 and $\rho<0.05$ significance level. The result is not surprising because as explained by Adidam, Banerjee, and Shukla (2012) the effectiveness of the CI depends exclusively on the implicit knowledge and expertise aptitude in linking the event, perceptions and data into trends in the organisation's decision making process (Grant, 2010). Furthermore conceptual competency requires the abilities to diagnose issues, analyse thoughts and match new concepts with the current knowledge and capabilities (Baron & Ensley, 2006; DeTienne & Chandler, 2004). The result in Table 4 also confirms that strategic competency has a positive relationship on competitive intelligence with ß value at 0.16, t-value equals to 1.80 and $\rho<0.05$ significance level. Strategic competency also requires the ability to plan for the short and long term and to anticipate issues (Ashurst, Cragg, & Herring, 2011).

Technical competency also has a positive relationship on competitive intelligence with ß value at 0.255, t-value equals to 2.975 and $\rho<0.01$ significance level. Thus, H3, H4 and H5 are supported. These findings are in agreement with Fai and Von Tunzelmann (2001) argument that organisations start with generic technologies and switch to incorporating new technologies to develop new processes and products innovation. Therefore, organisations with specific industrial technology and technical competency have a positive relationship with competitive intelligence. This line of reasoning is consistent with Sewdass and Toit (2014) which mentioned that to practise competitive intelligence, a person must be technically and scientifically competent to differentiate between information and intelligence. For CI to be successful, the core requirement is to be able to plan, collect, interpret, distribute and give feedback (Ranjan, 2009) which Muller (2002b) had identified as analytical skills and understanding scientific methodology. This agrees with Sarasvathy, Simon, and Lave (1998) study which has shown that entrepreneurs have these specific characteristics in information processing. Thus, strategic, conceptualisation and technical competencies play highly in determining the success of CI.

Table 4 also shows more pronouncedly that competitive intelligence is positively related to innovative performance with a ß value of 0.647, t-value equals to 15.737 and $\rho<0.01$ significance level. This is an interesting finding because there are evidences that lead to the conclusion that the use of CI leads to innovativeness, creation of ideas and increase in innovative performance (Hussein et al., 2011; Tanev & Bailetti, 2008). After all, Jaworski et al. (2002) has always emphasised the importance of CI in shaping strategic decision, new products and building the organisation. The outcome of H6 lends a strong support to the reasoning by Krücken-Pereira, Debiasi, and Abreu (2001) that competitive intelligence is a tool for innovation process. Competitive intelligence is designed to foresee customers actions, competitors and government strategy (Gilad, 1996). Competitive intelligence utilises and scans the whole competitive environment to gain insight into the market condition and is proven to support an organisation's business strategy, market penetration, and development and product innovation (Petri°or & Strãin, 2013). Therefore, competitive intelligence is a tool that assists in the development of an innovative performance of an organisation.

CONCLUSION AND IMPLICATION

The purpose of this study is to examine the relationships between entrepreneurial competencies, competitive intelligence and innovative performance. This study makes several contributions to the literature on Strategic Management. The result coincides with the Resource based theory that regardless of the size of the organisations, they are able to utilise the resources (tangible, intangible, competency and capability) to achieve competitive advantage. Secondly, while previous literature concentrates on large organisations and focus on the competitive intelligence process, this study extends the pool of literature by examining the SMEs entrepreneurial competencies that lead to competitive intelligence. It seems clear that this study acknowledges that entrepreneurial competencies and competitive intelligence are positively related to innovative performance but only *specific* entrepreneurial competencies exhibit significant relationship with competitive intelligence because these *specific* competencies (conceptual, strategic and technical) require high ability of scientific understanding. These *specific* competencies also demand high ability to link the past, the current and the future occurrences with depth understanding of the likelihood of the risks and the advantages from their decision making. Finally, the most important contribution of this study is that it conformed to the RBT that says that competitive/innovative advantage can be achieved more easily by exploiting internal rather than external factors and that this can be exploited from the bundle of valuable tangible and intangible resources of an organisation.

It is a common knowledge that entrepreneurship is indeed an important factor for a country's economic success. Although many studies postulate that entrepreneurial competencies impact on innovative performance, the intervening role of competitive intelligence does accelerate this impact. Therefore, the government agencies initiatives programmes should be more biased towards enhancing competitive intelligence behaviour. These training programmes need to be developed with the full comprehension of the cognitive psychology of the Malaysian SMEs and the programmes should be related to competitive intelligence and innovativeness rather than generic scientific and technological training programmes. Extra attention should be given to factors that relate to the championing of competency, such as cultural factors and prior experiences. Training programmes must be focusing on conceptual, strategic and technical competencies. For this reason, *specific* competencies (conceptual, strategic and technical) are essential in bringing about competitive intelligence. On the whole, the empirical evidence indicates that competitive intelligence boosts innovative performance. As a result, the model stipulates that competitive intelligence is a powerful tool that bridges the relationship between *specific* entrepreneurial competencies and innovative performance among SMEs.

REFERENCES

Abdullah, M. (1993). Konsep malu dan segan orang Melayu (Self-consciousness and the shame concept among Malay people). In A. H. Othman (Ed.), Psikologi Melayu (pp. 244-294). Kuala Lumpur, Malaysia: Dewan Bahasa dan Pustaka.

Abid, A. A., Rahim, M. M., & Scheepers, H. (2011). Experienced benefits and barriers of e-business technology adoption by SME suppliers. *BIMA Publishing Communications of the IBIMA*, *2011*, 1–11. doi:10.5171/2011.791778

Adidam, P. T., Banerjee, M., & Shukla, P. (2012). Competitive intelligence and firm's performance in emerging markets: An exploratory study in India. *Journal of Business and Industrial Marketing*, *27*(3), 242–254. doi:10.1108/08858621211207252

Ahmad, N. H. (2007). *A cross-cultural study of entrepreneurial competencies and entrepreneurial success in SMEs in Australia and Malaysia* [Ph.D Unpublished doctoral thesis]. University of Adelaide, Australia.

Ahmad, N. H., Abdul Halim, H., & Mohamed-Zainal, S. R. (2010). Is entrepreneurial competency the 'silver bullet' for SME success in a developing nation? *Interdisciplinary Journal of Contemporary Research in Business*, *2*(1), 217–236.

Ahmad, N. H., & Ramayah, T. (2010). Is entrepreneurial competency and business success relationship contingent upon business environment? A study of Malaysian SMEs. *International Journal of Entrepreneurial Behaviour & Research*, *16*(3), 182–203. doi:10.1108/13552551011042780

Aris, M. N. (2007). *SMEs: Building block for economic growth.* Retrieved from: http://www.statistics.gov.my/portal/images/stories/files/journal/smes.pdf

Ashurst, C., Cragg, P., & Herring, P. (2011). The role of IT competences in gaining value from e-business: An SME case study. *International Small Business Journal*, 1–19. doi:10.1177/0266242610375703

Asian Development Bank. (2014). *Asia SME finance monitor 2013.* Mandaluyong City, Philippines: ADB.

Bachman, R., & Schutt, R. K. (2014). *The practice of research in criminology and criminal justice* (5th ed.). Thousand Oaks, CA: Sage Publication, Inc.

Bakar, L. J. A., & Ahmad, H. (2010). Assessing the relationship between firm resources and product innovation performance: A resource-based view. *Business Process Management Journal*, *16*(3), 420–435. doi:10.1108/14637151011049430

Baron, R. A., & Ensley, M. D. (2006). Opportunity recognition as the detection of meaningful patterns: Evidence from comparisons of novice and experienced entrepreneurs. *Management Science*, *52*(9), 1331–1344. doi:10.1287/mnsc.1060.0538

Baron, R. A., & Markman, G. D. (2000). Beyond social capital: How social skills can enhance entrepreneurs' success. *The Academy of Management Executive*, *14*(1), 106–116.

Bartes, F. (2010). Competitive intelligence – Tool obtaining specific basic for strategic decision making TOP management firm. *Acta Universitatis Agriculturae et Silviculturae Mendelianae Brunensis*, *LVIII*(6), 43–50. doi:10.11118/actaun201058060043

Bergeron, P. (2000). Government approaches to foster competitive intelligence practice in SMEs: A comparative study of eight governments.*Proceedings of the ASIS Annual Meeting, 37*, 301-308.

Bernhardt, D. C. (1994). 'I want it fast, factual, actionable'—tailoring competitive intelligence to executives' needs. *Long Range Planning*, *27*(1), 12–24. doi:10.1016/0024-6301(94)90003-5

Bilton, C., & Cummings, S. (2010). *Creative strategy: Reconnecting business and innovation.* Chichester, UK: John Wiley & Sons.

Bird, B. (1995). Toward a theory of entrepreneurial competency. In J. A. Katz & R. H. Brockhaus (Eds.), *Advances in entrepreneurship, firm emergence, and growth* (Vol. 2, pp. 51–72). Greenwich, CT: JAI press.

Bose, R. (2008). Competitive intelligence process and tools for intelligence analysis. *Industrial Management & Data Systems*, *108*(4), 510–528. doi:10.1108/02635570810868362

Business, S. (2013). Govt targets SMEs contributing 40% to GDP by 2015. In The Staronline [Business news]. Kuala Lumpur, MY: Star Publications (M) Bhd.

Calof, J., & Dishman, P. L. (2002). The intelligence process: Front-end to strategic planning. working paper. University of Ottawa.

Chandler, G. N., & Jensen, E. (1992). The founder's self-assessed competence and venture performance. *Journal of Business Venturing*, *7*(3), 223–236.

Che-Ha, N., & Mohd-Said, S. (2012). Innovation practices by Malaysian SMEs. In *Malaysian SMEs in the new economy* (pp. 39–55). Cengage Learning.

Chen, O. (2012). Govts support SMEs against likely economic downturn. *China Daily Asia.* Retrieved from http://www.chinadailyasia.com/business/2012-03/10/content_72478.html

Chong, C. W., Chong, S. C., & Gan, G. C. (2011). The KM processes in Malaysian SMEs: An empirical validation. *Knowledge Management Research & Practice*, *9*(2), 185–196. doi:10.1057/kmrp.2011.8

Cornell University. INSEAD, & World Intellectual Property Organization. (2014). Global innovation index 2014 rankings. In S. Dutta, B. Lanvin, & S. Wunsch-Vincent (Eds.), The global innovation index 2014 -The human factor in innovation. Geneva, Switzerland: Authors.

Damanpour, F. (1996). Organizational complexity and innovation: Developing and testing multiple contingency models. *Management Science*, *42*(5), 693–716. doi:10.1287/mnsc.42.5.693

Dennis, W. J. Jr. (2003). Raising response rates in mail surveys of small business owners: Results of an experiment. *Journal of Small Business Management*, *41*(3), 278–295. doi:10.1111/1540-627X.00082

DeTienne, D. R., & Chandler, G. N. (2004). Opportunity identification and its role in the entrepreneurial classroom: A pedagogical approach and empirical test. *Academy of Management Learning & Education*, *3*(3), 242–257. doi:10.5465/AMLE.2004.14242103

Douglas, S. P., & Craig, C. S. (2007). Collaborative and iterative translation: An alternative approach to back translation. *Journal of International Marketing*, *15*(1), 30–43. doi:10.1509/jimk.15.1.030

Economic Planning Unit. (2001). *The third outline perspective plan 2001–2010.* Putrajaya, Malaysia: Prime Minister's Department.

Ecorys. (2012). *EU SMEs in 2012: At the crossroads. Annual report on small and medium-sized enterprises in the EU, 2011/12.* Rotterdam, The Netherlands: Ecorys Nederland BV. Retrieved from http://ec.europa.eu/enterprise/policies/sme/facts-figures-analysis/performance-review/files/supporting-documents/2012/annual-report_en.pdf

Edelman, L. F., Brush, C. G., & Manolova, T. S. (2002). The impact of human and organizational resources on small firm strategy. *Journal of Small Business and Enterprise Development*, *9*(3), 236–244. doi:10.1108/14626000210438562

Fai, F., & Von Tunzelmann, N. (2001). Industry-specific competencies and converging technological systems: Evidence from patents. *Structural Change and Economic Dynamics*, *12*(2), 141–170. doi:10.1016/S0954-349X(00)00035-7

Faul, F., Erdfelder, E., Buchner, A., & Lang, A.-G. (2009). Statistical power analyses using G*Power 3.1: Tests for correlation and regression analyses.[PubMed]. *Behavior Research Methods*, *41*(4), 1149–1160. doi:10.3758/BRM.41.4.1149

Fornell, C., & Larcker, D. (1987). A second generation of multivariate analysis: Classification of methods and implications for marketing research. *Revista de Marina (Valparaiso, Chile)*, *51*, 407–450.

Fuld, L. M. (2006). *The secret language of competitive intelligence*. New York: Crown Business.

Fuld, L. M. (2010). *The secret language of competitive intelligence: How to see through and stay ahead of business disruptions distortions, rumors and smoke screens*. Indianapolis, IN: Dog Ear Publishing.

Gates, B. (1999). *Business@ the speed of thought*. London: Penguin Books.

Gilad, B. (1996). *Business blindspots*. Chicago, IL: Probus Publishing Company.

Götz, O., Liehr-Gobbers, K., & Krafft, M. (2010). Evaluation of structural equation models using the partial least squares (PLS) approach. In V. E. Vinzi, W. W. Chin, J. Henseler, & H. Wang (Eds.), *Handbook of partial least squares* (pp. 691–711). Berlin: Springer; doi:10.1007/978-3-540-32827-8_30

Grant, R. M. (2010). *Contemporary strategy analysis* (7th ed.). John Wiley & Sons Ltd.

Guha, I. B. (2011). *Literature review of "Innovative financing for SMEs"*. Retrieved from http://www.smeworld.org/story/seminary-special-supplement/innovative-financing-for-smes.php

Hafeez, Shariff, & Lazim. (2013). Does innovation and relational learning influence SME performance? An empirical evidence from Pakistan. *Asian Social Science*, *9*(15), 204.

Hair, J. F., Black, W. C., Babin, B. J., & Anderson, R. E. (2010). *Multivariate data analysis*. Upper Saddle River, NJ: Pearson Education, Inc.

Hair, J. F., Hult, G., Tomas, M., Ringle, C. M., & Sarstedt, M. (2014). *A primer on partil least squares structural equation modeling (PLS-SEM)*. Los Angeles, CA: SAGE Publications, Inc.

Herring, J. P. (1992). The role of intelligence in formulating strategy.[PubMed]. *The Journal of Business Strategy*, *13*(5), 54–60. doi:10.1108/eb039518

Hilmi, M. F., Ramayah, T., Mustapha, Y., & Pawanchik, S. (2011, 5-6 December). *Exploring intellectual capital of Malaysian small and medium entrepreneurs*. Paper presented at the IEEE Colloquium on Humanities, Science and Engineering Research (CHUSER), Penang, Malaysia. doi:10.1109/CHUSER.2011.6163808

Hofstede, G. (1984). *Culture's consequences: International differences in work-related values* (2nd ed.). Beverly Hills, CA: SAGE Publications.

Hussein, R. D., Farzaneh, G., & Amiri, F. (2011). Analyzing the impact of competitive intelligence on innovation at scientific research centers In Isfahan science and technology town. *Interdisciplinary Journal of Contemporary Research in Business, 3*(5), 940–948.

International Monetary Fund. (2014). *Economic growth rates of advanced economies.*

Jabatan Perangkaan Malaysia. (2012). *Banci Ekonomi 2011 - Profil Perusahaan Kecil dan Sederhana.* Putrajaya, Malaysia: Jabatan Perangkaan Malaysia.

Janeway, W. H. (2012). *Doing capitalism in the innovation economy: markets, speculation and the state.* Cambridge, UK: Cambridge University Press; doi:10.1017/CBO9781139381550

Jaworski, B. J., Macinnis, D. J., & Kohli, A. K. (2002). Generating competitive intelligence in organizations. *Journal of Market-Focused Management, 5*(4), 279–307. doi:10.1023/B:JMFM.0000008071.19917.36

Johannessen, J.-A., Dolva, J. O., & Kolvereid, L. (1997). Perceived innovation success in the Russian market. *International Journal of Information Management, 17*(1), 13–20. doi:10.1016/S0268-4012(96)00039-4

Kahaner, L. (1996). *Competitive intelligence.* New York: Simon and Schuster.

Kayakutlu, G., & Büyüközkan, G. (2010). Effective supply value chain based on competence success. *Supply Chain Management: An International Journal, 15*(2), 129–138. doi:10.1108/13598541011028732

Kim, Y., & Ha, S. (2010). Innovation activities and innovation performances of SMEs: The Korean electronic parts industry 1990–19951). *Asian Journal of Technology Innovation, 18*(1), 125–160. doi: 10.1080/19761597.2010.9668685

Krücken-Pereira, L., Debiasi, F., & Abreu, A. F. (2001). Technological innovation and competitive intelligence: An interactive process. *READ - Electronic Journal of Directors, 7.*

Larivet, S. (2009). *Economic intelligence in small and medium business in France: A survey* Paper presented at the 3rd European Competitve Intelligence Symposium, Stockholm, Sweden.

Li, X. (2009). *Entrepreneurial competencies as an entrepreneurial distinctive: An examination of the competency approach in defining entrepreneurs (Master of Science).* Singapore: Singapore Management University.

Limpanitgul, T., & Robson, M. (2009). *Methodological considerations in a quantitative study examining the relationship between job attitudes and citizenship behaviors.* Paper presented at the 18th EDAMBA Summer Academy, Soreze, France.

Man, T. W. Y., Lau, T., & Chan, K. F. (2002). The competitiveness of small and medium enterprises: A conceptualization with focus on entrepreneurial competencies. *Journal of Business Venturing, 17*(2), 123–142. doi:10.1016/S0883-9026(00)00058-6

Man, T. W. Y., Lau, T., & Snape, E. (2008). Entrepreneurial competencies and the performance of small and medium enterprises: An investigation through a framework of competitiveness. *Journal of Small Business and Entrepreneurship, 21*(3), 257–276. doi:10.1080/08276331.2008.10593424

Mastor, K. A., Jin, P., & Cooper, M. (2000). Malay culture and personality: A big five perspective. *The American Behavioral Scientist*, *44*(1), 95–111. doi:10.1177/00027640021956116

McCrae, R. R., & Terracciano, A. (2005). Personality profiles of cultures: Aggregate personality traits. [PubMed]. *Journal of Personality and Social Psychology*, *89*(3), 407–425. doi:10.1037/0022-3514.89.3.407

Miller, S. (2001). *Competitive intelligence: An overview*. Alexandria, VA: Society of Competitive Intelligence Professionals.

MITI. (2006). *IMP3 Third industrial master plan 2006-2020 Malaysia -Towards global competitiveness*. Kuala Lumpur: Ministry of International Trade and Industry Malaysia.

Mohd-Said. (2012). SMEs in the Malaysian economy. In *Malaysian SMEs in the new economy* (pp. 1–17). Cengage Learning.

Morse, J. M., & Richards, L. (2002). *Read me first for a user's guide to qualitative research*. Sage.

Muller, M. L. (2002). *Managing competitive intelligence*. Johannesburg, South Africa: Knowres Publishing.

Muller. (2002b). *Creating intelligence*. Johannesburg, South Africa: Knowres Publishing.

National SME Development Council. (2014). *SME annual report 2013/2014 - Transitioning for productivity-led and innivation-driven growth*. Kuala Lumpur: SMECorp.

Nelson, R. R., & Winter, S. G. (2002). Evolutionary theorizing in economics. *The Journal of Economic Perspectives*, *16*(2), 23–46. doi:10.1257/0895330027247

PEMANDU. (2010). *Economic transformation programme. A roadmap for Malaysia*. Putrajaya, Malaysia: Performance Management and Delivery Unit, Prime Minister's Department.

Petrişor, I., & Străin, N. A. (2013). Approaches on the competitive intelligence. *USV Annals of Economics & Public Administration*, *13*(1), 100–109.

Podsakoff, P. M., MacKenzie, S. B., Lee, J.-Y., & Podsakoff, N. P. (2003). Common method biases in behavioral research: A critical review of the literature and recommended remedies.[PubMed]. *The Journal of Applied Psychology*, *88*(5), 879–903. doi:10.1037/0021-9010.88.5.879

Podsakoff, P. M., MacKenzie, S. B., & Podsakoff, N. P. (2012). Sources of method bias in social science research and recommendations on how to control it.[PubMed]. *Annual Review of Psychology*, *63*(1), 539–569. doi:10.1146/annurev-psych-120710-100452

Priporas, C. V., Gatsoris, L., & Zacharis, V. (2005). Competitive intelligence activity: Evidence from Greece. *Marketing Intelligence & Planning*, *23*(7), 659–669. doi:10.1108/02634500510630195

Quek, E. E. (2003). A survey of TQM practices in the Malaysian electrical and electronic industry. *Total Quality Management and Business Excellence*, *14*(1), 63–78. doi:10.1080/14783360309708

Ranjan, J. (2009). Business intelligence: Concepts, components, techniques and benefits. *Journal of Theoretical and Applied Information Technology*, *9*(1), 60–70.

Rogers, E. (2008). *Diffusion of innovations*. New York: Free Press.

Roitner, A. (2008). Competitive intelligence in Austria: An empirical study. *(Masters of Arts Unpublished)*. Wien University.

Saayman, A., Pienaar, J., de Pelsmacker, P., Viviers, W., Cuyvers, L., Muller, M.-L., & Jegers, M. (2008). Competitive intelligence: Construct exploration, validation and equivalence. *Aslib Proceedings*, *60*(4), 383–411. doi:10.1108/00012530810888006

Saleh, A. S., & Ndubisi, N. O. (2006). An evaluation of SME development in Malaysia. *International Review of Business Research Papers*, 2(1), 1–14.

Samad, N. A. (2007). *Positioning Malaysian SMEs in the global.* Paper presented at the Proceedings of Persidangan Kebangsaan IKS 2007, Kota Kinabalu, Malaysia.

Sarasvathy, D. K., Simon, H. A., & Lave, L. (1998). Perceiving and managing business risks: Differences between entrepreneurs and bankers. *Journal of Economic Behavior & Organization*, *33*(2), 207–225. doi:10.1016/S0167-2681(97)00092-9

Saunders, M., Lewis, P., & Thornhill, A. (2007). *Research methods for business students* (4th ed.). London: Prentice Hall.

Schumpeter, J. A. (1943). *Capitalism, socialism and democracy* (6th ed.). Routledge.

Sewdass, N., & Du Toit, A. (2014). Current state of competitive intelligence in South Africa. *International Journal of Information Management*, *34*(2), 185–190. doi:10.1016/j.ijinfomgt.2013.10.006

Shaker, S., & Gembicki, M. (1999). *The war room guide to competitive intelligence.* New York: McGraw-Hill.

Shukran, M. K., Sultana, M. R., & Rahman, F. (2011). *Moving towards the knowledge-based economy: A comparative study between Malaysia and South Korea.* Paper presented at the Research and Innovation in Information Systems (ICRIIS), 2011 International Conference.

Singapore Government. (2013). *SPRING Singapore. Performance indicators.* Retrieved from http://www.spring.gov.sg/aboutus/pi/pages/performance-indicators.aspx#.Ud4_nkHI0-J

SMECorp Malaysia. (2013). *New definition of small and medium enterprises (SMEs).*

Strauss, A. C., & du Toit, A. S. A. (2010). Competitive intelligence skills needed to enhance South Africa's competitiveness. *Aslib Proceedings*, *62*(3), 302–320. doi:10.1108/00012531011046925

Sveiby, K. (1997). *The new organizational wealth: Managing and measuring knowledge-based assets.* San Francisco, CA: Berrett-Koehler Publishers.

Tanev, S., & Bailetti, T. (2008). Competitive intelligence information and innovation in small Canadian firms. *European Journal of Marketing*, *42*(7-8), 786–803. doi:10.1108/03090560810877150

Teece, D. J. (2010). Business models, business strategy and innovation. *Long Range Planning*, *43*(2), 172–194. doi:10.1016/j.lrp.2009.07.003

Ur-Rahman. (2013). SMEs can be engine of economic growth. *Pakistan's Premier Financial Daily.* Retrieved from http://www.brecorder.com/articles-a-letters/187/1181556/

Vedder, R. G., Vanecek, M. T., Guynes, C. S., & Cappel, J. J. (1999). CEO and CIO perspectives on competitive intelligence. *Communications of the ACM*, *42*(8), 108–116. doi:10.1145/310930.310982

Wadie, N. (2011). Competitive intelligence in Tunisian companies. *Journal of Enterprise Information Management*, *24*(1), 53–67. doi:10.1108/17410391111097429

World Economic Forum. (2014). *The global competitiveness report 2014 - 2015*. Geneva, Switzerland: World Economic Forum.

Wright, S., Pickton, D. W., & Callow, J. (2002). Competitive intelligence in UK firms: A typology. *Marketing Intelligence & Planning*, *20*(6), 349–360. doi:10.1108/02634500210445400

Zaugg, R., & Thom, N. (2003). Excellence through implicit competencies: Human resource management, organisational development, knowledge creation. *Journal of Change Management*, *3*(3), 199–212. doi:10.1080/714042541

Zylkifli, Kamal, & Char, Rushdan, Yasoa, & Hassan. (2010). Small and medium enterprises (SMES) competing in the global business environment: A case study of Malaysia. *International Business Research*, *3*(1), 66–75.

KEY TERMS AND DEFINITIONS

Competitive Intelligence: An ongoing, systematic evaluation of the external environment for opportunities, threats and developments that could have an impact on the organisation and influence reactive decision-making and follows specific steps from acquiring raw information to producing intelligence for decision makers in deciding action.

Entrepreneurial Competencies: The total ability package of an entrepreneur to perform the job role successfully which includes entrepreneurial, managerial responsibility, and technical functional role.

Innovation Economy: An economy that emulsifies knowledge, technology, entrepreneurship and innovation in order to accelerate productivity, which is the heart of economic growth.

Malaysia: A country that practices federal constitutional monarchy located in Southeast Asia.

SmartPLS: A statistical software with graphical user interface for variance-based structural equation modeling (SEM) using the partial least squares (PLS) method.

SMEs: Malaysian small medium enterprise is classified as micro, small or medium and depends on two criteria which are full-time employees and annual sales turnover as defined by the National SME Development Council.

Vision 2020: A blueprint for Malaysia to become a self-sufficient industrialised nation by the year 2020 encompasses economic prosperity, social well-being, educational world class, political stability, as well as psychological balance.

Chapter 4
Intercloud:
Delivering Innovative Cloud Services

Alexandru Aurel Costan
Technical University of Cluj-Napoca, Romania

Alexandru Radu
Technical University of Cluj-Napoca, Romania

Bogdan Iancu
Technical University of Cluj-Napoca, Romania

Adrian Peculea
Technical University of Cluj-Napoca, Romania

Petru Cosmin Rasa
Technical University of Cluj-Napoca, Romania

Vasile Teodor Dadarlat
Technical University of Cluj-Napoca, Romania

ABSTRACT

Cloud computing is considered the next step in the evolution of technological applications, services and products development, in the Future Internet and Internet of Things context. The permanent connection of users to the Internet has become widespread, so almost all available resources (data, software) can be placed on the Internet, shared between users - sometimes completely independently of others, and brought from the Internet to the user's computer on demand. The chapter highlights the benefits of using hybrid technology with a focus towards entrepreneurs and business innovators. Intercloud systems are built of heterogeneous services, offered by different providers of cloud computing. To emphasize the business opportunities offered by innovative Intercloud approaches, an architecture for the interconnection of cloud services offered by multiple cloud providers, with applications to e-health, is presented. The main concepts and business benefits of using Intercloud systems and an architecture for interconnecting cloud services from multiple cloud providers is described.

DOI: 10.4018/978-1-5225-0953-0.ch004

INTRODUCTION

Businesses bring value to a community. Entrepreneurs have a powerful impact on a local economy by creating innovations, employment and socio-economic benefits. Some of the key benefits entrepreneurs bring to an economy are (Kristos, 2014):

- Introducing innovations and inducing economic growth.
- Increased competition, thus improved performance and better quality of products.
- Positive employment effects in the short and long term.

Nevertheless, Kristos (Kristos, 2014) points out the negative effects on local economies:

- Negative employment effects in the medium term.
- Failure costs are sometimes paid by taxpayers (there is a substantial risk of failure associated with entrepreneurs and innovators).

Hugos and Hulitzk (2011) introduced the term of Cybernetic Economy to define the new industrial revolution - a service oriented economy, based on the Internet, mobile computing and digital media.

The Internet, with the associated applications and services, represents one of the fastest penetrating technology in the last century.

In Iancu and Iancu (2012) the vision of Future Internet is described as an all-inclusive range of computer networks combined into a global IT platform formed of seamless networks and networked smart things or objects. There are five concepts that need to be underlined: Internet of Things (IoT), Internet of People (IoP), Internet of Energy (IoE), Internet of Media (IoM), and Internet of Services (IoS).

By Internet of things (IoT) we understand an environment in which all information originating from different devices (computers, auto-vehicles, smart-phones, traffic lights and, any other thing that has a sensor) are accumulating and, applications (everything from Facebook social network to e-commerce platforms, from a fabrication system to traffic control system).

In brief, there is a need for data and a way to access them. Here, the Internet intervenes and interconnects the devices. Besides the two, there is a need for a system to work with the information, to process and analyze it. Usually, for this purpose software or automatic and, semi-automatic systems are created.

The Internet of things is referring to a wide array of devices that have incorporated short radius transmitters, used every day. Thus, a new way of communication between people and objects is created. The same concept describes a series of technologies used in Internet for the interaction between objects from the real world.

The advantages exposed by IoT technology are:

- The dynamic control of industry and day-to-day life.
- Improves the utilization ratio of the resources.
- Creates a closer relation between human and the environment.
- Flexible configuration.
- Universal transportation.

Internet of things (IoT) is an abstractization for the connection of different devices and hardware platforms to Internet using a unique IP address. Thus, an Arduino board connected to IoT offers the possibility to be controlled from distance and, his data to be processed in other development environment.

The concept IoT is based on the integration of the physical world with the virtual world, of the Internet, and thus can be defined as global infrastructure of networks that connects physical objects, virtual things and smart devices with capabilities of communication and activation via the Internet.

From the perspective of IoT numerous devices and objects will be connected to the Internet, each assuring data, information and services. Devices that provide these benefits can be objects as smartphone, tablet and digital cameras, but IoT involves more than that. Our daily environment, the house we live in, vehicles, garden devices connected via a gateway to the Internet, can become objects of the IoT (Figure 1).

IoT may be seen as a vast network consisting of network devices and computers connected through a series of intermediate technologies such as RFID, barcode, wired and wireless connections. International Telecommunication Union (ITU) IOT structured it in a 4-dimensional architecture (Tsai, Sun, & Balasooriya, 2010):

Figure 1. Internet of Things concept
Source: Vouk, 2008.

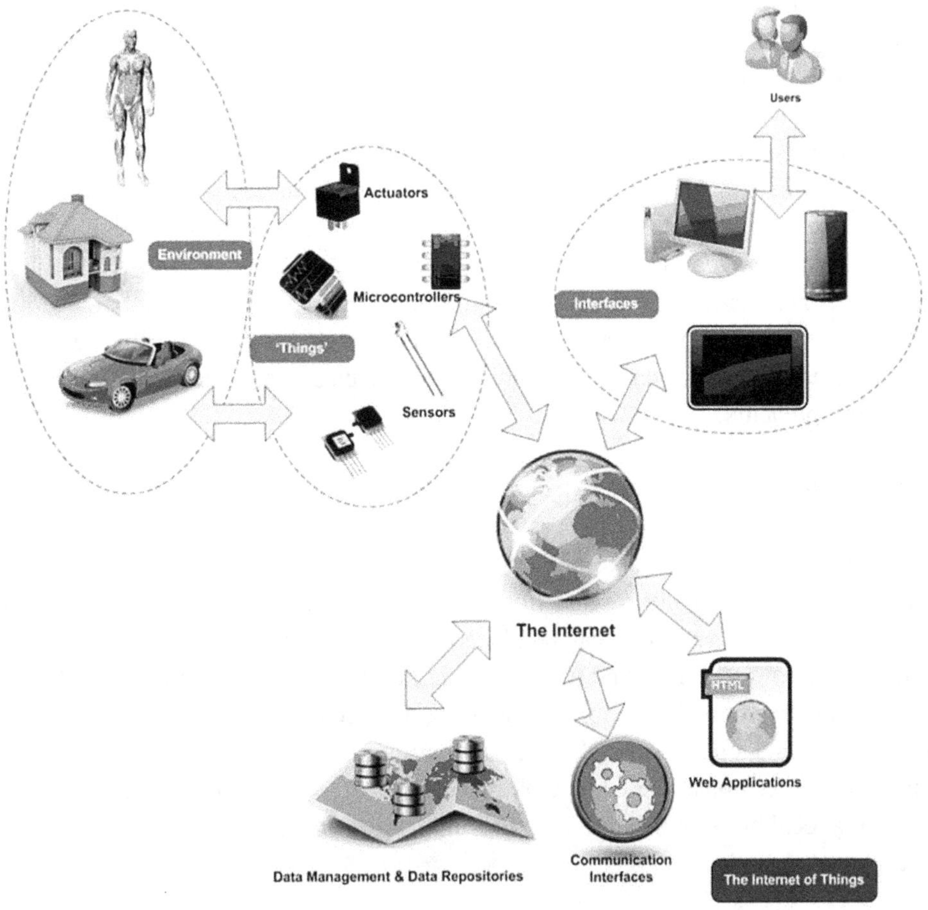

1. **Tagging Things:** Tracking and addressing objects by RFID in the IOT vision, the most important thing, because it allows the automatic identification of objects.
2. **Feeling Things:** Sensors act as devices that collect environmental data thus the real world is linked with the virtual world based on these data using communication protocols.
3. **Shrinking Things:** Minimizing devices and the use of nanotechnology facilitated the possibility of really small things connected communicating their status e.g. water quality monitoring.
4. **Thinking Things:** By incorporating intelligence into devices and sensors, an intelligent control network can be formed through which devices can communicate their condition via Internet.

Cloud computing refers to renting hardware and software virtual resources, over the Internet. Through this service, clients will not acquire physical servers to install different software applications; rather they will receive processing capabilities and virtual storage that can be access online. Cloud computing is defined as a private data center. The cloud can be hosted by any company, manufacturer or service organization. In simple terms, the cloud means storing and accessing data and applications over the Internet instead of accessing it through personal hard drives.

The Intercloud scenario is based on the key concept that any cloud by itself does not have infinite natural resources or stable geographic location (Bernstein, 2009). Interconnecting different cloud providers aims at offering enhanced services for business owners. A distributed architecture for interconnecting existing cloud services, from different cloud providers (e.g. Amazon, AT&T, Cisco), will provide a scalable and flexible infrastructure. Creating networks between cloud systems have many advantages: the user has a higher number of services available than if using a single cloud; the number of services and options available for storage is higher; desktop applications can use the service and store information or data belonging to any cloud with the help of a single service.

The chapter will be structured as follows. The first part consists of a theoretical study of the concepts related to the Internet of things (IoT), hybrid cloud services communication, and the business opportunities brought by the Intercloud scenario. The second part of the chapter will exemplify the design of a prototype solution for interconnecting and using Intercloud services. To highlight the innovative use of cloud and Intercloud services, an application for e-health data acquisition and device monitoring is described. Nowadays, business managers need to have ITC related technical skills, besides managerial skills. Thus, the focus of the chapter is on technical and business issues, pointing out the benefits and drawbacks for entrepreneurs.

Cloud Computing

Cloud computing is considered the next step in the evolution of technological applications, services and products development. The permanent connection of users to the Internet has become widespread, so almost all available resources (data, software) can be placed on the Internet, shared between users- sometimes completely independently of others, and brought from the Internet to the user's computer on demand.

Cloud computing has a highly flexible service-oriented architecture, a reduced set of technologies to be installed on the client's system and. Cloud computing is based on virtualized resources. Typical cloud computing providers make available standard commercial applications; the user has access to them only by means of a local browser because both the application and the user's own data are hosted in the cloud, on the service provider's server.

The definition of cloud computing is based on the five attributes of each cloud service (Höfer & Karagiannis, 2011):

- **Sharing Resources:** Unlike previous providers how were providing dedicated resources; cloud computing is based on a business model in which the same resource is shared by multiple users.
- **High Scalability:** Resource needs may change depending on user requirements.
- **Elasticity:** Users can quickly increase or decrease the amount of resources allocated to their tasks. This is particularly important because other users can then use the freed resources.
- **Pay as You Go:** Cloud users pay only for the resources they use and only for the actual time of use.
- **Self-Provisioning of Resources:** Users are able to manage existing resources and to add new resources, connect with external servers, etc.

IT companies are hoping to attract their clients in their cloud-type solutions. However, the obscure policy of prices and diverse errors are pushing their clients away. In this moment, very few services are offered without the additional option of cloud-storing. Be it pictures, videos, MP3s or, even entire sites, the cloud technology is considered the most important of all modern storing solutions. Buying software packages from shops is a legacy for big companies like Microsoft or Adobe, forcing the users to use the online services.

In June 2011, a study realized by Version One (Version One, 2011) demonstrated that 91% of the programming and IT experts do not know what Cloud Computing represents and two thirds of the financial experts have a clear vision of the concept, a fact that indicates the nature of this technology (Cauveri). In September 2011, a study realized by the Aberdeen Group concluded that disciplined companies have realized an average increase of 68% in IT costs due to using the services provided by Cloud Computing and, a decrease of only 10% of the costs for energy of the data center (Gibbs, 2012). Chris Capossela, Senior Vice-President of Microsoft mentioned that Cloud Computing help companies to reduce their costs of infrastructure with up to 50%. The percentages presented above do not appear to bring a substantial financial and resources economy, but reported to companies with giant data centers and that offer a multitude of services, the economies are impressive. In 2015 there were an estimated 18 million cloud computing jobs globally, a number that is forecast to increase in the next years (Jones, 2016).

Rossa and Blumenstein (2015) describe Cloud computing technologies as enablers for SMEs, by providing worldwide access, connectivity and collaboration between SMEs and by lowering IT&C costs. Nonetheless, the global market of Cloud technologies can represent a potential threat for SMEs. SMEs have to adapt even faster to an ever changing environment, due to global based competitors, added on top of local competition, thus creating a massive pressure on the companies. In this context Cloud-based business models and IT solution need to be adopted.

Cloud Architectural Layers

Cloud Computing can be viewed as a collection of services that can be presented similar to a stratified architecture of Cloud Computing, as in the Figure 2.

Figure 2. Stratified architecture of Cloud Computing

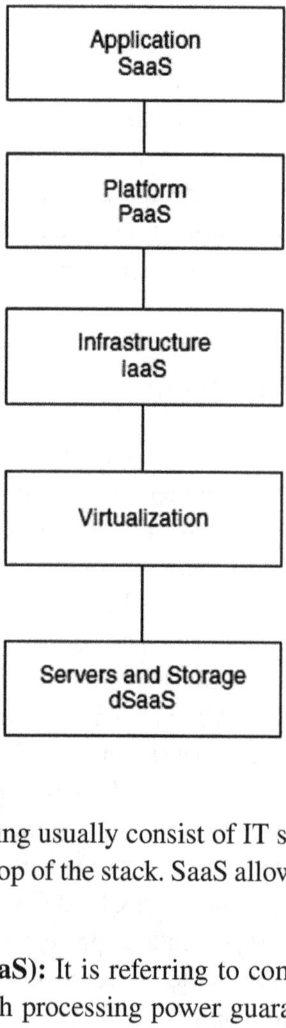

Services offered by Cloud Computing usually consist of IT services, mentioned as SaaS (*Software-as-a-Service*), which is situated at the top of the stack. SaaS allow the client to use the applications from distance from cloud:

- **Infrastructure-as-a-Service (IaaS):** It is referring to computer resources as a service. This includes virtualized computers with processing power guaranteed and broadband used for storing and accessing the internet.
- **Platform-as-a-Service (PaaS):** It is similar to IaaS but includes even operating systems and services needed for a certain application. In other words, PaaS is IaaS with a specialized stack for a certain application. PaaS supplies a development environment (IDE – Integrated Development Environment), including data security, backup and restore, storing of applications and architecture.
- **Data-Storage-as-a-Service (dSaaS):** Offers storage space that the client uses, including broadband requirements for storing.

There are three categories of cloud services, as illustrated in Figure 3. Figure 3 presents SaaS cloud service, if the entire application runs in cloud. The client has a simple browser to access the application. A known example of SaaS cloud service is salesforce.com

In Figure 4 two types of cloud services are illustrated. The first two figures represent the case in which application is running on client; however, it is accessing the functions and useful services situated in cloud. One example of application that runs locally at the client and is accessing cloud services is the iTunes application of the Apple Company.

Figure 3. Cloud services

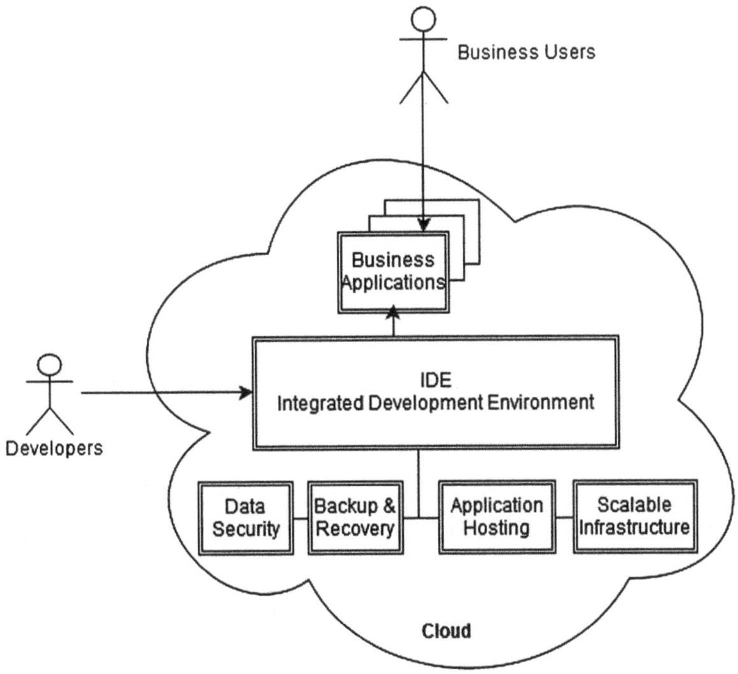

Figure 4. Cloud services types

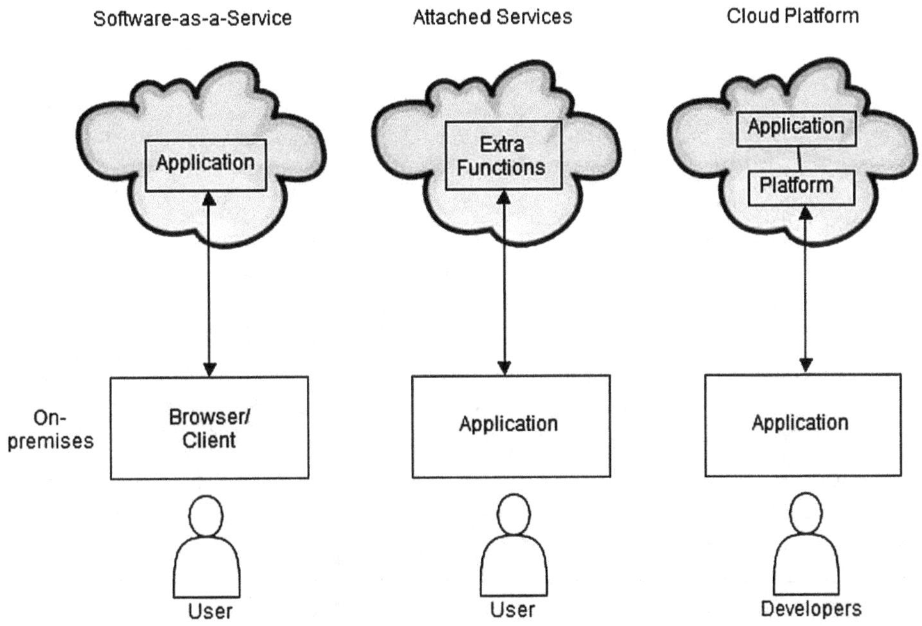

The desktop application brings the music, in which time the cloud service is used to buy new audio and video content. One example, at company level, of this type of cloud service is Microsoft Exchange HostedServices. Locally, the exchange server uses services from the cloud, including spam-filtering, archives and other functions. In the end, in the right side of the Figure 4 a cloud platform for the creation of applications is presented, which is used by programmers (developers). Application developers have created a new SaaS application using cloud platform.

Cloud Computing Types

There are four types of cloud services: public cloud, private cloud, hybrid cloud and, cloud developed for community. They are represented in Figure 5.

In the *public cloud* (external cloud) computer resources are made public on the Internet by web applications and web services, so a public cloud can be viewed as an additional service. Public clouds are used by different clients with different needs, but these applications are all stored on the same cloud server, on the same storing systems, using the same network. In the case of public cloud services, the data are stored and protected in the data centers of the most powerful cloud providers, like Microsoft or Google.

- The *private cloud* (internal cloud) is owned by an organization and it is designed to be used only by the organization.
- A *hybrid cloud* environment combines many different models of private and public clouds. Hybrid clouds are introducing the functionality of determining distribution way of application, both for public and private clouds.

Figure 5. Types of cloud

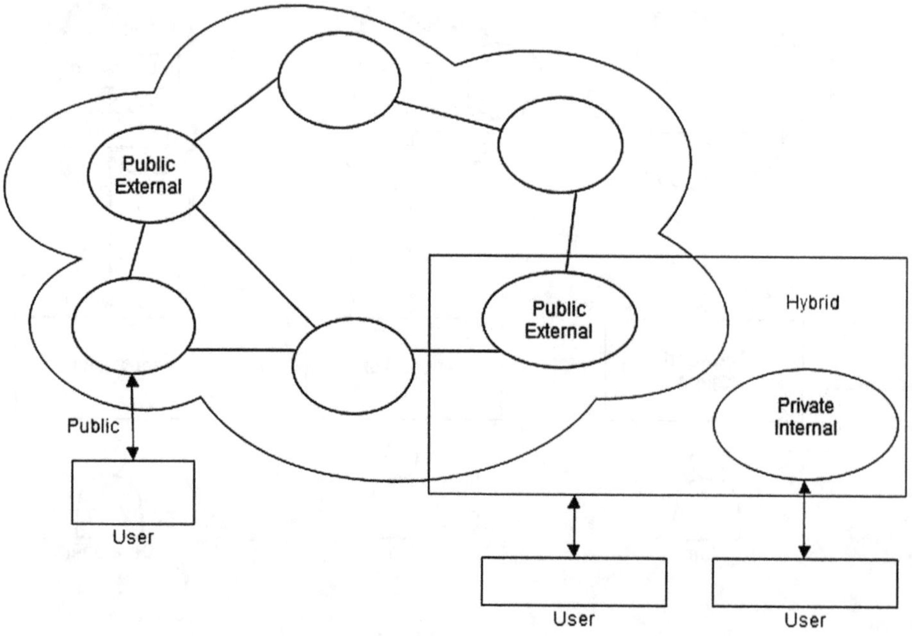

- A *community cloud* is an effort of collaboration in which the infrastructure is shared between multiple organizations of a society with the same preoccupations (security, conformity, competence), administered either internally or by a third party and stored internally or externally. The costs are distributed to a small number of users, so only a small part of the cloud savings are realized (Briscoe & Marinos, 2009).

To understand the entire functionality of the Cloud Computing system, Jangra (2011) describe the four major actors that play an important role in executing different cloud activities and functions:

1. A *client* can request services, which can be one of the three main cloud services: application as a service (Software-as-a-Service), platform as a service (Platform-as-a-Service) or infrastructure as a service (Infrastructure-as-a-service), from cloud brokers or directly from cloud providers.
2. A *cloud services provider* is an organization or an entity that is responsible for providing cloud services to clients. For providing services, every cloud provider implements the cloud system in conformity with the four implemented models: private cloud, public cloud, community cloud or hybrid cloud.
3. A *cloud auditor* can evaluate the independent value of cloud services regarding the impact on the integrity, on the performance and on the control of security.
4. A *cloud broker* manages the providing of services for the clients because the integration of the cloud services can be too complex. A cloud broker can increase the quality of the offered services by increasing their capacity and by, also integrating one or more services in other services to create new services.

To fully understand the entrepreneurial value creation in the cloud, Bhat and Shroff (Bhat&Shroff, 2014) used the business model concept – a highly used concept in IT innovations, to study this aspect. Using the V4 Business Model – value-proposition, value-network, value-architecture, and value-finance, the study explored how technology start-ups leverage cloud computing for value creation. They have found that start-ups derive value from two aspects of cloud: access to affordable IT infrastructure for product development and the ability to deliver products and services by making cloud part of their business model.

Intercloud

Many challenges have been raised by enterprise applications, challenges that can affect cloud technology development and use by organizations for outsourcing applications with sensitive content. One of these problems is Intercloud interoperability.

Inter-Operativity in Cloud Technologies and Intercloud

Today, technology represents a new and promising paradigm that is developing in fast pace in the industry. Cloud services have become appreciated in different fields, having the capacity to participate in the development of the organizations, especially in the case of the small and medium ones, that are contributing to the effort of global development in a significantly way, creating work opportunities and valuable services. There are some justifications behind the scenes of the recent use of cloud services in diverse fields, among the main advantages being the small costs of the software applications and, of the technical infrastructure; payment possibility in accordance with the use; high flexibility and measurability and, simplified requirements regarding the training of personnel.

Cloud technology has at its basis a number of well researched concepts and technologies, like distributive calculus, calculation grid or virtualization. The great innovation of cloud technology consists in the way services are delivered to clients. The National Standardization and Technology Institute (NSTI) proposed a definition of cloud technology largely accepted. According to this, cloud technology represents a model of convenient access, on demand, to a calculation resources base, commonly configured that can be provided in a small amount of time and, released after some management efforts and interaction with minimum services provider (Mell & Grance, 2009).

There are numerous challenges regarding research in cloud technology field. Cloud affects the evolution of the externalization process for applications, by heterogeneous organizations.

Recent innovations indicate that regarding the challenges associated with the principle of inter-operability, inter-cloud represents a key element in the future evolution of cloud technology. The concept of Intercloud technology is based on the limited number of calculation resources of a cloud services provider. However, the majority of cloud systems do not present any problems regarding the principle of inter-operability, while the specific standards of the cloud environment do not create support for cloud inter-operability. In addition, the time necessary for their elaboration can span over the years. Thus, a more intensified research is needed to increase the functionality, so that a unified global collaboration is possible for cloud technology.

The new problems associated to enterprise-type application have raised numerous challenges to which no proper attention was gave yet, challenges that can affect the evolution of cloud technology and the use of it by originations with the purpose of externalizing applications with sensitive content. One of these problems is the problem of Intercloud inter-operability.

The concept of Intercloud had become known at the beginning of 2009 (Bernstein, 2009). This is based on the principle according to which every cloud has limited calculation resources in a certain geographical region. The Intercloud technology is aiming at the inter-operability of different cloud calculation instances, every cloud having the possibility to access the calculation resources of any other in the network. Inter-operability is the condition regarding cloud environments with the purpose of reducing the producing/measuring costs related to improvement of components. The clients need to have the possibility to access, leave and, inter-change different clouds, depending on their needs, without being obstructed to select another provider. In addition, the providers must have the possibility to collaborate to find new providers, whose services are of a better quality. Intercloud network, available in present, is connecting only different cloud systems, the providers having different strategies regarding the interactions between applications/clients and clouds. Feldhaus (Feldhaus, 2015) presents the challenges regarding inter-operability in cloud technology field on short, as follows:

- A number of standards are available regarding cloud technology formulated by different parts;
- OCF (Open Grid Forum) standards are only partially applicable or even inapplicable at the cloud level;
- It is necessary to elaborate a strategy of binding different standards regarding cloud technology/ programming interfaces of the applications;
- It is necessary to improve the inter-operational character of the existent implementations regarding the programming interfaces of the applications;
- It is necessary an improvement of the verification sequences of the combined inter-operability;

- It is essential to analyze the problems associated with existent specifications and, of the implementation procedure.

Today, different organizations like IEEE seek to formulate certain standards and programming interfaces of the adequate applications (Grozev & Buyya, 2012). Intercloud future networks will expand the necessary function to permit the interactions between different cloud services.

Applications to Entrepreneurial Business

Srinivasan (Srinivasan, 2014) states the success of a new technology can be measured by its acceptance degree among the designated clients. At this point, cloud computing has spanned over the initial business target (large companies and SMEs) and has been accepted by millions of simple users.

The available global infrastructure is allowing an even faster adoption rate for cloud computing solution. Cloud service providers deliver reliable computing resources using their complex and secure infrastructure. Thus, businesses – of different size, can grow without the need of investing in computing and communication infrastructure. Business owners only need to invest a small amount of their capital for creating and managing internal IT departments. Having the companies' data backed-up online and available anywhere, anytime allow business owners the flexibility and agility they need to rapidly adapt to the new global market.

Cloud computing - a technological innovation, has been widely used for mobile commerce (m-commerce – accessing of information wirelessly on the move) and electronic commerce (e-commerce – allowing business transactions to be conducted electronically) (Ratten, 2013), lowering the entry barriers for entrepreneurs, thus allowing the development of new business. Ratten (Ratten, 2013) has shown that successful adoption of cloud computing can be achieved by focusing marketing campaigns on the emotional value attached to technological innovations. Furthermore, value can be created for entrepreneurial technology service providers by customizing marketing strategies, considering that marketing is one of the most important determinants of an individual's behavior.

In Kavis (2014) the three main cloud computing paradigms are described: Software-as-a-service (SaaS), Platform-as-a-service (PaaS) and Infrastructure-as-a-service (IaaS). The SaaS concept assumes that a software provider hosts and distributes applications. Some of the best known examples of SaaS applications are Google Apps, Salesforce or SuccessFactors. A PaaS is a development environment where a customer can create and develop applications on a provider's computing environment. Some examples of such popular platforms are Google Compute Engine, AWS Elastic Beanstalk or Microsoft Azure. In the IaaS model, a company can rent a data center environment. The most popular infrastructures are Amazon EC2, Windows Azure Virtual Machines or Rackspace Cloud.

Cloud computing is considered a disruptive technology for traditional businesses and a business innovation. Cloud adoption is changing the business world with remarkable speed. New business opportunities arise every day. Alternative career opportunities are opened for youth business initiatives all over the globe. By retraining young people to work in the IT sector, the low rate of employment can be overcome. Several careers of the future are now days derived from the cloud initiatives: cyber-security specialists, cloud and networking engineers or cloud sales consultants.

Three major evolutionary forces contributed to the large-scale adoption of cloud computing (Hugos & Hulitzky, 2011):

- Software companies benefited from widespread availability of the Internet.
- Innovative software companies changed the market.
- Large companies (such as Amazon.com, Google, IBM, and Microsoft) provided access to their massive data centers.

The evolution of Cloud Computing has influenced the way companies view and define productivity. Today, cloud solutions offer a flexible way of working, achieving decreases in costs by transition of servers, of their maintenance and, of their productive flows.

To understand what Cloud Computing is, it is very important to understand the way in which this calculation model has evolved in time. Like Alvin Toffler notes in his famous book, *The ThirdWave (Bantam, 1980),* the civilization evolves in waves: the first wave being represented by the agricultural societies, the second by industrial age and, the third by the information age.

In every wave, there are a couple of important sub-waves. In the post-industrial, information age, in which we live, many consider that we live at the beginning of the so-called Cloud Computing age. Those sub-waves are represented in Figure 6. It began with mainframe computers, it evolved to mini-computers, then to PCs; after that appeared client-server models and network protocols, and now we are entering the Cloud Computing age.

Another point of view illustrates that Cloud Computing is in itself a logical evolution of the computer. Figure 7 represents Cloud Computing and cloud services providers (CSP – Cloud Service Providers), as extensions of the internet services providers (ISP – Internet Service Providers).

Cloud Computing is a model that evolves rapidly, with new aspects and capabilities announced frequently. Simple examples of Cloud Computing are Yahoo! mail, Gmail or Hotmail. There is no need for a software or a server to use them. All the user needs to use them is an internet connection and it can begin to send e-mails. The management server and software are both situated in the cloud (Internet)

Figure 6. Sub-waves in information era
Source: Mather, Kumaraswamy and Latiif (2009).

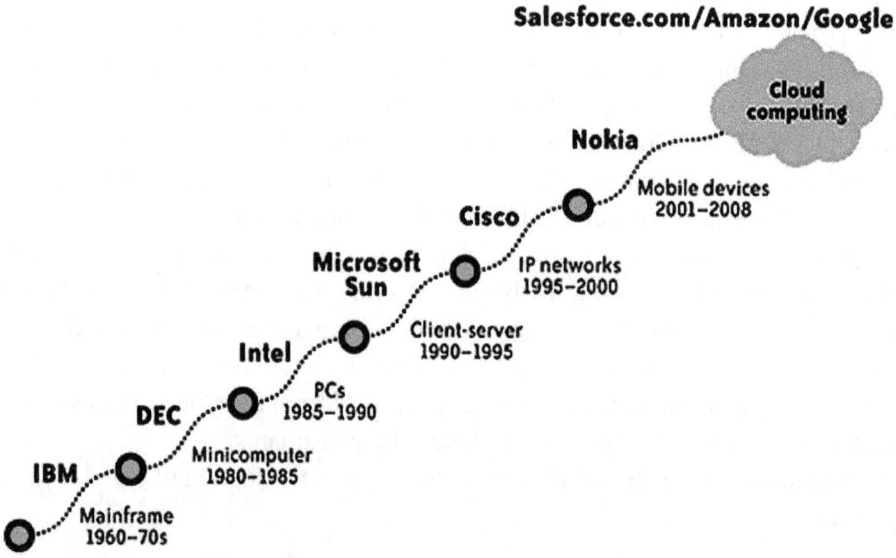

Figure 7. Cloud computing evolution
Source: Mather, Kumaraswamy & Latiif (2009).

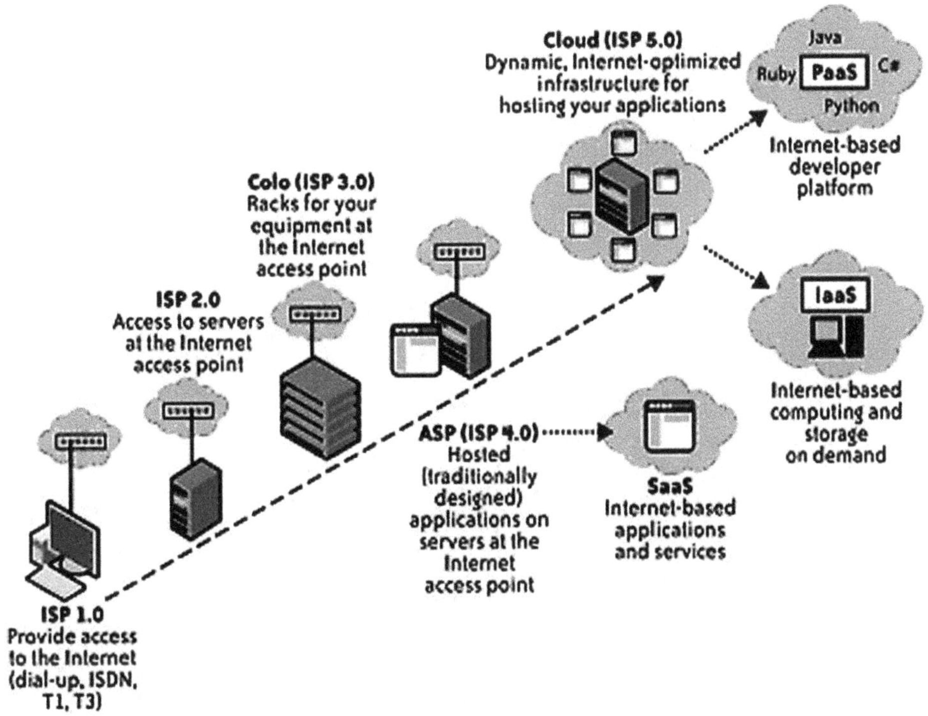

and are completely managed by the cloud service provider (Yahoo, Google etc.). The client uses only the software and he can rejoice from the offered benefits, without having an interest in administration policy behind the service.

Cloud Computing Business Opportunities

Cloud Computing it is not much of a technology, because it is a combination of multiple preexistent technologies. These technologies had been developed differently and in different contexts and were not conceived like a coercive system. However, these technologies had come together to create a technical frame for what is to be known as Cloud Computing. New developments in the areas of processors, virtualization technology, disk storing, broadband network connections and fast inexpensive servers were combined to make the cloud a more convincing solution.

Cloud computing businesses are currently focused on three major directions:

1. **Cloud Access Devices:** The palette of cloud access devices have expanded itself in the last years. Personal computers, company, computer networks, mobile phones, portable personalized devices (including refrigerators) are all on-line.
2. **Browsers and Small Clients:** Users of many types of devices can access applications and information from everywhere, using a browser. It is true that the browsers become more and more sophisticated. Applications dedicated to the professional medium, like SAP and Oracle, can be accessed

Figure 8. Relevant technologies in cloud
Source: Pamula, Jairam, and Rajesh (2014).

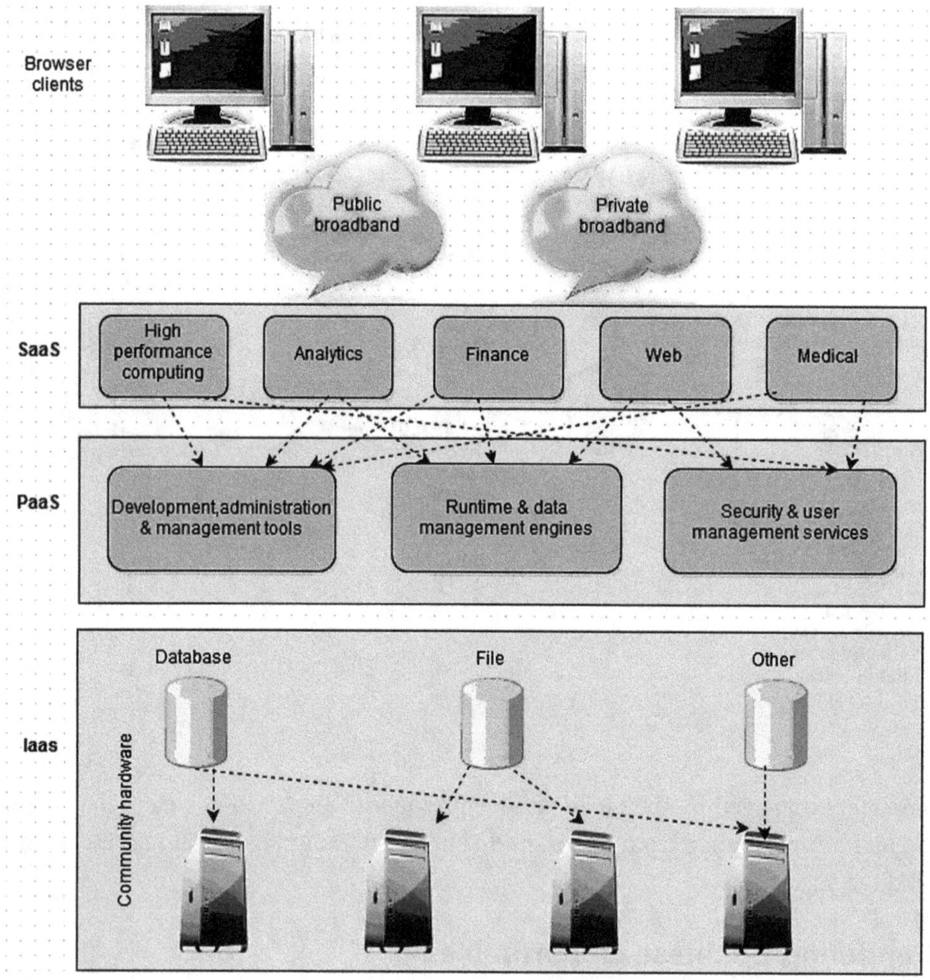

by a browser interface. The general population has become more familiar with the function of the browser and can use a discreet application, where the context is intuitive without necessitating user guides or any other instructions.

3. **Data Centers and Server Farms:** The services based on cloud need a big calculation capacity and are stored in data centers and server farms. These data centers distribute server "farms" stretching over many interconnected locations providing calculation power and services delivering capabilities.

A series of today examples illustrate the flexibility and scalability of the cloud power. For example, Google has connected a great number of cheap servers to offer an extraordinary flexibility and power. Elastic Compute Cloud (EC2), offered by Amazon, offers virtualization in the data center to create a giant number of virtual instances for the services that are required.

Companies operate in unpredictable, more volatile business environments that in the last decade. Srinivasan (Srinivasan, 2014) points out, that from an entrepreneurship point of view, the rapid adoption of cloud computing has eliminated the entry barrier for many startups and not only, by providing an easy to use, reliable and secure IT infrastructure. Hugos and Hulitzk (Hugos & Hulitzk, 2011) present the vision of cloud-based models for business organizations, which allow for a variable cost operating model to be applied.

USING INTERCLOUD TO DEVELOP INNOVATIVE APPLICATIONS

E-Health Application Scenario

This scenario aims at presenting an architecture for the interconnection of cloud services coming from multiple cloud providers. The goal is to highlight the advantages of using hybrid technology composed of heterogeneous services offered by various providers of cloud computing. Due to the large growth of the concept of cloud computing and the emergence on the market of more solutions offered as services, market and the developers of software and hardware solutions trend is to use existing services, trying to link them to get the best quality in comparison with the implementation time and production costs.

Recently the concept of Intercloud has had a significant increase, being very well promoted and supported by software manufacturers in the field of cloud. The concept of cloud computing also offers a number of basic facilities that all developers use to develop services. After the cloud computing revolution, much more specialized dedicated and performing solutions appeared on the market on various fields. Based on these existing services already well developed and mature, all at the same time at an extremely affordable price, the new challenge in the IT world is to be able to interconnect existing services in order to build new IT products. The concept of creating software from scratch is transformed by the existing market into the concept of creating modular and subdivided into components architecture, each component of the system being implemented as a service.

To exemplify the inter-communication between cloud services, the scenarios describes a system that retrieves data from a series of sensors and sends them in real time to a cloud service specialized in the distribution of data. This cloud service is recorded as consumers, a mobile application that listens to and displays the received data and a dedicated cloud service that offers the possibility of processing and data storage. According to the results obtained from the processing of the data received from distribution service, the processing cloud service sends specific messages to a service specialized in the notification of mobile devices remotely.

In the following, a prototype is presented, which will help to understand how the Intercloud concept can be used to design and build an entire e-health monitoring system.

For an easier understanding, the entire system can be divided in three major subsystems. The acquisition subsystem is made using the Arduino Yun development board, which is responsible for environmental data acquisition and sending them in the cloud. To monitor the entire system, view the data, receive notification and take some decisions according to the values received a smartphone will be used. The most important subsystem is the Intercloud, which is used for data management, send push-notifications, which means that establish the connection between the others subsystems.

For the proper functionality of the system, before being used, it should be configured, so the functionality of the system can be divided in this way:

1. First part is the one where the system is configured and its components are synchronized.
2. Second part is represented by acquisition and data management.

Cloud Services Communication Architecture

The communication architecture is a distributed architecture based on services consisting of hardware and software components. The hardware - Arduino board and a mobile device, is dealing with retrieving and displaying the values, and the software is used for data processing and data storage, business logic of events and notification messages.

A low cost device - Arduino Yun Development Board, is used for monitoring and interpretation of received signals from the sensors. Another purpose of it is to pass through the HTTP protocol data packets that contain information concerning the value of the e-health parameter at any time using the integrated Wi-Fi module.

The packages are picked up by the Firebase service which purpose is the management of events and signals transmission in real time to the mobile device. This cloud service was chosen because it is specialized for the distribution of events in real time having notable performance in this area.

Mobile Microsoft Azure Service is dedicated for storing and processing large data that are steadily increasing. Azure platform capability is to provide computational resources needed for a larger amount of data. The purpose of this component is to store data regarding the evolution of the sensor values in time and to process it's for setting notification decisions. The last role of this module is to send specific messages to the notification service regarding decisions taken after the processing of the data.

Figure 9. Cloud services communication architecture

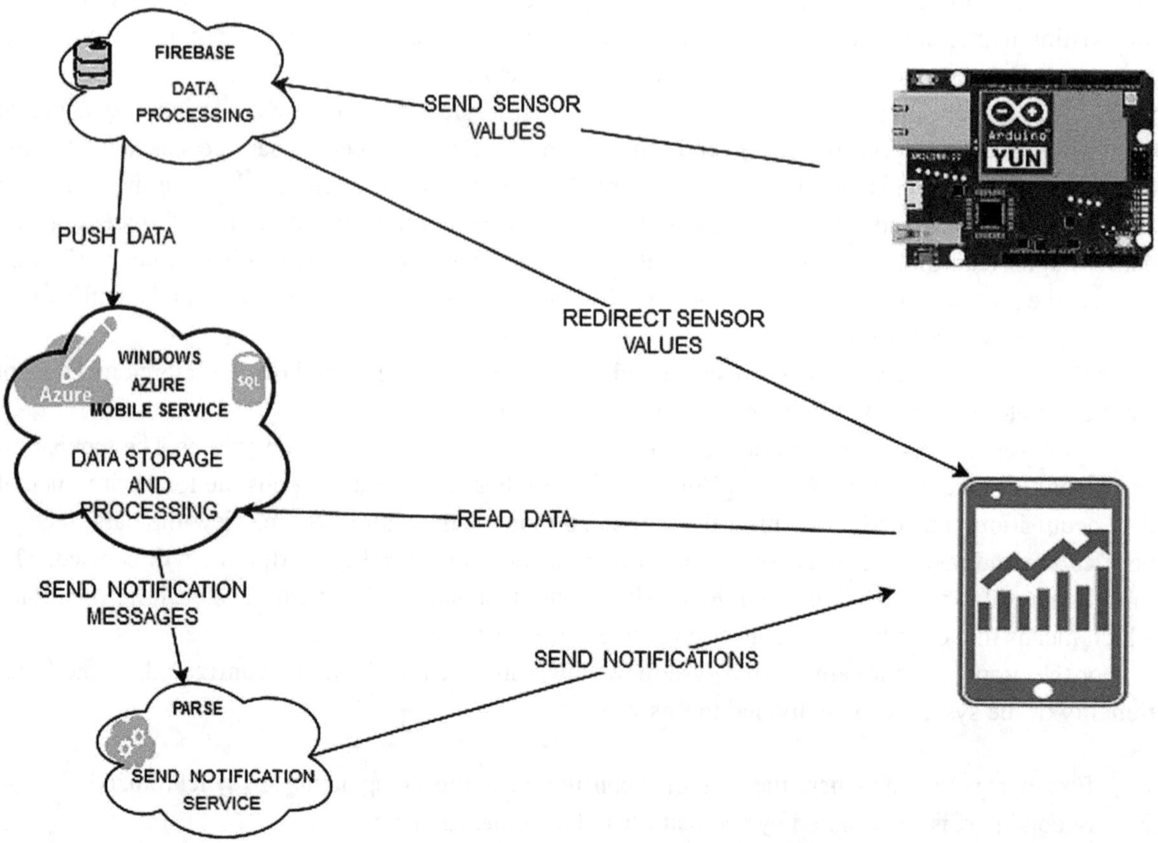

Parse Cloud service is specialized on the management of notification remote messages. Its role is to retrieve specific messages from mobile service and send them further on the channel of one or more subscribers to notification message. In our case the subscribers of the channels are mobile device users.

Another hardware device is the smartphone, which will show the sensors' values and a graphic concerning its evolution over time. This mobile device will retrieve data for graph provided by service provider Azure, will get real-time values of the Firebase service and will receive notifications using service provided by Parse.

Interconnecting Clouds

The role of this subchapter is to explain the functional principles of cloud services used for Intercloud system deployment and to describe how their capabilities were used to develop an innovative application. From the business point of view, the Intercloud scenario bring added value, with reduced costs due to fast development and deployment time.

Firebase Architecture

In this architecture, the Firebase is situated between the server and the clients. Servers can connect to the Firebase and can interact with data as another client. The level of security and rules of language permit total access to the Firebase data on the server. Thanks to this service, the server can listen to any data change made by clients and respond or react immediately. Firebase processes all changes in real time and uses an API- REST that can be accessed from any environment that provides the ability to send HTTP requests. Another feature of the architecture that uses the Firebase is the ability to have synchronization on both sides. On the client side, we need a library that offers support for typical situations such as refreshing the window, accessing data in times when it is offline or synchronizing data when a device is reconnected to the service.

Windows Azure Structure

Windows Azure Mobile Service is a backend solution for mobile applications where structured storage system in the cloud capabilities are offered. As well as methods for operating on Firebase database CRUD (create, read, update and delete), are based on the requests of type REST. Windows Azure service facilitates the storage of data in the cloud. When creating a Windows Azure mobile service, automatically an associated database SQL Azure cloud-site infrastructure is created.

The service provides support for secure operations of writing and reading of data from remote applications by using the REST methods on both sides and using JSON formats.

Using the Windows Azure platform, we have support for database management in the creation of new tables, searching and controlling data access permissions. Other capabilities offered by Azure Portal are monitoring traffic data and use of the database or possibility to change the resources allocated to the service cloud according to usability by offering scalability to the system.

On the Microsoft Windows Azure is configured a virtual machine where is installed an Arduino compiler. With this compiler, according to the settings selected by the user, is generated the program which will be loaded on the development board firmware.

Configuring Notification Service

Cloud Parse service is specialized on sending remote notifications. To configure this service, just like at the Firebase, it is necessary to create an application on the Parse platform where the resources needed for the management of notifications are offered. After registering the application, the platform generates a token which must be saved on each device that wants to have access to notifications.

For each device it is created a channel of communication between it and the Push Notification service. In addition, Parse has the capability to notify all subscribed devices on all channels or creating a group to which the message will be send.

The E-health prototype that monitors a patient pulse parameter exemplifies the design and usage of Intercloud services, as a valid business model in e-health applications. While many Intercloud initiatives are centered on resource management, this scenario focuses on the design and development of a services centered architecture. Rapid deployment and integration, scalability and versatile use of mobile devices represent the strong point of the prototype.

CONCLUSION

Through time, technology evolved rapidly, from the concept of mainframe to the concept of cloud. Cloud is one of the technologies that are being used more and more frequently due to their performances. In the last decades the number of scientists and companies that are using distributed computing to remain competitive on market increased significantly. Furthermore, the amount of data generated by mobile devices and social networks increased significantly. Because the costs of operating personal super-computers are high, not all companies and research groups are affording to pay the high prices generated by the acquisition of the infrastructure needed to run their personal applications and experiments. Thus, they need to count on resources from external sources.

In the beginning, clusters and grids were those who attracted the attention of the academic and economic areas, but they had some limitations. The new technology of cloud-computing is promising all needed resources, at any moment, based on the model "pay how much you use". Accessing it should be like accessing any other utility such as water, electricity etc. However, to have the necessary infrastructure at your needs it is not sufficient for all the clients, because some of them do not have the knowledge to profit from the advantages of the new technologies.

Cloud computing is considered the next step in the evolution of technological applications, services and products development. Furthermore, the Intercloud paradigm allows interconnecting different cloud providers with the aim of offering enhanced services for business owners. The permanent connection of users to the Internet has become widespread, so almost all available resources (data, software) can be placed on the Internet, shared between users- sometimes completely independently of others, and brought from the Internet to the user's computer on demand.

Forbes magazine (Columbus, 2016) estimates a $127 billion Worldwide Public IT Cloud Service Revenue and a $256 billion in Managed Services. All major cloud computing providers (Amazon, Google, Microsoft etc.) are estimating significant business growth on this sector.

The chapter highlights the benefits of using hybrid technology consisting of heterogeneous services, offered by different providers of cloud computing. Furthermore, to emphasize the business opportunity of innovative Intercloud approaches an architecture for the interconnection of cloud services offered by multiple cloud providers, with applications to e-health is presented.

Nevertheless, the explosive evolution of cloud services requests is affecting cloud computing providers. The demand for cloud computing must be met by substantial investments in both infrastructure and novel storage technologies. Intercloud paradigm can be a solution for cloud computing providers to keep up with the ever increasing customers' demands and also to provide innovative solutions. Intercloud can also be seen as a new business model, from an entrepreneurial perspective.

When taking into consideration location, IT&C companies tend to be more flexible. Intercloud solutions also transcend location concerns. Businesses can use and combine different cloud technologies, and allows them not to have location constrains, due to the online nature of this technologies. Although this is a major advantage for entrepreneurs, it is also a drawback for local communities. A company can easily relocate, affecting employment rates and local budgets (Curran, Lynn, O'Gorman, 2016).

Some key feature, are making Intercloud the new ITC market niche: flexibility (integration of different Cloud solution from different providers – not bound to a single provider), low cost (due to Cloud computing paradigm "pay as you go"), easy to use (based on the Saas paradigm). Entrepreneurs can develop new products by combining the power of several cloud computing providers

In conclusion, the chapter presents the main concepts and business benefits (but also several drawbacks) of using Intercloud systems and an architecture for interconnecting cloud services from multiple cloud providers is described. The developed prototype highlights the benefits of using hybrid technology and heterogeneous services offered by different cloud computing providers in the field of e-health applications.

REFERENCES

Ahmed, M., Chowdhury, A.S.M.R., Ahmed, M., & Rafee, M.M.H. (2012). An advanced survey on cloud computing and state-of-the-art research issues. *International Journal of Computer Science Issues, 9*(1).

Bernstein, D. (2009). The Intercloud: Cloud Interoperability at Internet Scale. *Sixth IFIP International Conference on Network and Parallel Computing.*

Bhat, J. M., & Shroff, B. P. (2014). Entrepreneurial Value Creation in the Cloud: Exploring the Value Dimensions of the Business Model. In TDIT 2014, IFIP AICT 429, (pp. 301–310).

Briscoe, G., & Marinos, A. (2009). Digital ecosystems in the clouds: Towards community cloud computing. *3rd IEEE International Conference Digital Ecosystems and Technologies, DEST '09.* doi:10.1109/DEST.2009.5276725

Cauveri, A. (2015). *Literature survey on distributed Computing.* Retrieved August 12, 2015, from http://www.academia.edu/8298875/Literature_survey_on_distributed_computing

Columbus, L. (2016). *Roundup Of Cloud Computing Forecasts And Market Estimates.* Retrieved April 21, 2016, from http://www.forbes.com/sites/louiscolumbus/2016/03/13/roundup-of-cloud-computing-forecasts-and-market-estimates-2016/#4e794d7474b0

Curran, D., Lynn, T., & O'Gorman, C. (2016). The Role of Personal Factors in the Location Decision of Software Services Start-up Firms. *European Planning Studies, 24*(3), 551–567. doi:10.1080/09654313.2015.1046369

Feldhaus, F. (2015). *Cloud Interoperability*. Retrieved August 4, 2015, from https://www.ogf.org/OGF34/materials/2417/Cloud+Interoperability+BoF.pdf

Gibbs, S. (2012). Cloud Computing. *International Journal of Innovative Research in Engineering & Science*, *1*(1), 10–17. PMID:22397113

Grozev, N., & Buyya, R. (2012). Inter-Cloud architectures and application brokering: Taxonomy and survey. *Software, Practice & Experience*, 369–390.

Höfer C.N., & Karagiannis G. (2011). Cloud computing services: taxonomy and comparison. *Journal of Internet Services and Applications*, 81-94.

Hugos, M., & Hulitzky, D. (2011). *Business in the Cloud. What every business needs to know about cloud computing*. John Wiley & Sons.

Iancu, I., & Iancu, B. (2012). Future Internet: challenges and opportunities in advertising. In *PR Trend. New media challenges and perspectives International Conference*. Babeş-Bolyai University and Hanns Seidel Foundation.

Jangra, A., & Bala, R. (2011). Spectrum Of Cloud Computing Architecture: Adoption And Avoidance Issues. *International Journal of Computing and Business Research*. Retrieved August 2, 2015, from http://www.researchmanuscripts.com/PapersVol2N2/IJCBRVOL2N2P13.pdf

Jones, E. (2016). *The top five in-demand cloud skills for 2016*, Retrieved May 12, 2016 from http://www.cloudcomputing-news.net/news/2016/feb/16/top-five-demand-cloud-skills-2016/

Kavis, M. (2014). *Architecting the Cloud: Design Decisions for Cloud Computing Service Models (SaaS, PaaS, and IaaS)*. John Wiley & Sons. doi:10.1002/9781118691779

Kritikos, A. (2014). Entrepreneurs and their impact on jobs and economic growth. *IZA World of Labor*, *8*. doi:10.15185/izawol.8

Mather, T., Kumaraswamy, S., & Latiif, S. (2009). *Cloud Security and Privacy: An Enterprise Perspective on Risks and Compliance (Theory in Practice)*. O'Reilly Media.

Mell, P., & Grance, T. (2009). *NIST Definition of Cloud Computing Version 15*. NIST.

Pamula N. B., Jairam K., & Rajesh B. (2014). Cache-Aside Approach For Cloud Design Pattern. *International Journal of Computer Science and Information Technologies*, 1423-1426.

Ratten, V. (2013). Cloud computing: A social cognitive perspective of ethics, entrepreneurship, technology marketing, computer self-efficacy and outcome expectancy on behavioral intentions. *Australasian Marketing Journal*, *21*(3), 137–146. doi:10.1016/j.ausmj.2013.02.008

Srinivasan, S. (2014). *Cloud Computing Basics*. SpringerBriefs in Electrical and Computer Engineering. DOI 10.1007/978-1-4614-7699-3_1

Tsai, W. T., Sun, X., & Balasooriya, J. (2010). Service-Oriented Cloud Computing Architecture. In *Proceedings of the 2010 Seventh International Conference on Information Technology: New Generations*.

Vouk, M. (2008). Cloud Computing – Issues, Research and Implementations. *Journal of Computing and Information Technology*, 235-246.

Chapter 5
Use of New Innovative Technologies in Business by All Age Groups

Stelian Stancu
Bucharest University of Economic Studies, Romania

Laura Elly Naghi
Bucharest University of Economic Studies, Romania

Constanta-Nicoleta Bodea
Bucharest University of Economic Studies, Romania

Oana Madalina Popescu
Bucharest University of Economic Studies, Romania

Alina Neamtu
Bucharest University of Economic Studies, Romania

ABSTRACT

This paper makes a foray into the new innovative business technologies by presenting several important aspects of the Digital Economy, Business models, Entrepreneurship, Information and Communication Technologies (ICT), ICT4D, Innovation Technology. The second part of the paper consists of an analysis of new innovative technologies results in business, using Global Entrepreneurship Monitor (GEM), thus presenting the conceptual framework GEM, How GEM Defines Entrepreneurship, Measures of Entrepreneurial Activity, Entrepreneurial Aspirations, Entrepreneurial Attitudes. Based on this analysis, empirical results of new innovative business technologies in Romania are configured, highlighting the key indicators for Romania (2015), Entrepreneurial Activity rates in efficiency-driven EU Countries in 2014 (% of population aged between 18-64 years), distribution of entrepreneurs by gender, age, education and household income in Romania 2014 (%), both individually and overall. The paper concludes with key findings and proposals for the following questions to be explored in future research.

DOI: 10.4018/978-1-5225-0953-0.ch005

INTRODUCTION

With early trade, entrepreneurship played a central role in creating and distributing wealth (Schumpeter, 1934), whereas entrepreneurs are those who perform innovation in order to create wealth. Schumpeter (1939) defined an entrepreneur based on his activities, and not as entrepreneurship. Innovation and technological advances have been closely related to entrepreneurship; until the second part of the twentieth century, literature has focused on the role that technology had in entrepreneurship (Utterback, 1971; Bhupatiraju et al., 2012).

More recently, in the last part of the twentieth century, both the Internet and the rapid penetration of personal computers have brought a new wave of studies on the impact of Information and Communication Technologies (ICTs) and innovation enterprises (Fichman, 1999; Hitt & Brynjolfsson, 1996; Barnes et al., 2004).

In the late 1990 and the first part of 2000, academia debated about whether ICT should be considered a top line predictor variable for value creation or a moderator variable bottom line (Hitt & Brynjolfsson, 1996; Oliner & Sichel, 2000; Hamilton & Asundi, 2008). Researchers followed the growing trend of new online companies (Lockett and Brown, 2000), but the terms e-Entrepreneurship (Matlay, 2004) and digital entrepreneurship have been used to refer to a new discipline only at the end of the last decade (Zhu et al., 2006; Hull et al., 2007).

In the last ten years, data networks and ubiquitous wireless smart-phones (Berman, 2012) have accelerated the number of new online businesses and have created the environment for selling products/digital services online (Taylor & Murphy, 2004; Barnes et al., 2004; Lockett & Brown, 2000; Wall et al., 2007; Kollmann, 2006; Asghari & Gedeon, 2010).

According to European Commission, digital entrepreneurship has five dimensions: strengthening innovation and the ICT digital marketing sector; strengthening digital infrastructure and simplifying the business process; facilitating the access to digital investment growth; the need for digital literacy and e-leadership by promoting digital literacy through education and training; the need for an entrepreneurial culture.

NEW INNOVATIVE TECHNOLOGIES IN BUSINESS

Digital Economy

Digital Ventures are defined as new digital enterprises that trade and operate exclusively online. Digital economy introduced by Tapscott (1996) is the digital e-business enterprise or a qualitatively different living being in a different environment, according to Barr (2001).

In a purely digital context, concepts such as contribution and marginal effects have a greater impact; therefore, economic activity should be measured differently (Brynjolfsson & Kahin, 2000; Gopal et al., 2003).

In literature, both e-commerce and e-business terms are commonly interchangeably used and are considered to be a business transaction that is executed electronically (Wall et al., 2007).

Business Models

Business model were used in literature in different ways by association to the scope. Timmers (1998) defined a business model as following:

1. An architecture for products, services and information flows, including the description of various business actors and their roles;
2. A description of the potential benefits to different business; and
3. A description of income sources.

Osterwalder et al. (2002) and Osterwalder and Pigneur (2002) presented a framework to explain how the elements of a e-Business model reach a deep level of detail on the conceptualization terms, components and relationships between them (Weill & Woerner, 2013; Berman, 2012).

Wall et al. (2007) made a comprehensive review, expanding the classification of e-business models offered by Timmers (1998) and Rappa (2003). In fact, both Sako (2012) and Doganova and Eyquem-Renault (2009) admit that the business model itself can be a source of innovation. Doganova and Eyquem-Renault (2009) also describes how business models were defined and how they were used by technology entrepreneurs.

Entrepreneurship

The word "entrepreneurial" is currently widely used. However, it is still challenging to provide a clear and specific definition of entrepreneurship due to the multitude of definitions that rarely agree with each other on certain essential characteristics of entrepreneurship. Davidsson (2004) lists at least seven definitions. This is no surprise, taking into account that entrepreneurship is a new field of study.

According to Shane and Venkataraman (2000), who gave one of the most cited definitions, "entrepreneurship is concerned with discovering and exploiting profitable opportunities." However, it is necessary to distinguish entrepreneurship from research entrepreneurship.

Davidsson (2004) suggests that entrepreneurship is defined as "a behavior that drive competitive marketing process, while the study of entrepreneurship is structured around the idea of the emergence of new business ventures."

The research provides a definition based on previous literature (Shane and Venkataraman, 2000; Zahra and Dess, 2001, Gartner, 1988; Venkataraman, 1997). It is also possible to observe the permanent development of entrepreneurship in time. For example, the definition of Shane and Venkataraman in 2000 was rewritten by Shane in 2003.

In the book "The General Theory of entrepreneurship", Shane removes the profit: "We define an entrepreneurship opportunity as a situation where a person can create a new framework by recombining the resources so that the entrepreneur believes it will lead to a profit". He adds that "readers should note that entrepreneurial opportunities are not necessarily profitable" (Shane, 2003).

Dealing with all available definitions - from Schumpeter to Gartner - Low and McMillan (1988), "the problem with these definitions is that even though each captures an aspect of entrepreneurship, none captures the whole picture".

In contrast with Shane and Venkataraman (2000), even if Davidsson does not use the word "opportunity," it is impossible to deny that research entrepreneurship "implies learning opportunities and sources, discovering, evaluation and exploitation of opportunities" (Shane & Venkataraman, 2000).

Casson (1982) defined entrepreneurial opportunities as those situations in which new goods, services, materials and methods of organization can be introduced and sold below the cost of production. However, not all opportunities are entrepreneurial opportunities.

The premise is "finding new means of closing relationships (Shane & Venkataraman, 2000). In the same way that not all opportunities are profitable, not all industries are equally profitable. If an entrepreneur randomly started a business in some industries, that person would be more likely to have a very fast growing private company or public (Shane, 2003, p. 18). Therefore, there is a relationship between the specific characteristics of industry and entrepreneurial opportunities.

In fact, Shane and Venkataraman's definition is more like a suggested boundary of the scientific field of entrepreneurship, rather than a definition of the term. This distinction is further enhanced by Davidsson (2004) to how the new association designs future goods and services and subsequently develops them, by whom and with what consequences (Low & McMillan, 1988).

Information and Communication Technologies (ICT)

ICTs can be considered raw materials for Digital Ventures. Bhupatiraju et al. (2012) stated that in the 1960s and 1970s innovation [INN] and scientific and technological studies [STS] were partially overlapped. Bhupatiraju et al. (2012) explains also that technology literature has evolved itself as a discipline with a cumulative knowledge.

Since the early 2000s, several papers have paid attention to the relationship between technology, innovation and entrepreneurship (Doganova & Eyquem-Renault, 2009; Hamilton & Asundi, 2008; Hindle & Yencken, 2004; Groen et al., 2008; Miller & Garnsey, 2000).

Emergent ICTs were chosen empirically based on their use of high technology industry jargon for the past two years: Cloud Computing and Web 2.0. (Ojala & Tyrvainen, 2011; Barnes et al. 2012; Lee et al., 2008).

The National Institute of Standards and Technology (NIST) definition of cloud computing is: "a model for enabling convenient on-demand network access to a shared pool of configurable computing resources that can be rapidly provisioned and released with minimal management effort or service provider interaction" (Jansen & Grance, 2011).

Barnes et al. (2012) explains that Web 2.0 can be better understood as a range of [ICT] internet-based applications. Barnes et al. (2012) studied how SMEs are using Web 2.0. Brynjolfsson and Hee Oh (2012) discuss the challenges to measure the economic value of Web 2.0 services, which are apparently presented as a free producer/consumer.

ICT4D

The potential positive rate, higher for SMEs, of adopting electronic trading could be a real help for the global economy (Boateng et al., 2008; Martha, 2004; Hinson et al., 2008), but it is generally true for ICT adoption in general (Middleton & Byus, 2011).

ICT4D authors include recommendations to be implemented by governments, non-governmental organizations (NGOs) and the private sector in order to improve the adoption of ICT and e-commerce SMEs (Kenny, 2003; Knight, 2011; Ngwenyama & Morawczynski, 2009; Nair et al., 2005; Hitt & Brynjolfsson, 1996).

Based on ICT4D literature, it could be believed that digital ventures have a positive socio-economic impact in developing countries (Thewarapperuma, 2013; Facet, 2011; Carayannis & von Zedtwitz, 2005).

Innovation Technology

Drucker (2002) offered a definition of innovation that captured the essence of such a relationship: "Innovation is the specific function of entrepreneurship, whether in an existing business, a public service institution or a new venture, started by a lone individual in the family kitchen".

From Bhupatiraju et al. (2012), Figure 1 shows an extended timeline of authors that are relevant for Technology Innovation.

Early authors (Cantillon, 1755; Marshall, 1890) considered innovation as a change of technology that offered a creative solution to a particular need, with results in wealth creation and economic growth.

Rogers (1962) has taken further the concept of innovation by providing a diffusion frame that today is still a reference for researchers in a broad spectrum of disciplines (Zhu et al., 2006; Rahman et al., 2013; Maertens & Barrett, 2013). Rogers (1962) proposed that there is a defined process, called diffusion, whereby people learn about, and decide whether to adopt innovations.

Rogers (1962) identifies the following characteristics of innovation that influence the adoption rate: relative advantage, compatibility, complexity, process ability, and observability.

Later, Utterback (1971) suggested a three-step process of technical innovation: idea generation, problem solving and implementation, possibly followed by diffusion.

Kamien and Schwartz (1982) proposed a theory of market structure and innovation where innovation was also considered a contextually targeted marketing company. Similarly, Tornatzky and Fleischer (1990) presented an empirical innovation framework called contextual Technology Organization Environment (TOE), composed of: context of external environment, the context of technological and organizational context. Acs and Audretsch (1990) refocused previous research innovation and have strengthened innovation and entrepreneurship for SMEs. Fichman (1999) focused on ICT diffusion and assimilation theories, and provided a framework to classify their key buildings.

Drucker (2002), highlights that there is even a discipline approach as innovative elements. In 2008, Wang yield new insights entrepreneurial orientation, learning orientation, and firm performance, for 2014 Autio, et al., address essential aspects of Entrepreneurial innovation: the Importance of context.

Figure 1. Literature timeline for technology innovation
Source: Adapted Quinones, G. et al., 2013 and contributing authors

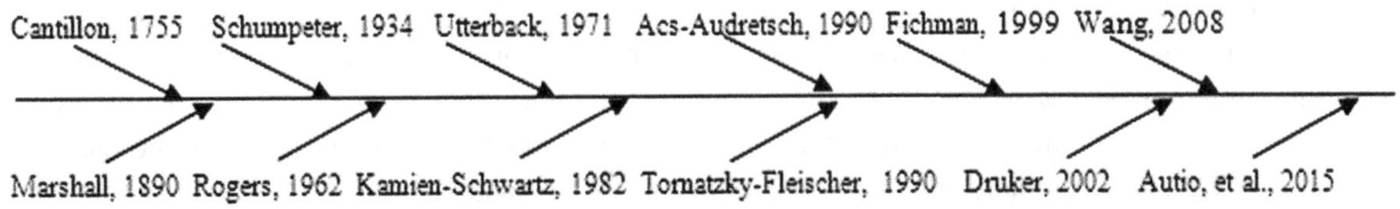

Table 1. Differences between incremental, radical, and disruptive innovations

Radical Innovations	Incremental Innovations	Disruptive Innovations
Discontinuous (with or without predecessor; substantial, non-linear improvement)	Continuous (linear improvement in the value received by customers)	Inferior performance, cheaper, fast improving
Based on new technology	Based on old technology	Leading customer rejection
Leads to new dominant design	Dominant design unchanged	Performance overshooting of established technology
Can lead to paradigm shift	Does not lead to paradigm shift	Lower profits until a new business model is found
Involves great uncertainty	Involves low uncertainty	Emerging market success (in isolated niches)
Entire new set of performance features	Feature improvements	Intersecting trajectories lead to invasion of the incumbent's market;
Need for re-education, new organization and skills	Existing organization and qualifications are sufficient	First-mover advantages
Attributed to chance, not to necessity; might be influenced by R&D policy	Result of rational response, of necessity	
Driven by technology push (important in early phase of technology)	Driven by market pull (important in late phase of technology)	
To achieve economic long-term goals	To achieve economic short-term goals	

Source: Quinones, G. et al., 2013. adapted from (Latzer, 2009).

Foster and Heeks (2013) explains how the Systems of Innovation (SoI) and Technology Diffusion frameworks can be used to study new forms of innovation in developing countries by conceptualizing what they define as including innovation.

It is generally accepted that subsidizing ICT technology and innovation has a positive effect on productivity (Crespi & Zuniga, 2012; Daude, 2010; Lastres & Cassiolato, 2003; Hilbert, 2010).

DIGITAL ENTREPRENEURSHIP

A type of entrepreneurial activity is digital entrepreneurial, defined as a new opportunity to create business generated by ICT - internet, mobile technologies, social computing and digital platforms (Davidson & Vaast, 2010; Oestreicher-Singer & Zalmanson, 2013; O'Reilly, 2007; Onetti, Zucchella, Jones, & McDougall-Covin, 2012).

Despite the development of small digital businesses, Javalgi, Todd, Johnston and Granot (2012, p. 743) notes that so far academic investigation only skimmed the surface, compared to how the small Internet firms develop and grow. ICT needs to trigger their entrepreneurial activities and infrastructure that support interactions with stakeholders (Lusch & Nambisan, 2015).

Digital entrepreneurship is the practice to continue new opportunities with risk presented by new media and Internet (Davidson & Vaast, 2010, p. 8). It is similar to traditional entrepreneurship, since the digital associations aimed at generating a financial return and entered directly in the economic field, such as creating a new company or a marketing innovations (Davidson & Vaast, 2010, p. 2). Some or all of entrepreneurial association occurs digitally instead of more traditional formats (Hair, Wetsch, Hull, Perotti & Hung, 2012, p. 3, Hair et al., 2012).

Digital entrepreneurial ventures range from large established companies that develop hardware, software and network technologies to small start-up firms that use information and communication technologies (ICT) to carry out their business activities (Rosenbloom et al., 1993, p. 462). This includes associations working in the ICT sector (e.g. online accounting, Wi-Fi-hotspot, backup phone, software artificial intelligence) (Vasilchenko & Morrish, 2011), sites for social calculating (e.g. for consumption of music) (Oestreicher-Singer & Zalmanson, 2013) and small mobile enabled by the internet (e.g. classification/cataloging information specialist, travel/recruitment agencies and the sale of market) (Javalgi et al., 2012).

Digital Entrepreneurship and Local Context

The previous discussion shows that, extending ICT as a resource operant shaping digital entrepreneurship, ICT entrepreneurs depends on the skills/competencies, orientation and liabilities as determinants of entrepreneurship complementary digital market (Ngoasong, 2015).

These factors should obviously be considered in relation to the local context in which the digital entrepreneur operates. Local context is being used to refer to the context of 1) technology and 2) geographic context, which both require contextual influences on the digital entrepreneur (e.g. Autio et al 2014).

Mills and Pawson (2012) examines how entrepreneurs training in the ICT can sense their experiences of start-up by describing their approach to motivation, risk taking and self-identity in the context in which they operate. In this case, contextual influences include entrepreneurs making choices in pursuing opportunities and challenges associated with dealing.

Country study include geographic context while context, technology include ICT industry in which the entity operates. In addition to country and industry, Gill and Larson (2014, p. 522) suggests that local context is also a factor for development of entrepreneurial high-tech identity.

Proposition about Digital Entrepreneurship in Emerging Economies

Proposition 1: In emerging economies, entrepreneurs who either have a formal ICT training or a strong passion for ICT are more likely to create successful digital businesses than their peers.

Proposition 2: In emerging economies, digital entrepreneurs with limited capital for start-up are more likely to succeed by creating digital businesses that rely on ICT infrastructure to manage interactions between key stakeholders within their enterprise value chain.

Proposition 3: Digital entrepreneurs with proper market orientation, who are able to implement innovative digital solutions for issues affecting their society are more likely to succeed than their peers in creating digital businesses.

Proposition 4: Emerging markets, knowledge and skills to meet the challenges associated with geographic digital entrepreneurship increase the probability of successfully launch and operate a digital enterprise.

Proposition 5: In emerging markets, entrepreneurs who are able to handle the responsibility by outsourcing parts of their digital businesses that require physical interaction with key stakeholders are more likely to reduce their operational costs than their peers.

E-ENTREPRENEURSHIP

In the second half of the 1990s and early 2000s, the emergence of companies like Netscape, Amazon, Google and eBay has triggered a wave of studies focused on specific challenges and opportunities of operating a business solely on the Internet (Timmers 1998; Afuah & Tucci, 2000; Souitaris & Cohen, 2003; Barnes, Hinton & Mieczkowska, 2004; Pateli & Giaglis, 2003).

However, Matlay (2004) was the first to use e-entrepreneurship to refer specifically to the SME phenomenon created to trade exclusively on the internet. Other terms, such as Internet Entrepreneurship (Batjargal, 2005), and Digital Entrepreneurship (Hull et al., 2007) were used with a similar meaning.

e-Entrepreneurship is still in the development of theoretical frameworks that are mainly based on the disciplines around as shown in Figure 2.

In Table 2, entrepreneurial innovation is seen as involving disruption of existing industries and developing new processes and multi-level stakeholders, several actors and multiple contexts which are different entrepreneurial ecosystems (Isenberg, 2010).

Figure 2. Literature thematic categories around e-entrepreneurship and their relationships
Source: Quinones, G. et al., 2013.

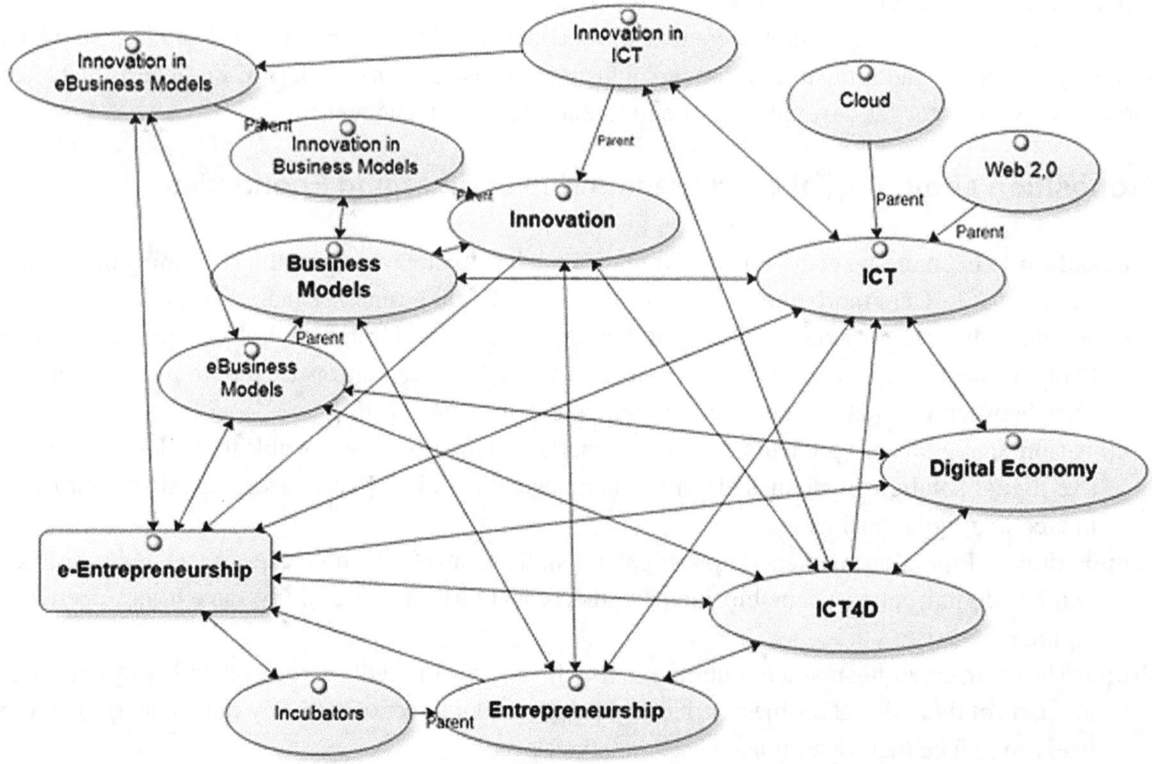

Table 2. National innovation systems, entrepreneurship, and entrepreneurial innovation compared

	National Innovation System	Entrepreneurship	Entrepreneurial Innovation
Definition of innovation	Narrow view: focus on technological and science based innovation	All actions that lead to the formation of a new organization?	Focus on radical innovation
Innovation driver	R&D and technology, invention	Entrepreneurial cognition and learning, opportunity recognition and creation	Co-creation and evolution with ecosystem
Role of the Individual agent	Not considered or expected to appear automatically	Central, but analysis usual in vacuum and focuses usually on individuals (often does not recognize team character of much entrepreneurship)	Individuals/teams operating within a context that includes social, institutional, industrial, organizational, temporal and spatial networks
Context	Institution as actors	Generally ignored but emerging agenda	Agency within a multi-dimensional context
Mechanisms	Top down (i.e. Government policy) and complex set of interaction	Bottom up (i.e. individual entrepreneurs), decentralized, non-linear processes; social networks; resource orchestration	Multi-level and multi-actor processes
Mode, organizational forms	University, research labs, tech transfer, large corp	Start-ups, corporate entrepreneurship	Entrepreneurial firms embedded in networks
Perspective	Actions of macro-level actors	Opportunities and motivation	Interaction between entrepreneur and ecosystem
Unit of analysis	Institutional context	The individual; the firm	Multiple units of analysis: Context and individual
Policy emphasis	To foster R&D, provide subsidy programs,, transfer/ bridging policies	Fostering firm creation and growth	To foster development of entrepreneurial ecosystems
Financing	Emphasis public support for knowledge production and technology transfer?	Government entrepreneurship programs, incubators, subsidies to BA and VC sector	Multilevel, multi-dimensional and public-private funds and funds of funds

Source: Autio, E. et al., 2014.

ANALYSIS RESULTS OF NEW INNOVATIVE TECHNOLOGIES IN BUSINESS USING GLOBAL ENTREPRENEURSHIP MONITOR (GEM)

GEM Conceptual Framework

GEM began in 1999 as a joint project between Babson College (USA) and London Business School (UK) with aim to investigate why some countries are more entrepreneurial than others. 17 years later, GEM became the richest resource of information on this subject, publishing a series of global, national and special topic reports on an annual basis.

GEM conceptual framework describes the multiple features of entrepreneurship, recognizing the responsible conduct proactive, innovative and risk individuals, always in interaction with the environment. In this regard, GEM survey was conceptualized as the interdependence between entrepreneurship and economic development, in order to:

- Discover the factors that encourage or hinder entrepreneurial activity in terms of social values, personal attributes and entrepreneurial ecosystem;
- Provide a platform for assessing the extent to which entrepreneurship influences economic growth within individual economies;
- Discover policy implications for strengthening the entrepreneurial capacity in the economy.

The social, cultural, political and economic fields were represented by the national framework conditions, which impact the evolution of society through three phases of economic development and the framework conditions for entrepreneurs.

The latter consists of: Finance business, government policies, entrepreneurship programs of government, entrepreneurship education, research and development (R & D) transfer, business infrastructure and legal, internal market dynamics and regulation of entry, physical infrastructure and the rules cultural and social.

Social values related to entrepreneurship include how society values entrepreneurship as a good career choice, whether entrepreneurs have high social status and, as far as the media are positive entrepreneurship in an economy.

Individual attributes include demographic characteristics (sex, age, etc.), self-perception (perceived capabilities, perceived opportunities, fear of failure) and the reasons for starting a business (ie, necessity versus opportunity).

Entrepreneurial activity includes several phases of business (nascent, new business, established business, discontinuation), the potential impact (creation of jobs, innovation, internationalization) and the activity type (Total Early-stage Entrepreneurial Activity (TEA), Social Entrepreneurial Activity (SEA) or Employee Entrepreneurial Activity (EEA)). Operational definitions of business phases and entrepreneurial characteristics are shown in Figure 3.

GEM survey is based on primary data collection through a study on adult population of at least 2,000 randomly selected adults (18-64 years) in each economy. In addition, national teams collect the opinions of experts on entrepreneurship ecosystem components via National Expert Survey (NES).

How GEM Defines Entrepreneurship

While entrepreneurship is a multifaceted phenomenon with many different meanings and definitions, GEM defines entrepreneurship as any attempt at new business or new venture creation, such as self-employment, a new business organization or expanding an existing business by an individual, a team of individuals, or an established business.

While GEM rather narrow entrepreneurship view as business activity, we need an overview of what business activity is supposed to be. For example, unlike many official records of new business activity, GEM definition is not limited to newly registered businesses.

GEM adopts the perspective of professional entrepreneurship, even if it seems beyond individuals officially registered as a freelancer. Entrepreneurship can also be seen in terms of behavior, by identifying employees within organizations who act entrepreneurially (also known as intrapreneurship or corporate entrepreneurship).

For years, GEM has focused on combining stage phase before starting a new business (entrepreneurship training) and the stage immediately after starting a new business (-management firm that owns us). Taken together, this phase is defined as "early-stage entrepreneurial activity (TEA).

Figure 3. The GEM conceptual framework
Source: Singer, M.H. et al., 2015.

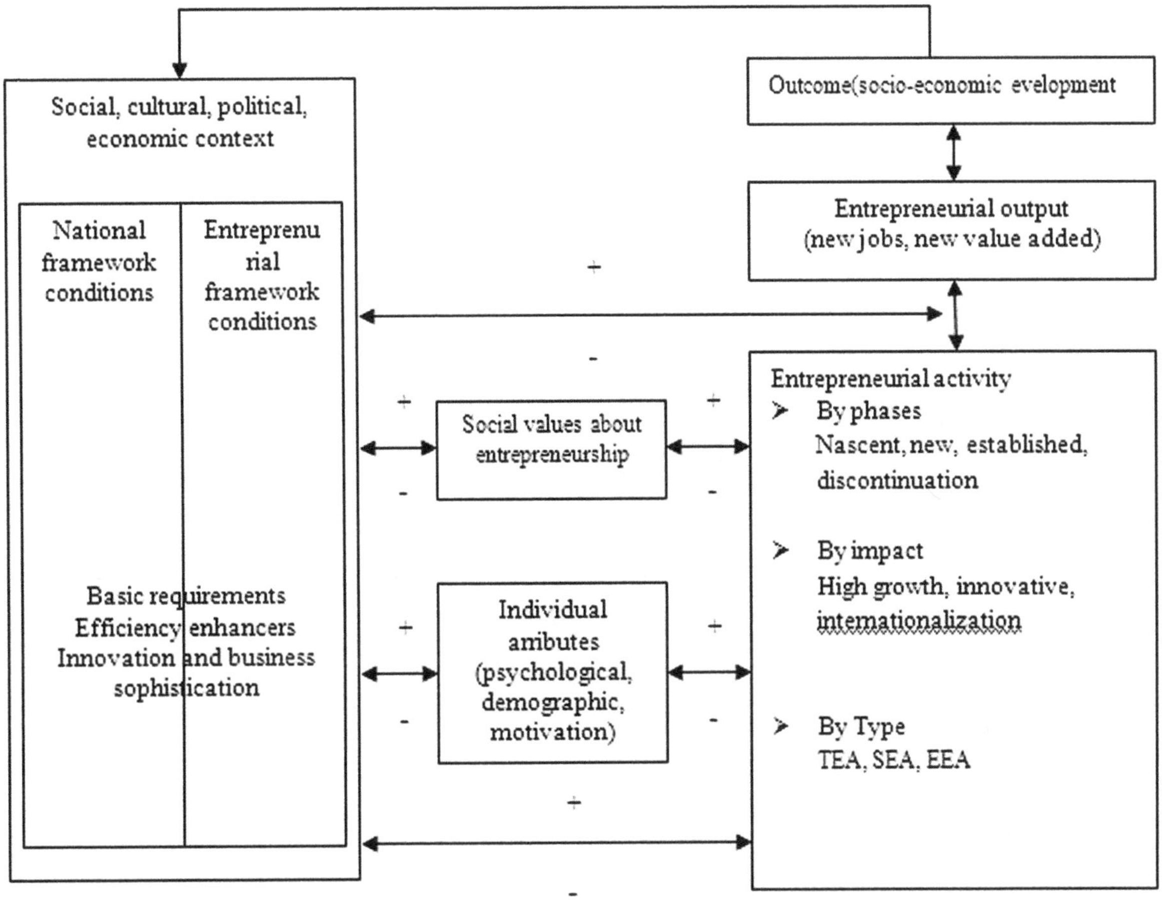

In addition, people with entrepreneurial attitudes - individuals involved as owner-managers in companies- could lead to entrepreneurial activity. These categories that distinguish different stages of entrepreneurship are derived from GEM questionnaire. Figure 4 shows how individuals taking part in Population Survey Adult labeled as entrepreneurs budding owner-managers of new firms and owner-managers of established businesses that depend on the answers to questions specific GEM, which are likely recurrent.

Measures

Entrepreneurial Activity

- **Established Business Ownership Rate (*Estbbuyy*):** Percentage of 18-64 population who are currently owner-manager of an established business, namely, ownership and management of operating businesses that pay wages, salary or other payments to owners for more than 42 months.

Figure 4. Model of Business phases and entrepreneurship characteristics represented in GEM
Source: Singer, M.H. et al., 2015.

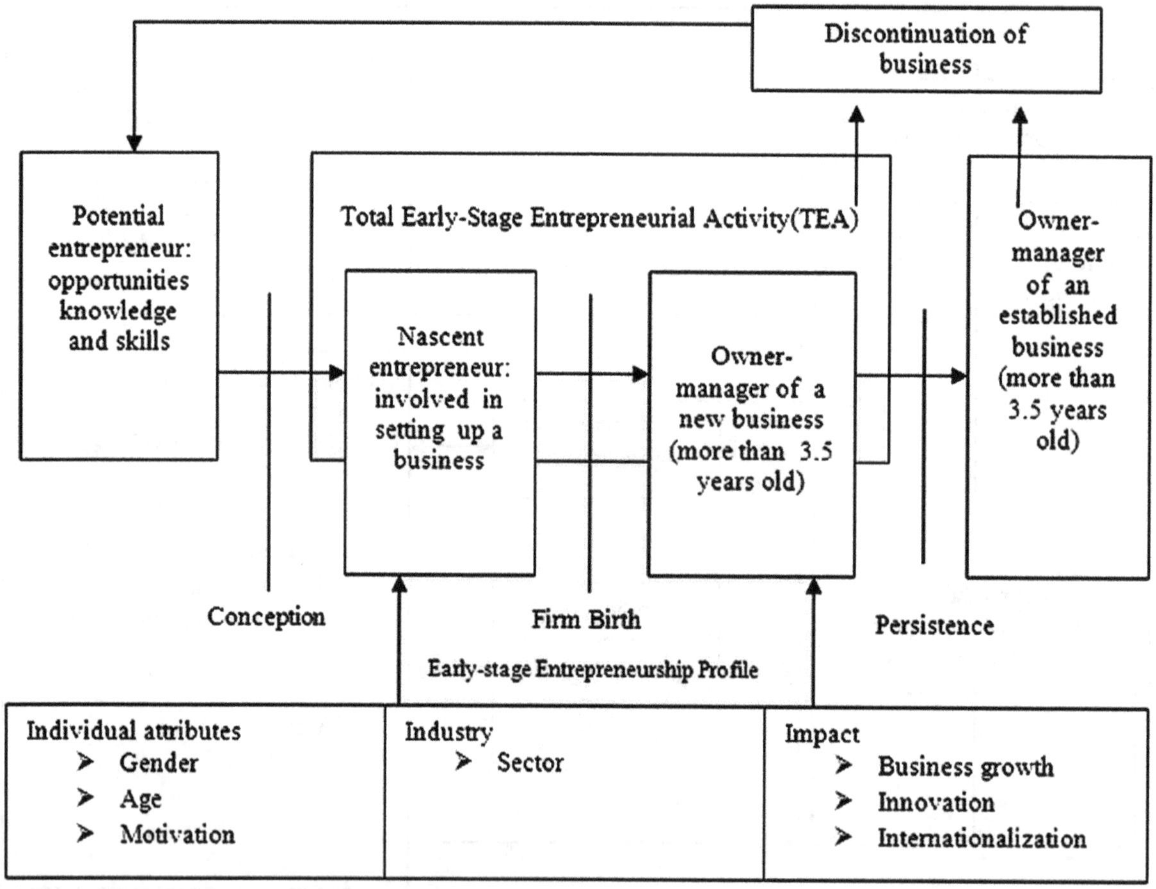

- **Improvement-Driven Opportunity Entrepreneurial Activity - Relative Prevalence (*Teayyido*):** Percentage of those involved in TEA:
  - Claims to be driven by opportunity, as opposed to finding no other option for work; and
  - Indicating the main factor to be involved in this opportunity is being independent or increasing their income, rather than just maintaining their income.
- **Informal Investors Rate (*Busangyy*):** Percentage of 18-64 population who have provided personal funds for a new business started by someone else in the past three years.
- **Nascent Entrepreneurship Rate (*Suboanyy*):** Percentage of 18-64 population who are currently undergoing training an entrepreneur, that is actively involved in setting up a business you will own or co-own; this business has not paid salaries or other payments to the owners of more than three months.
- **Necessity-Driven Entrepreneurial Activity: Relative Prevalence (*Teanec_p*):** Percentage of those involved in TEA, which are involved in entrepreneurship because they had no other option for work.
- **New Business Ownership Rate (*Babybuyy*):** Percentage of 18-64 population who are currently owner-manager of a new business, namely, holding and managing a business operation that has

paid salaries, wages or any other payments the owners for more than three months but not more than 42 months.

- **Total Early-Stage Entrepreneurial Activity (TEA) (*Teayy*):** Percentage of 18-64 population who are either a nascent entrepreneur or owner-manager of a new business (as defined above).
- **Total Early-Stage Entrepreneurial Activity for Male Working Age Population (*Teayymal*):** Percentage male 18-64 population who are either a nascent entrepreneur or owner-manager of a new business (as defined above).
- **Total Early-Stage Entrepreneurial Activity for Female Working Age Population (*Teayyfem*):** Percentage female 18-64 population who are either a nascent entrepreneur or owner-manager of a new business (as defined above).

Entrepreneurial Aspirations

- **Growth Expectation Early-Stage Entrepreneurial Activity: Relative Prevalence (*TEAyyjg5*):** Percentage of TEA who expect to employ at least five employees five years from now.
- **New Product Early-Stage Entrepreneurial Activity (*TEAyynwp*):** Percentage of TEA who indicate that their product or service is new to at least some customers.
- **International Orientation Early-Stage Entrepreneurial Activity (*TEAyyint*):** Percentage of TEA who indicate that at least 25% of the customers come from other countries.

Entrepreneurial Attitudes

- **Entrepreneurial Intention (*Futsupno*):** Percentage of 18-64 population (individuals involved in any stage of entrepreneurial activity excluded) who intend to start a business within three years.
- **Entrepreneurship as Desirable Career Choice (*Nbgoodyy*):** Percentage of 18-64 population who agree with the statement that in their country, most people consider starting a business as a desirable career choice.
- **Fear of Failure Rate (*Frfailop*):** Percentage of 18-64 population with positive perceived opportunities who indicate that fear of failure would prevent them from setting up a business.
- **High Status Successful Entrepreneurship (*Nbstatyy*):** Percentage of 18-64 population who agree with the statement that in their country, successful entrepreneurs receive high status.
- **Know Startup Entrepreneur Rate (*Knoentyy*):** Percentage of 18-64 population who personally know someone who started a business in the past two years.
- **Media Attention for Entrepreneurship (*Nbmediyy*):** Percentage of 18-64 population who agree with the statement that in their country, you will often see stories in the public media about successful new businesses.
- **Perceived Capabilities (*Suskilyy*):** Percentage of 18-64 population who believe to have the required skills and knowledge to start a business.
- **Perceived Opportunities (*Opportyy*):** Percentage of 18-64 who see good opportunities to start a firm in the area where they.

EMPIRICAL RESULTS OF NEW INNOVATIVE TECHNOLOGIES IN BUSINESS IN ROMANIA USING GEM

The level of early stage entrepreneurial established by the analysis is one of the largest in the region of Central and Eastern Europe. The proportion of those involved in business activities in Romania has increased in the last three years due to an improvement of business owners (whose activity is older than 42 months).

In the adult population, there are nearly two and a half more motivated entrepreneurs in early stage opportunity (those pursuing an opportunity) than those motivated by necessity (those with no other option for work).

Romanians seem to have confidence in their abilities to start a new business and also in their ability to identify opportunities. Despite this, they still fear of failure.

The rate of those who intend to start a new business within three years is the largest in the region. Early-stage entrepreneurs in Romania have high expectations in terms of the number of people to be hired, but low ambitions in terms of innovation.

The Romanian typical early stage entrepreneur is male, aged 25-44 years, with an upper secondary level of education and household income in the upper third of the population.

Good internal market dynamics, commercial and legal infrastructure and physical infrastructure positively affect the business environment in Romania. Among the biggest constraints are: taxes and regulations; entrepreneurship education at primary and secondary levels, and financial resources availability:

- **The Agency for Implementing Projects and Programmes for SMEs:** There are a number of government programmes aiming to promote entrepreneurship in Romania. For example, the Agency for Implementing Projects and Programmes for SMEs has spearheaded initiatives to improve entrepreneurial abilities and skills among young entrepreneurs.
- **Trends Over Time:** In recent years, the number of Romanian adults intending to start a business has increased.
- **Challenges for the Future:** Key challenges for the development of Romanian entrepreneurship are: ensuring that government policies provide real support to entrepreneurs; promoting entrepreneurial education (predominantly in primary and secondary schools); providing easy access to different sources of finance, including equity, government funding, loans, business angels and IPOs.

It can be concluded that Romania is detached in advance by Intentional Entrepreneurial indicator, while for the other indicators, the other countries comparatively surveyed register small differences.

In terms of Entrepreneurs distribution by gender in Romania 2014 (%), masculine gender is superior to the female gender for all indicators: Potential, Intentional, Nascent, Young, TEA or EB. The results are presented in Figure 6.

For all indicators, age distribution of Romanian Entrepreneurs in 2014 (%) reveals a superior positioning for groups 25-34 and 35 - 44, while the age groups 18-24, 45-54 and 55-64 have similar values, but obviously lower than the first two. The results are displayed according to Figure 7.

Distribution of entrepreneurs by education in Romania, 2014 (%) reveals a superior positioning for SSD - Some secondary degree and SD - Secondary degree, while EL - educational groups, respectively GE - Graduate experience have similar values, but obviously lower than the first two groups, regardless of the chosen indicator. The results are presented in Figure 8.

Figure 5. Entrepreneurial activity rates in efficiency-driven EU countries, 2014 (% of population aged between 18-64 years)
Source: GEM, Adult Population Survey, 2014 and authors' calculations.

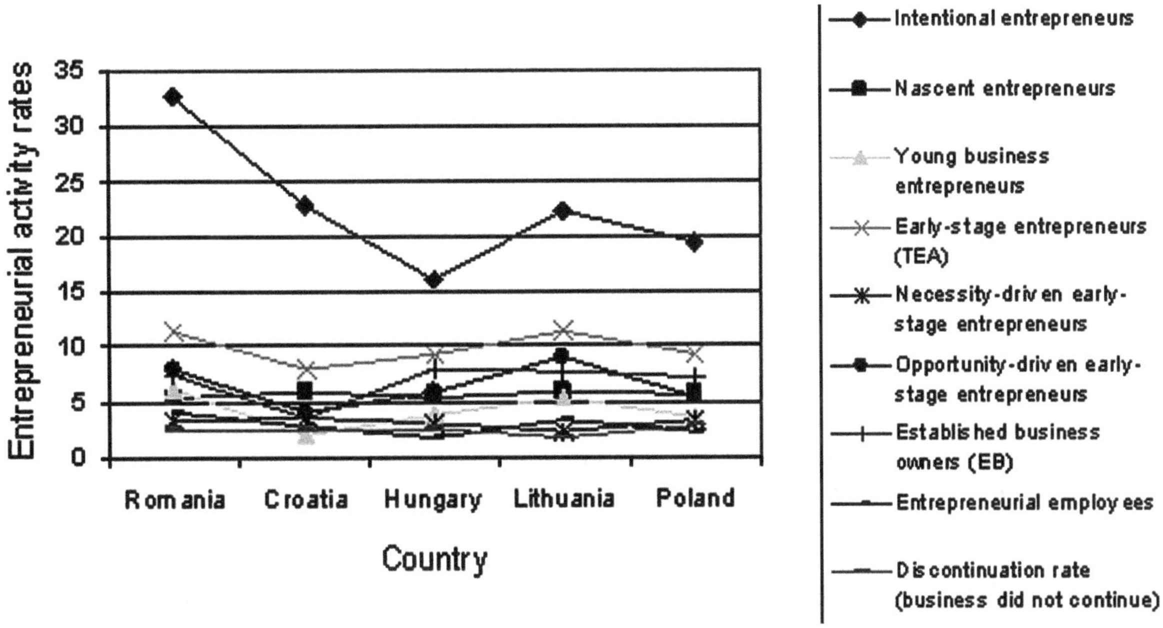

Table 3. Key indicators for Romania (2015)

Symbol Indicator	Indicator	Value (%)	Rank/60
TEA	Total early-stage Entrepreneurial Activity: percentage of 18-64 population who are either a nascent entrepreneur or owner-manager of a new business	11,3	30
Established Business Ownership	Percentage of 18-64 population who are currently an owner-manager of an established business, i.e., owning and managing a running business that has paid salaries, wages, or any other payments to the owners for more than 42 months	7,6	26
Perceived Opportunities	Percentage of 18-64 population (individuals involved in any stage of entrepreneurial activity excluded) who see good opportunities to start a firm in the area where they live	32	45
Perceived Capabilities	Percentage of 18-64 population (individuals involved in any stage of entrepreneurial activity excluded) who believe they have the required skills and knowledge to start a business	48	35
Entrepreneurial Intention	Percentage of 18-64 population (individuals involved in any stage of entrepreneurial activity excluded) who are latent entrepreneurs and who intend to start a business within three years	32	16
Fear of Failure	Percentage of 18-64 population (individuals involved in any stage of entrepreneurial activity excluded) who indicate that fear of failure would prevent them from setting up a business	41	42

Source: http://www.gemconsortium.org/country-profile/103 and authors' calculations.

We can say that Romania's position varies between position 16 and position 45, based on the key indicators.

Figure 6. Distribution of entrepreneurs by gender in Romania, 2014 (%)
Source: GEM, Adult Population Survey, 2014 and authors' calculations.

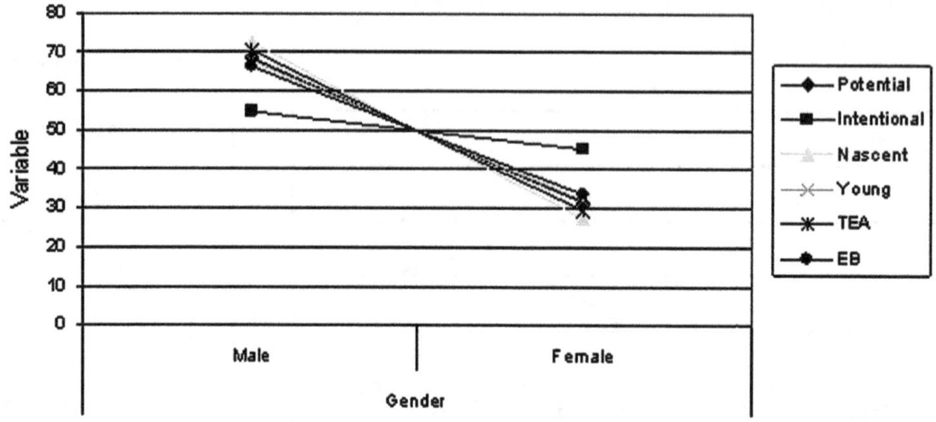

Figure 7. Distribution of entrepreneurs by age in Romania, 2014 (%)
Source: GEM, Adult Population Survey, 2014 and authors' calculations.

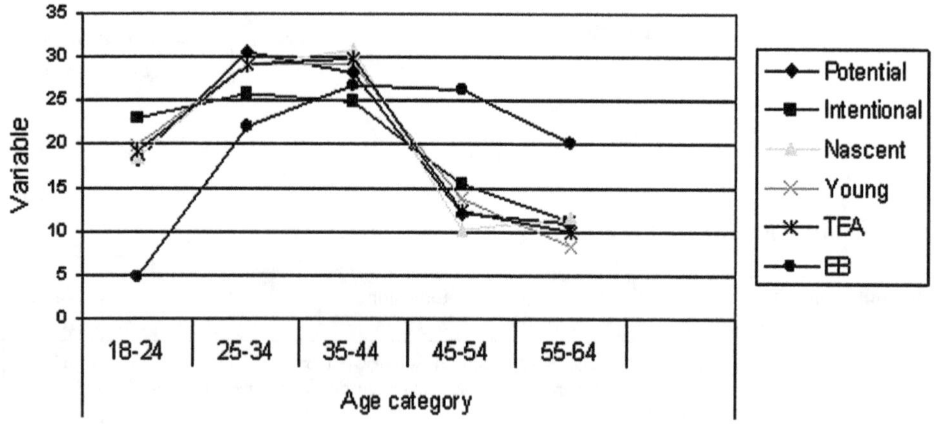

Figure 8. Distribution of entrepreneurs by education in Romania,2014 (%)
Source: GEM, Adult Population Survey, 2014 and authors' calculations. EL - Educational level; SSD - Some secondary degree; SD - Secondary degree; PSD - Post-secondary degree; GE - Graduate experience.

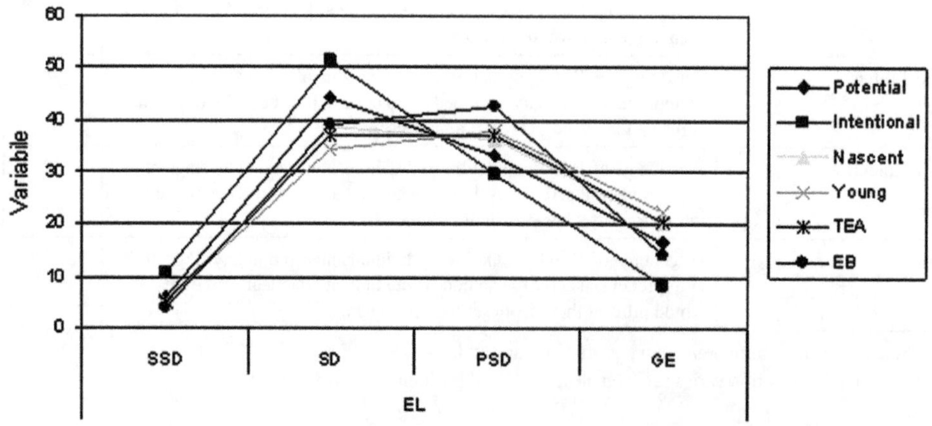

Figure 9. Distribution of entrepreneurs by household income in Romania,2014 (%)
Source: GEM, Adult Population Survey, 2014 and authors' calculations. HI - Household income; L 33% - Lowest 33% tile; M 33% - Middle 33% tile; U 33% - Upper 33% tile.

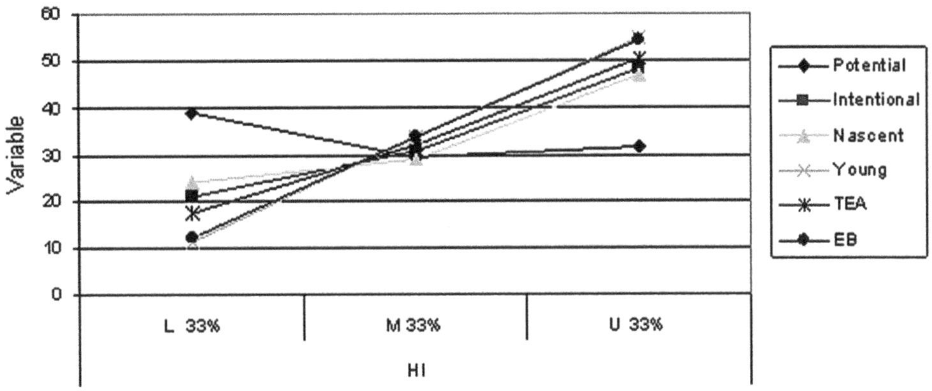

Distribution of Entrepreneurs by household income in Romania 2014 reveals an increasing as it progresses from Group revenue L 33% - 33% Lowest tile, through M 33% - 33% Middle tile, the group came U 33% - 33% Upper tile. (%), for all indicators. The results are presented graphically in Figure 9.

Figure 10 presents the aggregate distribution of entrepreneurs by gender, age, education and household income in Romania, 2014 (%).

CONCLUSION

Digital entrepreneurship in emerging economies needs developing training theory by international business scientists.

Innovation and entrepreneurship studies appear to be closely related and and those concepts seem to converge in a handful of theories that have become the cornerstone of contemporary researchers.

Theories focused on Business Models and Digital Economy are new and scarce defined; however, there is already a solid research on e-Business Models, where these two areas converge.

Theoretical frameworks used in such disciplines should inspire any future research on this new field. However, the lack of generally accepted theories of e-Entrepreneurship still allows creating new constructions that enrich existing frameworks. Therefore, it seems that an empirical approach would be better suited to fill the gaps.

Regarding the total distribution of entrepreneurs by gender, age, education and household income in Romania, 2014 (%), regardless of the chosen indicator - Potential, Intentional, Nascent, Young, TEA or EB - results are presented above.

Figure 10. Distribution of entrepreneurs by gender, age, education and household income in Romania, 2014 (%)
Source: GEM, Adult Population Survey, 2014 and authors' calculations. G – Gender; M – Male; F – Female; AC - Age category.

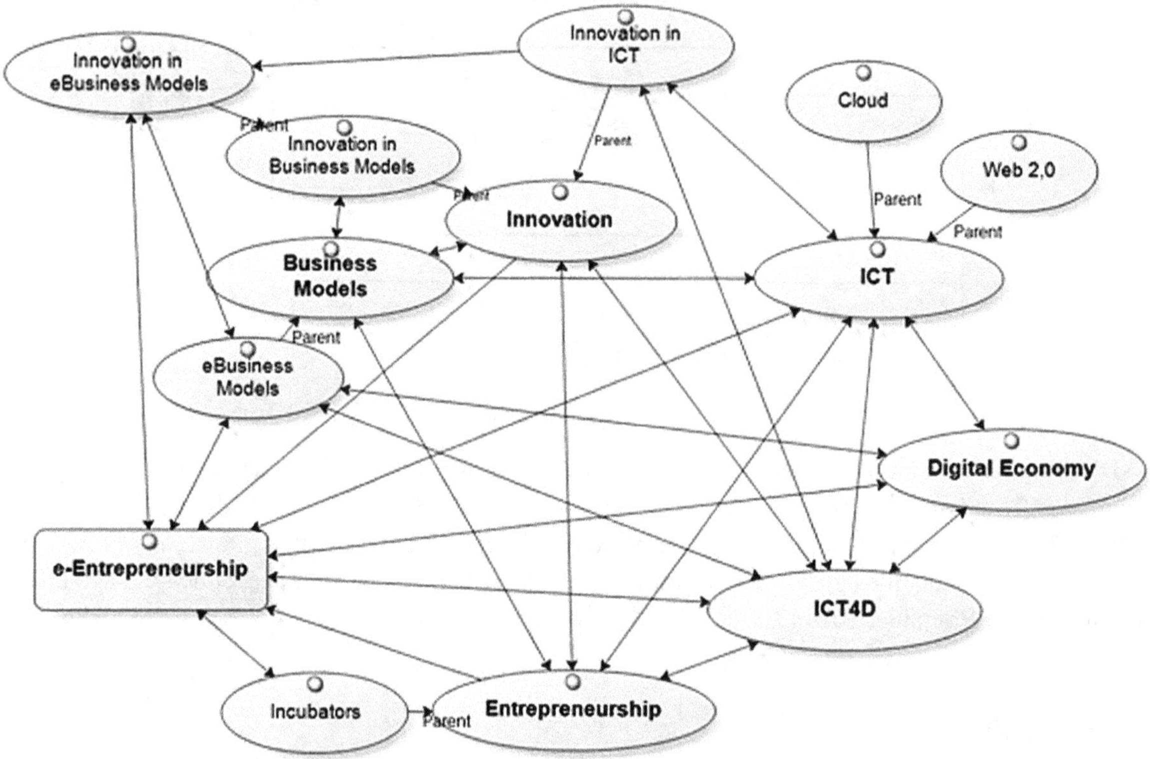

FUTURE RESEARCH DIRECTIONS

Finally, the following questions are proposed to be explored in the future research:

- Are current frameworks on e-Entrepreneurship Digital Ventures applicable in the context of developing economies, and especially in Romania? Given the expiry of the research literature based empirically on e-Entrepreneurship in developing countries and the experience gained from studies of e-commerce adoption for SMEs, it is anticipated that current theoretical frameworks have to be adapted or extended to better understand e-Entrepreneurship in this specific context.
- If Digital Ventures in Romania are internationalized, how will they compete with their international counterparts? Current frameworks seem to converge on the concept of internationalization when SMEs implement an e-business model. As Arenius et al. (2005) indicates, when SMEs are implementing an e-Business, they automatically enter in a global market.
- How are ICTs being used by Digital Ventures in Romania? Both the Frontier Technology proposed by Caselli and Coleman (2006) and low rates of adoption and diffusion of ICT in countries found by researchers in developing ICT4D suggest that digital combination in emerging countries may adopt ICT in a different way.

- Do incubators play a decisive role in the creation and survival Digital Ventures in Romania? As discussed here, incubators can be an excellent case study of Digital Ventures in Romania.

REFERENCES

Acs, Z. J., & Audretsch, D. B. (1990). *Innovation and Small Firms*. The MIT Press.

Afuah, A., & Tucci, C. L. (2000). *Internet business models and strategies: Text and cases*. McGraw Hill Higher Education.

Aliyu, M. S., Rogo, H. B., & Mahmood, R. (2015). Knowledge Management, Entrepreneurial Orientation and Firm Performance: The Role of Organizational Culture. Asian Social Science, 11(23). 10.5539/ass.v11n23p140

Asghari, R., & Gedeon, S. (2010). Significance and Impact of Internet on the Entrepreneurial Process: E-Entrepreneurship and Completely Digital Entrepreneurship.*Proceedings of the 4th European Conference on Innovation and Entrepreneurship, Academic Conferences Limited*, 70.

Autio, E., Kenney, M., Mustar, P., Siegel, D., & Wright, M. (2014). Entrepreneurial innovation: The importance of context. *Research Policy*, *43*(7), 1097–1108. doi:10.1016/j.respol.2014.01.015

Barnes, D., Clear, F., Romano, D., Harindranath, G., Harris, L., & Rae, A. (2012). Web 2.0 and micro-businesses: An exploratory investigation. *Journal of Small Business and Enterprise Development*, *19*(4), 687–711. doi:10.1108/14626001211277479

Barnes, D., Hinton, M., & Mieczkowska, S. (2004). Avoiding the fate of the dotbombs: Lessons from three surviving dotcom start-ups. *Journal of Small Business and Enterprise Development*, *11*(3), 329–337. doi:10.1108/14626000410551582

Barr, C. J. (2001). The Digital Enterprise: How to Reshape Your Business for a Connected World. *Journal of the American Society for Information Science and Technology*, *52*(12), 1084–1085. doi:10.1002/asi.1167

Batjargal, B. (2005). *Internet entrepreneurship*. Networks and performance of internet ventures in China.

Berman, S. J. (2012). Digital transformation: Opportunities to create new business models. *Strategy and Leadership*, *40*(2), 16–24. doi:10.1108/10878571211209314

Bhupatiraju, S., Nomaler, Ö., Triulzi, G., & Verspagen, B. (2012). Knowledge flows – Analyzing the core literature of innovation, entrepreneurship and science and technology studies. *Research Policy*, *41*(7), 1205–1218. doi:10.1016/j.respol.2012.03.011

Boateng, R., Heeks, R., Molla, A., & Hinson, R. (2008). E-commerce and socio-economic development: Conceptualizing the link. *Internet Research*, *18*(5), 562–594. doi:10.1108/10662240810912783

Brynjolfsson, E., & Hee Oh, J. (2012). T*he Attention Economy: Measuring the Value of Free Digital Services on the Internet.Thirty Third International Conference on Information Systems*, Orlando, FL.

Brynjolfsson, E., & Kahin, B. (2000). *Understanding the digital economy: data, tools and research.* Massachusetts, USA: The MIT Press.

Cantillon, R. (1755). *Essay on the Nature of Trade in General*. London: Macmillan.

Carayannis, E. G., & Von Zedtwitz, M. (2005). Architecting gloCal (global–local), real-virtual incubator networks (G-RVINs) as catalysts and accelerators of entrepreneurship in transitioning and developing economies: Lessons learned and best practices from current development and business incubation practices. *Technovation*, *25*(2), 95–110. doi:10.1016/S0166-4972(03)00072-5

Casson, M. (1982). The Entrepreneur. Totowa, NJ: Barnes & Noble Books.

Crespi, G., & Zuñiga, P. (2012). Innovation and productivity: evidence from six Latin American countries. World Development, 40, 273–290.

Daude, C. (2010). *Innovation, productivity and economic development in Latin America and the Caribbean*. OECD Publishing. doi:10.1787/5kmlcz254421-en

Davidson, E., & Vaast, E. (2010). *Digital Entrepreneurship and its Sociomaterial Enactment*. Paper presented at 43rd Hawaii International Conference on System Sciences (HICSS).

Davidsson, P. (2004). *Researching Entrepreneurship*. New York: Springer.

Doganova, L., & Eyquem-Renault, M. (2009). What do business models do?: Innovation devices in technology entrepreneurship. *Research Policy*, *38*(10), 1559–1570. doi:10.1016/j.respol.2009.08.002

Drucker, P. F. (2002). The discipline of innovation. *Harvard Business Review*, *80*, 95–104. PMID:12195923

Facet, T. (2011). Lessons on business virtual incubation services. In INFODEV (Ed.), Information for Development Program. The Netherlands: infoDev.

Fichman, R. G. (1999). The Diffusion and Assimilation of Information Technology Innovations. In Framing the Domains of IT Management: Projecting the Future…Through the Past. Cincinnati, OH: Pinnaflex Educational Resources, Inc.

Foster, C. & Heeks, R. (2013). Conceptualising Inclusive Innovation: Modifying Systems of Innovation Frameworks to Understand Diffusion of New Technology to Low-Income Consumers. European Journal of Development Research, 25, 333–355.

Gartner, W. B. (1988, Spring). Who is an entrepreneur is the wrong question. *American Small Business Journal*.

Gartner, W. B. (1990). What are we talking about when we talk about entrepreneurship? *Journal of Business Venturing*, *5*(1), 15–28. doi:10.1016/0883-9026(90)90023-M

Gem Consortium. (2014). *The level of early-stage entrepreneurship has staibilisied – and is one of the highest in the Central and Eastern Europe region*. Retrieved from: http://www.gemconsortium.org/country-profile/103

Gem Consortium. (n.d.). Retrieved from: http://www.gemconsortium.org/

Gill, R., & Larson, G. S. (2014). Making the ideal (local) entrepreneur: Place and the regional development of high-tech entrepreneurial identity. *Human Relations*, *67*(5), 519–542. doi:10.1177/0018726713496829

Gopal, R. D., Ramesh, R., & Whinston, A. B. (2003). Microproducts in a Digital Economy: Trading Small, Gaining Large. *International Journal of Electronic Commerce*, *8*, 9–29.

Groen, A., Oakey, R. P., & Cook, G. (2008). *New Technology-based Firms in the New Millennium* (Vol. 6). Bingley, UK: Emerald Group Publishing.

Hair, N., Wetsch, L. R., Hull, C. E., Perotti, V., & Hung, Y.-T. C. (2012). Market Orientation in Digital Entrepreneurship: Advantages and Challenges in A Web 2.0 Networked World. *International Journal of Innovation and Technology Management*, *9*(6), 1250045. doi:10.1142/S0219877012500459

Hamilton, L. C., & Asundi, R. (2008). Technology usage and innovation: Its effect on the profitability of SMEs. *Management Research News*, *31*(11), 830–845. doi:10.1108/01409170810913033

Hargadon, A., & Kenney, M. (2012). Misguided policy? Following venture capital into clean technology. *California Management Review*, *54*(2), 118–139. doi:10.1525/cmr.2012.54.2.118

Helfat, C.E., Finkelstein, S., Mitchell, W., Peteraf, M., Singh, H., Teece, D. & Winter, S. G. (2009). *Dynamic capabilities: Understanding strategic change in organizations*. Wiley.com.

Hilbert, M. (2010). When is Cheap, Cheap Enough to Bridge the Digital Divide? Modeling Income Related Structural Challenges of Technology Diffusion in Latin America. *World Development*, *38*(5), 756–770. doi:10.1016/j.worlddev.2009.11.019

Hindle, K., & Yencken, J. (2004). Public research commercialisation, entrepreneurship and new technology based firms: An integrated model. *Technovation*, *24*(10), 793–803. doi:10.1016/S0166-4972(03)00023-3

Hinson, R., Boateng, R., & Olav Jull, S. (2008). E-business financing: Preliminary insights from a developing economy context. *Journal of Information. Communication & Ethics in Society*, *6*(3), 196–215. doi:10.1108/14779960810916219

Hitt, L. M., & Brynjolfsson, E. (1996). Productivity, Business Profitability, and Consumer Surplus: Three Different Measures of Information Technology Value. *Management Information Systems Quarterly*, *20*(2), 121–142. doi:10.2307/249475

Hull, C. E. K., Hung, Y.-T. C., Hair, N., Perotti, V., & DeMartino, R. (2007). Taking advantage of digital opportunities: A typology of digital entrepreneurship. *International Journal of Networking and Virtual Organisations*, *4*(3), 290–303. doi:10.1504/IJNVO.2007.015166

Isenberg, D. (2010). The big idea: How to start an entrepreneurial revolution. *Harvard Business Review*, *88*(6), 40–50.

Jansen, W. & Grance, T. (2011). *Guidelines on security and privacy in public cloud computing*. NIST Special Publication, 800-144.

Javalgi, R. G., Todd, P. R., Johnston, W. J., & Granot, E. (2012). Entrepreneurship, muddling through, and Indian Internet-enabled SMEs. *Journal of Business Research*, *65*(6), 740–744. doi:10.1016/j.jbusres.2010.12.010

Kamien, M. I., & Schwartz, N. (1982). *Market structure and innovation*. New York: Cambridge University Press.

Kenny, C. (2003). The Internet and economic growth in less-developed countries: A case of managing expectations? *Oxford Development Studies*, *31*(1), 99–113. doi:10.1080/1360081032000047212

Knight, P. T. (2011). e-Development in Brasil. *Journal of the Knowledge Economy*, *2*(1), 77–116. doi:10.1007/s13132-010-0024-5

Kollmann, T. (2006). What is e-entrepreneurship?–fundamentals of company founding in the net economy. *International Journal of Technology Management*, *33*(4), 322–340. doi:10.1504/IJTM.2006.009247

Lastres, H. M. M., & Cassiolato, J. E. (2003). *Systems of innovation and development from a South American perspective: a contribution to Globelics*. Paper presented at the First Globelics Workshop on Innovation Systems. Aalborg University, Aalborg, Denmark.

Lee, S.-H., Dewester, D., & Park, S. R. (2008). Web 2.0 and opportunities for small businesses. *Service Business*, *2*(4), 335–345. doi:10.1007/s11628-008-0043-5

Lockett, N. J., & Brown, D. H. (2000). eClusters: The potential for the emergence of digital enterprise communities enabled by one or more intermediaries in SMEs. *Knowledge and Process Management*, *7*(3), 196–206. doi:10.1002/1099-1441(200007/09)7:3<196::AID-KPM85>3.0.CO;2-7

Low, M. B., & McMillan, I. C. (1988). Entrepreneurship: Past research and future challenges. *Journal of Management*, 14.

Lusch, R., & Nambisan, S. (2015). Service Innovation: A Service-Dominant Logic Perspective. *Management Information Systems Quarterly*, *39*(1), 155–175.

Maertens, A., & Barrett, C. B. (2013). Measuring Social Networks' Effects on Agricultural Technology Adoption. *American Journal of Agricultural Economics*, *95*(2), 353–359. doi:10.1093/ajae/aas049

Manafi, M., & Subamanian, I. D. (2015). The role of perceived justice in the relationship between human resource management practices and knowledge sharing: A study of Malaysian Universities Lecturers. *Asian Social Science*, *11*(12), 131–143. doi:10.5539/ass.v11n12p131

Marshall, A. (1890). *Principles of Economics*. London: MacMillan.

Martha, G. M. (2004). Institutions and the Adoption of Electronic Commerce in Mexico. *Electronic Commerce Research*, *4*(3), 201–219. doi:10.1023/B:ELEC.0000027980.16492.af

Matlay, H. (2004). E-entrepreneurship and small e-business development: Towards a comparative research agenda. *Journal of Small Business and Enterprise Development*, *11*(3), 408–414. doi:10.1108/14626000410551663

Middleton, K. L., & Byus, K. (2011). Information and communications technology adoption and use in small and medium businesses. *Management Research Review*, *34*(1), 98–110. doi:10.1108/01409171111096496

Miller, D., & Garnsey, E. (2000). Entrepreneurs and technology diffusion: How diffusion research can benefit from a greater understanding of entrepreneurship. *Technology in Society*, *22*(4), 445–465. doi:10.1016/S0160-791X(00)00021-X

Mills, C., & Pawson, K. (2012). Integrating motivation, risk-taking and self-identity: A typology of ICT enterprise development narratives. *International Small Business Journal, 30*(5), 584606. doi:10.1177/0266242610390594

Nair, M., Kuppusamy, M., & Davison, R. (2005). A Longitudinal Study on the Global Digital Divide Problem: Strategies to Close Cross-Country Digital Gap. *The Business Review, Cambridge, 4*, 315–326.

Ngoasong, M. Z. (2015). *Digital Entrepreneurship in Emerging Economies: The role of ICTs and local context.42nd AIB-UKI Conference*, Manchester, UK.

Ngwenyama, O., & Morawczynski, O. (2009). Factors affecting ICT expansion in emerging economies: An analysis of ICT infrastructure expansion in five Latin American countries. *Information Technology for Development, 15*(4), 237–258. doi:10.1002/itdj.20128

O'Reilly, T. (2007). What is web 2.0: Design patterns and business models for the next generation of software. *Communications & Strategies*, 17.

Oestreicher-Singer, G., & Zalmanson, L. (2013). Content or Community? A Digital Business Strategy for Content Providers in the Social Age. *Management Information Systems Quarterly, 37*(2), 591–616.

Ojala, A., & Tyrvainen, P. (2011). Developing Cloud Business Models: A Case Study on Cloud Gaming. *Software, IEEE, 28*(4), 42–47. doi:10.1109/MS.2011.51

Oliner, S. D., & Sichel, D. E. (2000). The Resurgence of Growth in the Late 1990s: Is Information Technology the Story? *The Journal of Economic Perspectives, 14*(4), 3–22. doi:10.1257/jep.14.4.3

Onetti, A., Zucchella, A., Jones, M., & McDougall-Covin, P. (2012). Internationalization, innovation and entrepreneurship: Business models for new technology-based firms. *Journal of Management & Governance., 16*(3), 337–368. doi:10.1007/s10997-010-9154-1

Osterwalder, A., Lagha, S. B., & Pigneur, Y. (2002). *An ontology for developing e-business models.* IFIP DsiAge.

Osterwalder, A., & Pigneur, Y. (2002). *An e-business model ontology for modeling e-business.15th Bled Electronic Commerce Conference*, Bled, Slovenia.

Pateli, A., & Giaglis, G. (2003). *A framework for understanding and analysing e-business models.Bled Electronic Commerce Conference.*

Quinones, G., Nicholson, B., & Heeks, R. (2013). Positioning research on e-entrepreneurship in developing economies: a study of latin-american digital ventures, Paper for ICIER 2013, Rio de Janeiro.

Rahman, M., Hasan, M. R., & Floyd, D. (2013). Brand orientation as a strategy that influences the adoption of innovation in the bottom of the pyramid market. *Strategic Change, 22*(3-4), 225–239. doi:10.1002/jsc.1935

Rappa, M. (2003). *Business models on the web.* Available at http://digitalenterprise.org/models/models.html

Rogers, E. M. (1962). *Diffusion of innovations.* New York: The Free Press.

Sako, M. (2012). Business Models for Strategy and Innovation. *Communications of the ACM, 55*(7), 22–24. doi:10.1145/2209249.2209259

Schumpeter, J. A. (1934). *The Theory of Economic Development; an Inquiry into Profits, Capital, Credit, Interst, and the Business Cycle*. Cambridge, MA: Harvard University Press.

Schumpeter, J. A. (1939). *Business Cycles: A Theoretical, Historical, and Statistical Analysis of the Capitalist Process*. New York: McGraw-Hill.

Shane, S. A. (2003). *A general theory of entrepreneurship: the individual-opportunity nexus*. Cheltenham, UK: Edward Elgar. doi:10.4337/9781781007990

Shane, S. A., & Venkataraman, S. (2000). The Promise of Entrepreneurship as a Field of Research. *Academy of Management Review, 25*(1).

Souitaris, V., & Cohen, M. (2003). Internet-Business or Just Business?: Impact of 'Internet-Specific' Strategies on Venture Performance. *European Management Journal, 21*(4), 421–437. doi:10.1016/S0263-2373(03)00080-X

Tapscott, D. (1996). *The digital economy: Promise and peril in the age of networked intelligence*. McGraw-Hill New York.

Taylor, M., & Murphy, A. (2004). SMEs and e-business. *Journal of Small Business and Enterprise Development, 11*(3), 280–289. doi:10.1108/14626000410551546

Thewarapperuma, A. (2013). *East African virtual incubation pilot to launch in Nairobi, Kenya*. infoDev. Available: http://www.infodev.org/en/Article.968.html

Timmers, P. (1998). Business models for electronic markets. *Electronic Markets, 8*(2), 3–8. doi:10.1080/10196789800000016

Tornatzky, L., & Fleischer, M. (1990). *The Process of Technological Innovation*. Lexington.

Utterback, J. M. (1971). The Process of Technological Innovation within the Firm. *Academy of Management Journal, 14*(1), 75–88. doi:10.2307/254712

Vasilchenko, E., & Morrish, S. (2011). The Role of Entrepreneurial Networks in the Exploration and Exploitation of Internationalization Opportunities by Information and Communication Technology Firms. *Journal of International Marketing, 19*(4), 88–105. doi:10.1509/jim.10.0134

Venkataraman, S. (1997). *The distinctive domain of entrepreneurship research*: An 23 editor's perspective. In J. Katz & J. Brockhaus (Eds.), *Advances in Entrepreneurship, Firm Emergence, and Growth*. Greenwich, CT: JAI Press.

Wall, B., Jagdev, H., & Browne, J. (2007). A review of eBusiness and digital business—applications, models and trends. *Production Planning and Control, 18*(3), 239–260. doi:10.1080/09537280601127245

Wang, C. L. (2008). Entrepreneurial orientation, learning orientation, and firm performance. *Entrepreneurship Theory and Practice, 32*(4), 635–657. doi:10.1111/j.1540-6520.2008.00246.x

Weill, P., & Woerner, S. L. (2013). Optimizing Your Digital Business Model. *MIT Sloan Management Review*, *54*, 71–78.

Xenikou, A., & Simosi, A. (2006). Organizational culture and transformational leadership as predictors of business unit performance. *Journal of Managerial Psychology*, *21*(6), 566–579. doi:10.1108/02683940610684409

Zahra, S., & Dess, G. G. (2001). Entrepreneurship as a field of research: Encouraging dialogue and debate. *Academy of Management Review*, *26*(1), 8–10. doi:10.5465/AMR.2001.4011916

Zhu, K., Kraemer, K. L., & Xu, S. (2006). The Process of Innovation Assimilation by Firms in Different Countries: A Technology Diffusion Perspective on E-Business. *Management Science*, *52*(10), 1557–1576. doi:10.1287/mnsc.1050.0487

KEY TERMS AND DEFINITIONS

Adult Population Survey (APS): The APS is a comprehensive questionnaire, administered to a minimum of 2000 adults in each GEM country, designed to collect detailed information on the entrepreneurial activity, attitudes and aspirations of respondents.

Entrepreneurial Framework Conditions (EFCs): The nine conditions identified by GEM that enhance (or hinder) new business creation in a given country, and provide the basis of the NES.

Global Entrepreneurship Research Association (GERA): GERA is the umbrella organization that owns GEM, registered as a private limited not-for-profit company in the United Kingdom. It is funded by National Team fees, and the contributions of sponsoring institutions.

Individual-Level Data: A .sav data file, either APS or NES, containing all information for all interviews conducted.

National Expert Survey (NES): The NES is administered to 36 'experts' in each GEM country and collects data on the context in which entrepreneurship takes place in a country. It provides information about the nine aspects of a country's socio-economic milieu that are believed to have a significant impact on national entrepreneurship - the Entrepreneurship Framework Conditions (EFCs).

National-Level Data: A .sav data file, either APS or NES, containing GEM-calculated rates for a country or countries.

National Team: GEM is formed of a consortium of 'National Teams'. Each Team is led by a local university or academic institution. The team is the official national representative of the project - responsible for collecting GEM data in the country on an annual basis, producing a 'National Report' on their findings, and acting as the point of contact for GEM enquiries.

Request For Proposal (RFP) Package: The package, usually a .zip file, which includes all the documents which National Teams need to submit in order to request approval for data collection as well as the documents they will need to conduct the surveys.

Request For Proposal (RFP): This is a .pdf document issued by GEM, outlining the requirements for completing the APS and NES in a particular year. It is contained within the RFP package (see above).

Total Early-Stage Entrepreneurial Activity (TEA): GEM's most well-known index, representing the percentage of 18-64 population who are either a nascent entrepreneur or owner-manager of a new business.

Chapter 6
Entrepreneurship and Innovation in Latin American Family Firms:
The Case of México

Verónica Baños-Monroy
Tecnológico de Monterrey, Mexico

Edgar Ramírez-Solís
Tecnológico de Monterrey, Mexico

Lucia Rodríguez-Aceves
Tecnológico de Monterrey, Mexico

ABSTRACT

This chapter examines the role of entrepreneurship and innovation in the development of family businesses in Latin America. The socio-economic dynamics of such region is an interesting field for research, mainly because: it is a major manufacturing hub with growing importance in retail; it experienced a huge economical shift in the last ten years causing the growth of the middle class (an increase of 50 million people); family firms in the region are the main driver of growth and employment. The authors focused the analysis on Mexico due to its singular characteristics, making it a representative example of the region. Based on secondary sources, a characterization of the entrepreneurship and innovation in family owned and managed firms in the region is reviewed. Moreover, based on the Global Entrepreneurship Monitor database, the authors present some recent data related with entrepreneurship levels in the region and its impact in terms of innovation. Finally, public policy implications are provided in order to encourage entrepreneurship and innovation in family businesses.

DOI: 10.4018/978-1-5225-0953-0.ch006

INTRODUCTION

The growing interest in Latin American economic and social history began in the 1960's and led to the publication of many sectorial studies that became standard reference works. The study of family firms in the same region can be traced to the 1970´s. In the decades of the 80´s and 90´s the economic sectors in which family businesses operated became relevant, because the number of companies starting operations was different in emphasis in each country. For example, in Brazil, family firms involved in coffee, sugar and some specific manufacturing branches were predominantly located in Sao Paulo and Rio de Janeiro. In Colombia, the companies were working in sectors related to coffee, tobacco and banana production, particularly in the Antioquia region (Barbero, 2008).

Potential growth in Latin America has been estimated at between 3% and 4% annually over the past ten years, which is slightly below to the growth recorded during the most recent period of expansion, from 2004 to 2008. However, the slowdown has continued since 2010 as the international situation, which undoubtedly boosted economic activity in the past, has worsened, so the "new normal" for the region may well be lower than previously expected. Latin America's productivity in recent years has been disappointing compared with that of both OECD countries and emerging economies (OECD, 2014).

Moderate poverty fell from more than 40% in 2000 to less than 30% in 2010. This decline in poverty implies that 50 million Latin Americans saw improved living conditions in the first decade of the 21st century. The Latin American middle class grew very substantially: from 100 million people in 2000, to around 150 million in 2012. The emerging middle class differs, of course, from country to country, but there are a number of common threads. Middle class entrants are more educated than those they have left behind. They are also more likely to live in urban areas and to work in formal sector jobs. Middle class women are more likely to have fewer children and to participate in the labor force than women in the poor or vulnerable groups (Ferreira et al, 2013), despite microcredits received by the latter (Gámez-Gutiérrez & Saiz-Alvarez, 2012).

Stronger productivity would lead to more growth and would reduce the significant rates of inequality and poverty found in the region. Education and innovation reforms must ensure equal opportunities for access to a complete, high-quality cycle of education and a workforce with better skills: this can be achieved with better matches to the labor market. There must also be measures to promote formal employment. In Latin America and the Caribbean, 36% of companies operating in the formal economy struggle to find a properly trained workforce, compared to a global average of 21% per country and an OECD average of 15%. Latin American firms are 3 times more likely than South Asian firms, and 13 times more likely than Pacific-Asian firms to face serious operational problems due to a shortage of human capital (OECD, 2014).

Furthermore, Latin America is a major manufacturing hub with growing importance in retail, and is increasingly a supplier of services. Ford, Toyota, VW, General Motors and Renault amongst others, manufacture cars in the region while Rio de Janeiro is the support center for offshore oil exploration. The size and growth potential of the regional market has encouraged global brands such as Nestle, Unilever and Kraft to develop significant local production facilities encouraged by low labor costs. Embraer, a manufacturer of short/medium-haul aircraft, while not privately controlled, also has a presence in the region. It sells its aircraft worldwide and enjoys an excellent reputation for quality. Latin America is also a leader in financial services, suffering no significant impact from the credit crunch of 2008. Banks like Itau Unibanco and BTG Pactual, both privately controlled, have built strong franchises in retail/commercial and investment banking respectively.

Major Latin American economies experienced slow growth for decades before the boom of the 2000s; over 110 years, these economies grew at a slower rate than the United States. At the beginning of the 20th century, a simple average of the region's gross domestic product per capita was about 38 percent that of the United States. By 2012, that ratio was about 35 percent. A main player in this huge economical shift are the family firms. The top 10 family businesses generated approximately US$290 billion out of Latin America's US$5.3 trillion GDP in 2012 and employ almost one million people (E&Y, 2014).

As it was stated, the socio-economic dynamics of Latin America is an interesting field for research. Therefore, in order to go deeper in this chapter we focus our attention on entrepreneurship and innovation because they are factors influencing the social and economic changes in the region. The chapter is organized as follows: we start with an economic overview of Latin America family firms with focus in Mexico, due to this country offers representative characteristics within the Latin American context. Following that, we offer an overview of the current situation of some aspects surrounding entrepreneurship and innovation in Latin American family firms. We conclude explaining some challenges that family companies and policy makers face in this region.

OVERVIEW OF LATIN AMERICAN FAMILY FIRMS WITH FOCUS IN MEXICO

Family-owned or controlled firms are the leading form of business organization in Latin American countries, and this has been true for some time, even among large listed companies. Some of the oldest family businesses in Latin America are located in Mexico and Brazil. For example, we have the Ypióca Group founded in 1846 which is involved in the beverage industry and belongs to the Telles family. Also from Brazil we have two more examples: Rotermund S.A. Indústria E Comércio founded in 1877 and which delivers specialized design services and Klabin which was founded in 1890 by the Klabin family and which produces paper products. In Mexico, the oldest family business is Tardan Hermanos Sucesores S.A. which belongs to the Tardan family and which produces and sells hats and caps (E&Y, 2014).

In recent years, over 90 per cent of 33 of the largest businesses in Latin America are family-owned and managed (La Porta et al., 1999). In addition, family ownership in Latin America typically involves multiple generations of managers and superimposes generational hierarchy on managerial relations.

Family businesses generate 60% of Latin America's GDP and employ 70% of the workforce in the region. Regarding the generations leading those companies, we can say that 47% of them are managed by the first generation, 29% of the companies are managed by the second generation, 14% are managed jointly by the first and second generations at the same time and only 10% are led by the third or fourth generations (E&Y, 2014).

Family controlled businesses have been the driver of growth and employment in Latin America. With the exception of a handful of state controlled champions such as CODELCO in Chile, Petrobras in Brazil, PDVSA in Venezuela, as well as PEMEX in Mexico, the majority of economic activity is seeded and controlled by private family-owned companies or foreign multinationals, often in partnership with local family firms. There are, however, few businesses which can trace their family ownership back more than three generations, with the majority of businesses still in the founding or second generation. Diniz, De Moraes, De Andrade Faria and Marinho in Brazil; Matte and Said in Chile; Garza Lagüera, Zambrano, Azcárraga, Romo Garza, Saba, Arango, Peralta, and Salinas Pliego in Mexico and Perez Companc and Rocca in Argentina are all family names readily identifiable with prominent businesses in their home countries (Kirkman, 1999).

In Latin America, especially in Mexico, family firms are smaller and younger than their counterparts in the rest of the world; almost half of them are in fact first-generation controlled. Research shows that there is a lack of strategic planning and had centralized decision-making (Burgoa et al., 2013; Durán & San Martín, 2013). This situation implies a major problem for Latin American entrepreneurs because they need faster and more modern organizational systems in order to compete globally. However, only 15 years ago, it was unusual for a Latin American family company to have an international presence. Since then, a number of regional champions have internationalized their businesses in a series of well-planned opportunistic moves, and this is the case in the following Mexican firms: Bimbo, América Movil, Cemex and Televisa.

In Mexico, micro, small and medium sized (SMEs) enterprises represents 99.8% of the total economic units, contributes with 35.9% to the GDP and generates more than 74% of the jobs (INEGI, 2014). According to the SME Observatory (Observatorio Pyme), 65% of these companies are family businesses (CIPI, 2003). Therefore, family businesses represent the largest number of enterprises that currently exist in Mexico. Additionally, 60% of jobs are generated by FB and they contribute with half of the GDP; around 90% of the more than five million businesses are managed by a family. According to the National Institute of Statistics, Geography and Informatics (INEGI), 57% of the companies have been in existence for less than five years and these small, more recently-founded companies are made up by one family and several non-related workers. They also experience family conflicts in the course of their day-to-day operations and the new generations faced more challenges than the founders did (García Fuentes, 2015).

A totally different reality, in terms of size and age, is the one that characterizes big companies. In this regard, five of the top ten biggest companies in Mexico are family businesses: América Movil, Cemex, Femsa, Alfa and Grupo Bimbo. The other five companies in the top ten are Pemex and CFE (government companies); Walmart, GM and BBVA-Bancomer (Global Companies) (Expansion, 2015). More than 90% of the firms listed on the Mexican stock exchange (BMV) have a clear family representation in both capital and control (KPMG, 2013).

Unlike large businesses, there is not enough statistical information on medium and small family firms; there is no census to define the type of businesses (family and non-family) in Mexico and no official agency has registered firms in this sector to date (Avendaño, Kelly, Treviño & Madero, 2009).

The lack of accurate statistical information about family businesses in Mexico leaves us with many unanswered questions. This is because in Mexico there is not a legal entity recognized as a "family business" which, as in other countries, could be classified as an official entity like small or medium enterprises with rights and obligations. That is one of the reasons why many contemporary family businesses are working unofficially.

Now that the context of the FB in Latin America and México has been stated, in the next paragraphs we present some interesting facts about the entrepreneurship in the region, because as previous research suggests, entrepreneurship is a very important activity for a country´s competitiveness and growth and a significant source of social mobility (Amorós, Fernández & Tapia, 2012).

ENTREPRENEURSHIP IN LATIN AMERICA AND MEXICO

According to Miller (1983), entrepreneurship is a multidimensional concept that encompasses the firm's actions relating to product-market and technological innovation, risk taking and proactiveness. Entrepreneurship in the context of family firms has been referred to as corporate entrepreneurship (Zahra et al., 2000;

Kellermanns & Eddleston, 2006), entrepreneurial behavior (Kellermanns et al., 2008), entrepreneurial orientation (Naldi, Nordqvist, Sjöberg, & Wiklund, 2007), or entrepreneurial spirit (García-Tenorio & Sánchez, 2009). In relation to family firm performance associated with entrepreneurship, indicators as profitability, growth (employment, sales, profits), and others, such as cash-flow, have been used in those studies (Durán-Encalada, San Martin-Reyna, & Montiel-Campos, 2012). The effects of entrepreneurship on different performance metrics vary according to the length of time considered for analyzing these impacts, growth normally precedes profitability and some types of entrepreneurial initiatives may favor more clearly some performance measures instead of others (Wiklund & Shepherd, 2003).

From a practical perspective, entrepreneurship is a very important activity for a country´s competitiveness and growth and a significant source of social mobility (Amorós, Fernández & Tapia, 2012). New ventures have become an important aspect of countries´ economic growth, particularly in terms of their contributions to new job creation (Birch, 1987), therefore it is an elusive target of public policy. In terms of academic research, recently some authors begun to offer useful conceptual frameworks and some empirical evidence regarding the peculiarities of entrepreneurial family businesses (Habbershon, et al., 2010; Nordqvist & Melin, 2010). For example, Dyer (1986) highlighted "strong organizational cultures" as a general characteristic of family business. Moreover, Gupta and Levenburg (2010) studied the impact of the industry environment and national culture on organization's entrepreneurial capabilities.

Particularly in Latin America, while a paternalistic organizational culture may be associated with business activities, studies reveal that these patterns are often challenged (Lenartowicz & Johnson, 2003). In such region, family business cultures have to adapt rapidly to uncertainty and change with a high degree of flexibility and have often to consider initiating new business ventures (Hall, Melin, & Nordqvist, 2001). In the same vein, most entrepreneurial ventures receive support from family members both financially and physically (Astrachan, Zahra, & Sharma, 2003) as well as access to economic, market and social networks (Pistrui, Welsch, & Roberts, 1997). One explanation is that entrepreneurs in Latin America cannot rely on sophisticated regulatory systems or systems that provide financial support and venture capital.

In the particular case of Latin America, there are a number of cultural, social, and economic factors that have a negative influence on the entrepreneurial context compared with other regions (Kantis 2005). In fact, there is sufficient evidence showing that the number of growth-oriented and innovative ventures in the region is limited. In Latin America and other parts of the developing world, the entrepreneurship is commonly characterized as by "necessity", where entrepreneurs launched their business not by choice, but by the need to earn a living when other options, such as wage employment, are lacking. They often operate in the informal sector, have low levels of technology and earn a rate of return equivalent to that of a relatively modest salaried job (Lora & Castellani, 2014). In addition, Firms in LA tend to be smaller (in terms of the number of employees) at birth than firms in other regions at similar levels of development, and the growth process fails to compensate for the initial gap in employment (Lederman et al. 2014). Another salient feature of LA entrepreneurship is that new firms do not grow as much as firms in other regions and thus tend to remain small. Traditionally, Latin American societies are characterized as risk averse because of their heterogeneous economic, political and societal context.

In Mexico, one of the most consistent and distinctive characteristics of entrepreneurship is the connection with family business. Start-up capital most often comes from family. Besides, the staff, or at least executives or managers, are usually members of the family. Besides, one of the primary reasons for opening a new venture is to provide and independent living for one`s children (Young & Welsch, 1993). In Mexico, the low barriers to entry make the start-up phase for entrepreneurs relatively easy, nevertheless, the regulatory environment that is both, burdensome and inefficient, contributes to the business closing. Studies of Mexican business find that as companies hire employees, acquire fixed locations, and otherwise increase the scale of their businesses, profits do not rise proportionally (Brenes & Haar, 2012). Since the jumping from informal to formal is especially costly, entrepreneurs prefer to stay informal. In sum, in Mexico the most important challenges are the financing sources, the social distrust, the business informality and the traditional sectors of investment (Lederman et al. 2014).

As it was previously stated, in general in Latin America and therefore in Mexico, there is a lack of sufficient statistical data to improve the understanding of entrepreneurship in the region. For such reason, some useful initiatives have emerged in order provide information that can be useful for policy makers in order to improve the amount and quality of entrepreneurship, especially when focused on reducing social imbalances through social entrepreneurship and solidarity economics (Saiz-Alvarez, 2016).

The Global Entrepreneurship Monitor (GEM) has developed a rich understanding of global entrepreneurship for about 17 years. It is the largest survey of entrepreneurship in the world, covering 60 economies in 2015. GEM tracks rates of entrepreneurship across multiple phases and assess the characteristics, motivations and ambitions of entrepreneurs, and the attitudes societies have toward this activity. It is noteworthy that GEM takes a point-in-time snapshot of entrepreneurship and business activity, because the survey does not follow entrepreneurs over time. Nevertheless, in order to see how many new business progress to the established business phase, GEM provides a platform for the examination of the level of mature business activity relative to start-ups (Kelley, Singer & Herrington, 2015). Such platform consists of the comparison between two phases. The first phase is related with the total early stage entrepreneurial activity (TEA), which is defined by the nascent and new businesses having more than three months operating and less than 42 months. The second phase regards the established businesses, that are characterized for having more than 42 months operating. A high level of established business activity may therefore mean that a high level of start-up activity feeds the next phase and that there is an element of sustainability due to synergy among the feasibility of the businesses started, the abilities and ambitions of the entrepreneurs, and enabling factors in the environment. Based on the GEM database, we present in Figure 1, the evolution of the TEA and Established business for the past six years, comparing Latin America with Mexico.

TEA rates tend to be highest in the factor-driven economies, decreasing with higher levels of economic development. In fact, the average TEA rate in the factor-driven economies is over two and-a half times that of the innovation driven economies (Kelley, Singer & Herrington, 2015). At a regional level, TEA rates are typically highest in Africa and in Latin America and the Caribbean. Figure 1 shows a relatively steady behavior in the percentage of Latin American TEA, with a small increase from 18% in 2010 to 19.9% in 2015. In contrast, Mexico presents a growing and constant pattern from 9.6% in 2011 until 21% in 2015, with a total increase of more than 11% over the last five years. While economic development levels and regional location can explain similar patterns in entrepreneurial activity (Coduras, Cuervo-Arango, & Saiz-Alvarez, 2016), the variations exhibited across the GEM shows that other forces are in play. For example, regulatory environments, cultural values and so forth (Kelley, Singer, & Herrington, 2015).

Globally, established business ownership is highest among the factor-driven group, mainly because there is a larger base of people starting businesses. As it is shown in Figure 2, in Latin America there was a significant reduction in the percentage, from 9.3 in 2010 to 6.5% in 2011. After that year, a recovery in the percentage can be identified for the next four years, where 8.5% is the greater in 2015. In contrast, Mexico in 2010 had a deplorable percentage of established business of less than 1%, which

Figure 1. Total early-stage entrepreneurial activity (TEA) comparison
Source: Own elaboration with data from GEM database.

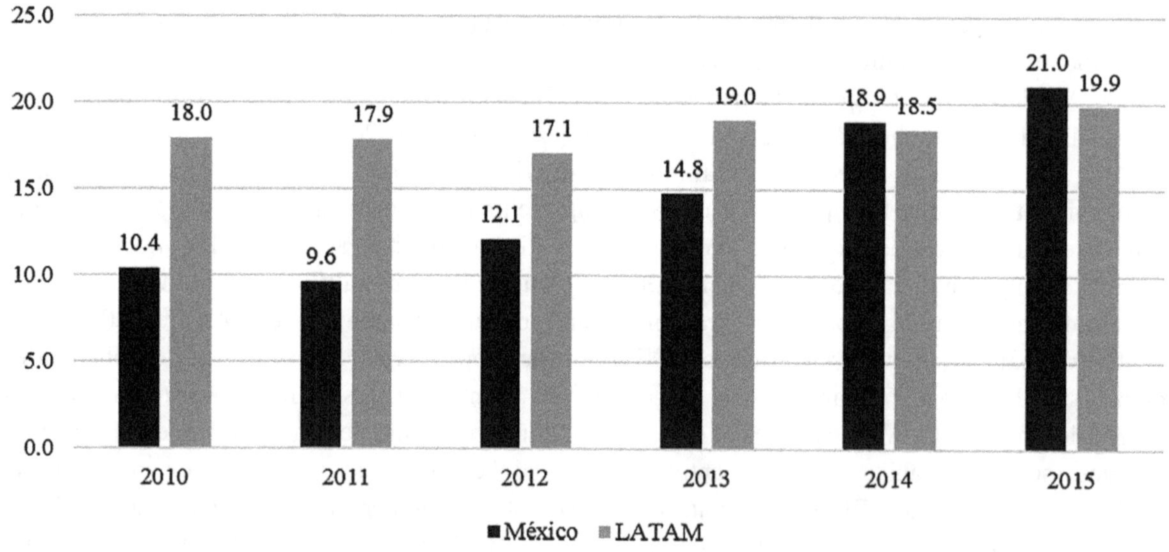

after two years became 4.7%. In the last three years an increasing pattern can be depicted, from 4.2% in 2013, until 6.9% in 2015.

Globally, relative to TEA, there are comparatively fewer established businesses in the factor- and efficiency-driven economies. In five economies of Latin America (i.e. Puerto Rico, Colombia, Peru, Mexico and Panama), established business rates are less than one-third the TEA rates for the last three years. In fact, in Latin America, 43% of the TEA survives more than 42 months, in Mexico, only 33% is going to make it. This shows a consistent imbalance between early-stage entrepreneurial activity and mature business activity, pointing to issues with sustainability of start-up efforts in these economies (Kelley, Singer, & Herrington, 2015).

In despite of the relevant function of entrepreneurship in job generation, persist the debate about the real impact that entrepreneurship has on competitiveness development and economic growth of the Countries (Acs & Storey, 2004; Van Stel et al. 2005: Acs & Amorós, 2008). One of the most prominent indicators of the impact of entrepreneurship is the innovation, in terms of the development of new products and services, as well as in the improvement of operational processes that allows companies to become more efficient and to increase their revenues. For such reason, in the next paragraphs we focus on that particular topic.

INNOVATION IN LATIN AMERICA AND MEXICO

On diverse literature, innovation refers to the introduction of the new products, diffusion of the new methods of production, new markets, discovery of alternative sources and changes in organization or management (Schumpeter, 1994, Dominguez, 2016). Generally speaking, innovation is a process built on different activities that result in a new product, expansion of a process, a specific market with local, national or international scope. As we know, there is not only one kind of innovation, it could be involved

Figure 2. Established business (more than 3.5 years of operation) comparison
Source: Own elaboration with data from GEM database.

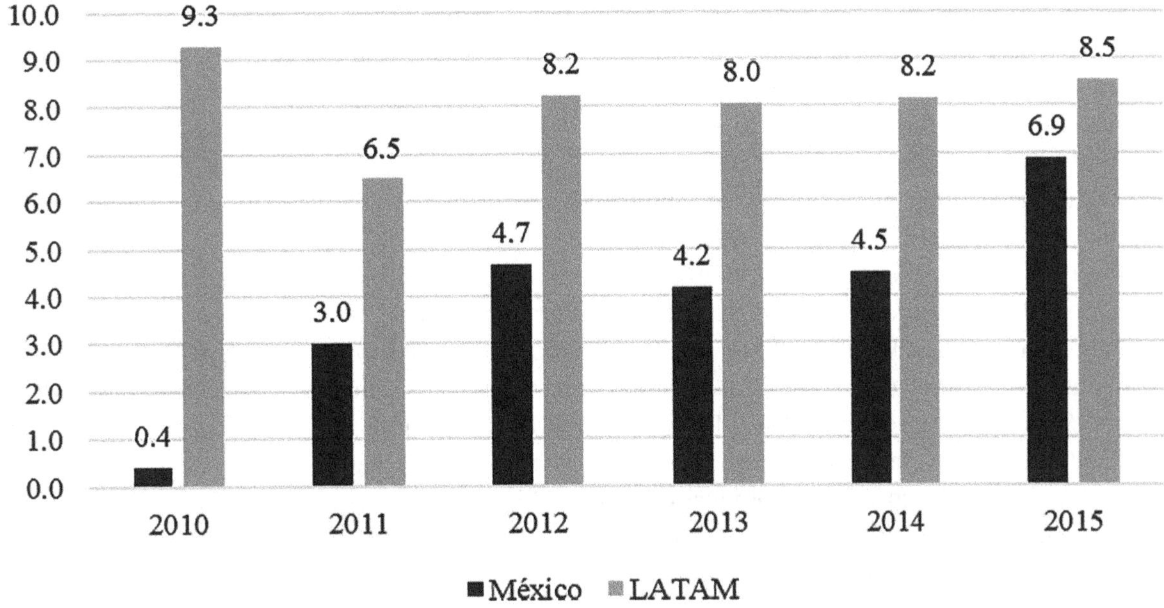

in the processes (which is the most common) or it can be included in the structural design (organizational chart) or even on the business model of the company, especially when there are mergers and acquisitions (Saiz-Alvarez & Colvin, 2015).

The definition of innovation capacity we found more accurate to our chapter is the one described by Lumpkin and Dess (1996) that states that innovation capacity is "a firm tendency to engage in and support new ideas, novelty, experimentation, and creative processes that may result in new products, services on technological processes" (p. 142). For the purpose of this chapter we considered innovation and innovation capacity as the same construct.

Empirical studies on innovating firms are based on surveys where the definition of innovation is related to the process that includes a set of scientific, technological, organizational, financial and commercial activities, as a result, these activities are supposed to create new products or improve them in technological terms; or use new technology or make an improvement in the manufacturing process (OECD, 1997).

Innovation and technology diffusion are considered to be the fundamentals of economic growth; it´s been said that these two factors together generate competitive advantages to companies as well as countries (Balwin & Johnson, 1996). Innovation usually has been related to a set of strategies, activities and attitudes that have an impact on performance and would be associated with higher performance and that the last result may be reflected in growth rate, generation of profits and participation in foreign markets (Lederman et al, 2014). The nature of innovation in a specific region of a country (especially in developing countries) that specializes in some economic activities is determined by structural factors like industrial concentration, the size of the local market and the cooperation among companies. Other important factors that also affect as we will see in paragraphs below are: financial sources, entrepreneurship spirit, scientific and technological infrastructure, systems of industrial protection and academic level. (Lederman et al, 2014)

As we will review in further paragraphs when we talk in detail about the factors that can or cannot help and promote innovation in Latin American countries, we could say now that to achieve development among regions it would be important to make public policies focused on the generation of firms that should have innovation as a premise, and the creation of an environment convenient to develop this kind of firms. The characteristics that identify a firm as an innovative one are not completely clear to the naked eye since these characteristics usually are not evident and their root is very diverse, such as: diversification in products and markets, research and development activities (R&D), design engineering and the acquisition of machinery and equipment, creation of new products and processes, including commercial strategies (Dominguez, 2016)

The types of innovation that triggers productivity growth in low- and middle-income countries differ from those that drive growth in high-income economies. Most firms in developing countries engage in activities not necessarily related to technology; they innovate by adopting and adapting products and production processes that have already been tested in countries that are already at the technology "state of the art" (Grosssman & Helpman, 1991; Dutz, 2007; Ayyagari, Demirgüc-Kunt, & Maksimovic, 20111). The evidence suggests that Latin America and the Caribbean (LAC) countries firms tend to score toward the bottom end of the spectrum in product innovation. Firms in East Asia and Eastern Europe tend to introduce new products more frequently, conduct more R&D, and obtain patents in the US more often than do firms in LAC. (Lederman et al. 2014).

After comparing innovation performance across regions, the findings were related to the factors that can potentially affect firm´s innovative potential. The focus was on four factors that have been shown to affect innovation: regulations, competition, access to finance and entrepreneurial skills (Lederman et al. 2014). Also the findings are related to the regulatory reforms that over the past 20 years improved the business environment of the region substantially, however although some benefits may still be extracted from improving the regulatory framework, regulation may no longer be the most severe obstacle to unleashing the private sector´s innovative potential; probably the most important challenge to address would be other aspects of the regulatory environment such as governance and uncompetitive practices.

Although there are serious technical challenges in measuring competition, LAC shows a pattern consistent with a few actors grabbing a large share of the market and having little incentive to innovate. The region exhibits high concentration both across and within markets: production remains less diversified than in other countries, and within sectors.

Regarding the human capital, the findings on the Lederman et al. (2014) study on the aspect of entrepreneurship, showed that entrepreneur´s technical and managerial background has been found to increase firm´s chances to success. Historical studies found a positive association between densities of engineers at the beginning of the 20th century and long-term economic growth (Maloney & Valencia-Caicedo, 2012). The analysis suggests that the region is still lagging along this dimension. Although cross-country comparisons remain a challenge in this field, most LAC countries have smaller percentages of science and engineering graduates than do countries in Eastern Europe or East Asia. The region also appears to have surprisingly few well-managed modern firms, as reflected by the poor quality of management practices in relatively large formal sector firms. (Lederman et al, 2014).

Among the four factors found to be the most important affecting innovation on the LAC region according to Lederman et al. (2014) study, the first factor to discuss is Regulations. Regulations are not bad by themselves since they can be implemented to solve market failures, prevent unfair practices and protect workers and consumers. How regulations are implemented and reinforced also matters. Risk-averse firms may be better off knowing the rules with certainty-even if they are costly-than facing uncertainty.

Higher uncertainty in the returns from innovation reduces the value of innovation, discouraging some firms from innovating.

The second factor to review is Competition. Innovation decisions are closed linked to profits, which depend crucially on the level of competition. The link between competition and innovation has been central in the policy debate in developing countries for many decades. There are two opposing points of view regarding this relation. Less competition can provide incentives for entrepreneurs to invest in innovation. This view has led many to argue in favor of stronger patent and licensing protection as a way to boost incentives to innovate. Such protection is equivalent to an increase in the value of innovation relative to the value of the status quo, which leads more firms to innovate (Lederman et al., 2014)

The third factor affecting innovation is Access to finance. Financial development can positively affect innovation by decreasing the fixed and variable costs of innovation by reducing financing costs. Financial development can thus have a positive effect on innovation. In particular, deepening financial development may have large effects on innovation if it expands available credit to firms that were previously financially constrained (Lederman et al., 2014)

The last factor affecting innovation according to this study is Entrepreneurial Skills. Entrepreneurial skills can affect both the number of innovations and the intensity of investment in innovation. To the extent that entrepreneurial skills increase the ability to introduce new products or improve existing technologies, they raise the value of innovation and thus the likelihood of engaging in innovation activities.

With a few exceptions, Latin American firms tend to engage less in innovation than firms in other parts of the world. On average, firms in the region are 20% less likely to have introduced a new product than their counterparts in Eastern Europe and Central Asia or in high-income countries. And the picture gets worst in the Caribbean, where the chance of introducing a new product is half that observed in Eastern Europe and Central Asia or high-income countries.

A second feature that distinguishes LAC from China and high-income countries on the continent is the preponderant role played by the public sector in R&D (Pagés-Serra, 2010) and this reflects the low level of private investment in innovation. Also to support the preceding statement, on the study of Lederman et al (2014) they found that on patenting activity of most LAC countries it remains low: no LAC country exhibits a level of patenting that equals that of a high-income country, and most countries have lower levels than their peers. Brazil, for instance, registered just 5 patents per million people between 2006 and 2010, half the number per capita of China (10) and only less than a quarter the number per capita in Bulgaria (22).

R&D and patenting as we saw are indicators of the intensity and quality of innovation, but they indicate only indirectly how firms perform in terms of process of innovation. But not all of the innovation depends directly on how firms design their own production processes. Also innovation has a lot to do in the way the companies are run by their management team.

Maloney and Sarrias (2012) conducted a research about the factors associated with the poor management practices of Latin American firms. Probably, part of the "management gap" that exists between the US and Latin American companies can be explained by firm characteristics. In the United States, midsized firms have a larger share of employees with a college degree, are larger, and are more likely to be multinational corporations than firms in Latin America. The proportion of family-owned business (which, on average, tend to be less well managed, less formal and less structured than publicly traded companies) is almost twice as large in Latin America (about 20%) as in the United States (about 10%). Furthermore, a large proportion of firms in Latin America are run by their founder (more than a third in Brazil) (Lederman et al., 2014).

Eddleston et al. (2008) stated that many family firms do not plan for the future or invest in developing their innovative capacity since these firms are often characterized as resistant to change (Hall et al., 2001), taking less risk (Naldi et al., 2007) and conservative in their strategic behavior (Sharma et al., 1997). And because innovation requires all these aspects, we could expect that innovation capacity is lower in family firms in contrast with non-family ones.

Furthermore, the culture of family businesses is internally focused and because of that knowledge and expertise reside within the family boundaries (Zahra et al., 2004). This internal orientation may stifle entrepreneurial behavior over time (Zahra et al., 2004) and consequently the innovation capacity.

On the contrary, some authors emphasize that family firms have a long term vision (Ward & Aronoff, 1991), because it is the ideal and dream of most family businesses to be passed on to later generations. This strengthens the idea of investing in the long run (Le Breton-Miller & Miller, 2006). This long term investment decreases the threat for liquidation (Dobrzynski, 1993) and makes it for family firms easier to pursue innovative strategies (Teece, 1992).

Although research about innovation capacity related to family is limited, a related domain, entrepreneurial behavior, is given a lot of attention in the family business literature (Kellermanns et al., 2008; Zahra, 2005; Zahra et al., 2004). Empirical research has shown that entrepreneurial activity is a common characteristic of many family firms (Zahra et al., 2004). Since innovation capacity is seen as a dimension of entrepreneurial behavior it could be said that this is something family businesses are investing in.

Innovation is a necessary condition for family firm's continuity; firms permanently have to find the ways to acknowledge and exploit new opportunities as well as refine existing resources in order to successfully grow and compete (Zahra, 2005). The decision for a family firm to start thinking itself as an innovative company can be very complex since family firms usually are trying to solve the competition between two major goals they have: on the one hand trying to cope with economic efficiency and obtaining constant revenues for the partners. On the other hand, trying to fulfill the social interests of the family members (Astrachan & Jaskiewicz, 2008; Chua et al 2003). For example, a business is expected to achieve financial success and at the same time the company has to be capable of managing the family needs of employment, identity and wealth.

The Global Entrepreneurship Monitor (GEM) collects some indicators that allows policy makers and researchers to assess the impact of the entrepreneurial activity. One of the indicators is the innovation, assessed as the extent in which entrepreneurs are introducing products or services that are new to some or all customers, and that are offered by few or no competitors. As it is expected, average innovation levels increase with development level, because of the higher levels of education, greater access to advanced technologies, and favorable conditions in the environment. In addition, many developed economies are characterized by crowded competitive spaces and markets accustomed to advanced solutions, therefore, entrepreneurs may need to introduce novel solutions in order to compete successfully (Kelley, Singer & Herrington, 2015). Using GEM database, we compared the number of entrepreneurs in Latin America and México which stated they offer new products to some customers, and that are offered by few or no competitors. As it is shown in Figure 3, the comparison takes into account the early stage entrepreneurial activity (TEA) and the established businesses (EB).

Almost 27% of entrepreneurs in TEA in Latin America offers innovative products or services. In contrast, in México, only 18% are offered this way. In the established businesses, the percentages are close in both regions, with a small difference of 2%. Such results could be explained because the only chance that new businesses have to enter into the market, is through the offering of new products or services to the customers. Perhaps, the established business already has loyal customers, which precludes the

Figure 3. New product market combination (customers and competitors)
Source: Own elaboration with data from GEM database.

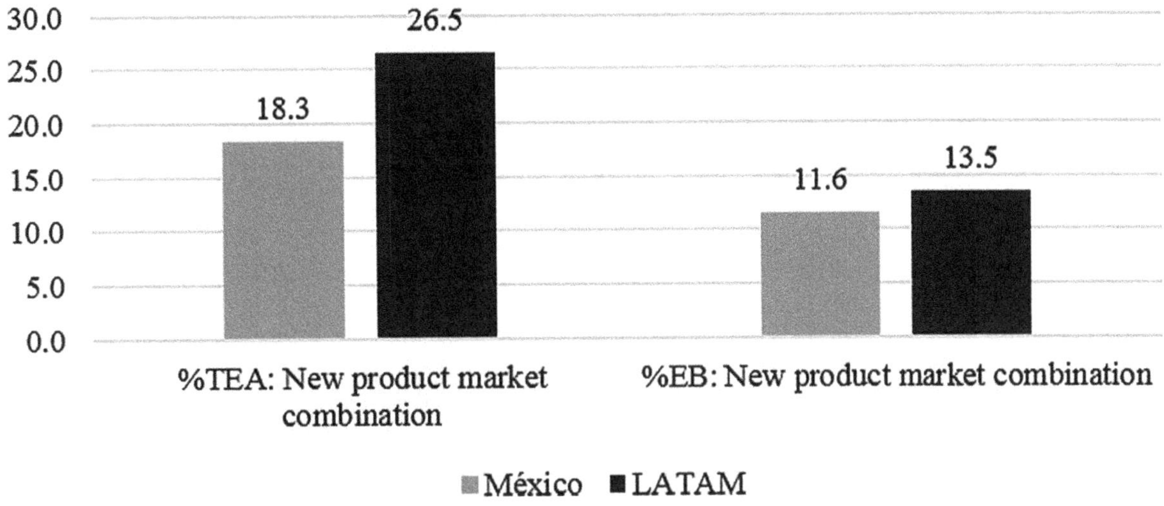

necessity to innovate. In the same vein, Figure 4 shows the evolution of the percentage of entrepreneurs in early stage who indicate that their product or service is new to at least some customers.

As it is shown in Figure 4, in 2015 in Mexico, only 31% of the early stage entrepreneurs confirms to offer new products to some of their clients. Such percentage increases to 46% in Latin America. The evolution of the indicator shows a relatively steady behavior for Latin America, and the opposite for

Figure 4. Percentage of entrepreneurs in early stage who indicated that their product or service is new to at least some customers
Source: Own elaboration with data from GEM database.

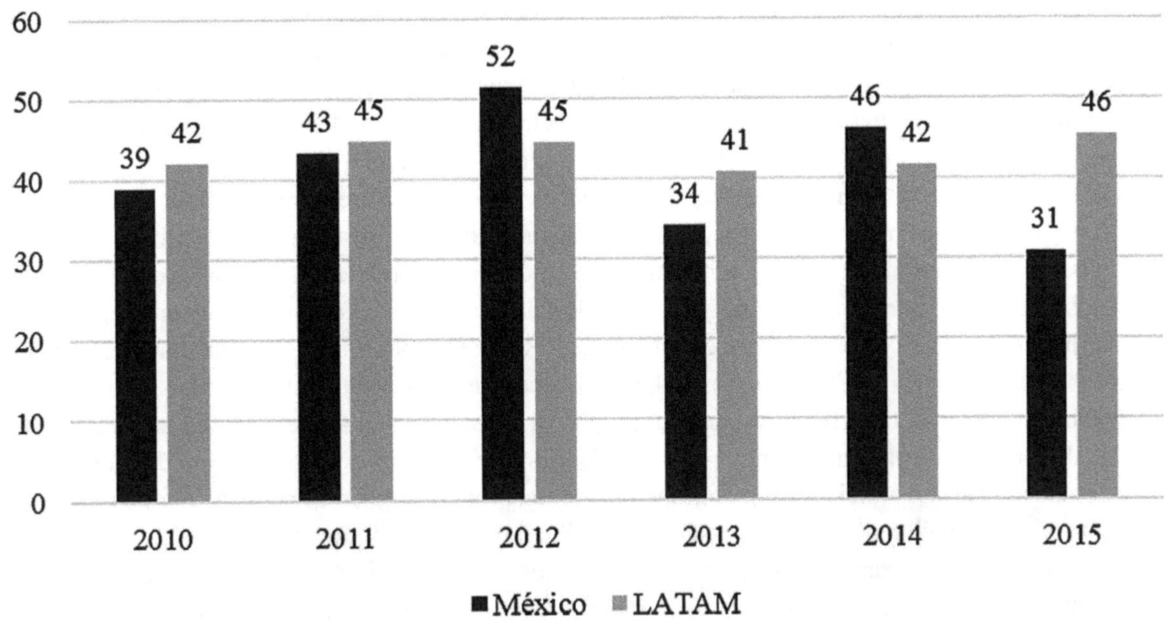

Mexico. Nevertheless, both percentages are relatively low when comparing with economies based-on innovation.

As it was previously stated, the landscape for Latin America and Mexico economic progress is promising but with a lot of challenges to overcome. FB are interconnected with entrepreneurship and innovation, and therefore can have an important influence on the development of the region. Therefore, policy makers have a lot of chances to encourage the creation of new ventures, owned and managed by families, and based on innovation in order to increase their chances to survive and grow. We conclude the chapter with some proposal for policy makers.

CONCLUSION

Creating an enabling environments for innovative entrepreneurship in family firms is difficult because it involves multiple policy areas that interact in complex ways. Of the usual factors named by many authors (i.e. inadequate protection of intellectual property rights and contracts, access to finance, competition, openness to international trade, and human capital) Latin America and Mexico appears to underperform in comparison to other regions.

Family firms considering whether to invest in research and development (R&D) or to develop new products or services, might have access to credit in economies where intellectual property rights are well established by law. Similarly, an institutional and legal environment in which contracts are unevenly enforced or economic transactions depend on informal enforcement mechanisms might limit access to finance for young firms and reduce risky investments in innovation. A long history of macroeconomic and financial instability might undercut both the incentives to innovate—by, for instance, widening the gap between downside risks and upside potential—and the availability of suitable financial services for firms.

A key consideration are costs of the necessary investments in innovation. Financial markets play a role in determining this cost. In some environments, family firms may be credit constrained or financial institutions may lend to them only at exorbitant interest rates, thus raising the costs of innovation investments. However, it is difficult to conclude that for the region as a whole, lack of access to finance is a main driver of the underperformance in innovation. Finance can come from various sources, not just from banks but also from venture capital, capital markets and of course from the family. We concluded that traditional financial markets do not meet the financing needs of small and young family firms, and this problem may affect many developing countries, not only countries in LAC.

The expected payoff from an investment in innovation affects an entrepreneur's decision to take the risk. It depends on the probability of discovering a profitable idea as well as on the ability of inventors to obtain benefits of their investments in innovation. The most directly relevant set of laws and institutions is arguably related to intellectual property rights (IPRs) and contractual certainty. Laws and regulations define the number and types of industries subject to IPRs, the number of international agreements on IPRs to which a country is a signatory, and the legal recourse available to patent holders in case of an alleged infringement of their IPRs. Intellectual property rights, however, are related to other legal areas. Hence, the establishment of intellectual property rights is inseparable from the quality and functionality of the broader legal and judicial system. Fixing intellectual property right regimes is in many cases likely to involve accompanying broader reforms to the legal and the judiciary frameworks.

Human capital for entrepreneurship and innovation needs also to be improved. Since at least the early 20th century, LAC has had a relatively low number of engineers per capita. In general, LAC countries have fewer engineers than the median country and fewer than expected given their level of development. This finding is a reality even for the region's relatively large middle- and high-income countries, such as Brazil, Chile, Colombia, and Mexico (World Bank, 2016). Indeed, if these limited data are reflective of the region as a whole, this constraint may be the single most important barrier to innovative entrepreneurship.

More research is needed before definitive conclusions can be reached, but some evidence suggests that the small rate of engineers in the population and excessive concentration of domestic markets in a few firms may be two of the major factors behind the region's innovation deficit. Intellectual property rights might also be important, but more research is definitely needed on this area.

Latin American universities have been focused into an emphasis on the humanities and law, as well as social, economic, and political fields. This tendency may constrain the ability to switch rapidly to educating more engineers. Such a switch would require very aggressive public policies in the education arena for almost all LAC countries. Young people may be attracted to fields of studies that are relevant to solve problems faced by their societies, which may explain why this region may have formed more macro than micro economists and sociologists. Given the progress of the region, there may be more incentives for students to enlist on scientific careers in the future.

Nowadays in Mexico for instance, there is a struggling between the Ministry of Education and the Teachers´s Union. For decades, the education in Mexico has been held by many different union´s leaders whose only purpose was having political power and that led the education field less and less developed. It is not a surprise that at least in this country, that many children remain without education and many of them that reach university do not have any motivation to study anything related to the engineering field since nobody has ever told them how to address this disciplines. Also it is not a surprise either that due to the lack of engineers on the field, the salaries for the ones to accomplish holding a diploma on almost any of the engineering disciplines also hold the highest salaries. Maybe that is one thing the teachers must tell the younger students in order to motivate them to study an engineer discipline.

In response to the complex issue of supporting environments for innovative entrepreneurship, policy makers are now beginning to recognize the merit of a more systems-based form of support for high growth entrepreneurship (Mason & Brown, 2013). Today, Policy makers across the OECD are now strongly focused on promoting high growth firms (HGFs) (OECD, 2013). The rationale for this focus is that HGFs are thought to drive productivity growth, create new employment, increase innovation and promote family business internationalization (OECD, 2013; Brown et al, 2014).

Based on Mason and Brown (2013), and considering the context of Latin America and Mexico, we present some suggestions for policy makers in the region:

- To provide assistance to family business ventures during the pre-start-up, start-up and early post-start-up stages through the provision of intensive support and mentoring (Roper & Hart, 2013);
- To fix intellectual property right regimes, involving broader reforms to the legal and the judiciary frameworks.
- To connect growth-oriented family owned SMEs, with sources of conventional expansion capital such as banks loans and newer former of finance such as crowdfunding, peer-to-peer lending and invoice-based finance (Collins et al, 2013).
- To build bridges between different entrepreneurial actors through the creation of communities of practice or entrepreneurial networks (professional networking organizations, entrepreneurship clubs).

- To promote a strong focus on enterprise education in schools, colleges and universities to encourage positive attitudes towards entrepreneurship based on innovation and technology, and the creation of entrepreneurial campuses (Mason & Brown, 2013).

FUTURE RESEARCH DIRECTIONS

Although we share the presumption that innovation is a driver of modern market economies around the world, a challenge for policy makers remains figuring out how to shape policies, including laws and institutions, to enhance the incidence of transformational entrepreneurship so that surviving entrepreneurs, in new generations of family firms, become true innovators. Therefore, future research lines should be focus on studying the behavioral additionality of R&D governmental subsidies. A better understanding of the changes in the behaviour of firms inducing a more efficient transformation of innovation inputs to innovation outputs (e.g. collaborations with other firms or universities, improving R&D management process) could be the key to increase the impact of such public support in developing countries.

REFERENCES

Acs, Z. J., & Amorós, J. E. (2008). Entrepreneurship and competitiveness dynamics in Latin America. *Small Business Economics*, *31*(3), 305–322. doi:10.1007/s11187-008-9133-y

Acs, Z. J., & Storey, D. J. (2004). Introduction: Entrepreneurship and economic development. *Regional Studies*, *38*(8), 871–877. doi:10.1080/0034340042000280901

Amorós, J. E., Fernández, C., & Tapia, J. (2012). Quantifying the relationship between entrepreneurship and competitiveness development stages in Latin America. *International Entrepreneurship Management*, *8*(3), 249–270. doi:10.1007/s11365-010-0165-9

Astrachan, J., & Jaskiewicz, P. (2008). Emotional returns and emotional costs in privately-held family businesses: Advancing traditional business valuation. *Family Business Review*, *21*(2), 139–150. doi:10.1111/j.1741-6248.2008.00115.x

Astrachan, J. H., Zahra, S. A., & Sharma, P. (2003). *Family-sponsored ventures.* Presented at the First Annual Global Entrepreneurship Symposium, New York.

Avendaño, J., Kelly, L., Treviño, R.N. & Madero, S. (2009). A Family Based Competitive Advantage: Handling Key Success Family Factors in Mexican Family Firms. *Cuadernos de Administración*, *22*(39), 191-212.

Ayyagari, M., Demirgüc-Kunt, A., & Maksimovic, V. (2011). Firm innovation in Emerging Markets: The Role of Finance, Governance and Competition. *Journal of Financial and Quantitative Analysis*, *46*(6), 1545–1580. doi:10.1017/S0022109011000378

Balwin, J.R., & Johnson. (1996). Business Strategies in more and less innovative firms in Canada. *Research Policy*, 25.

Barbero, M. I. (2008). Business History in Latin America: A Historiographical Perspective. *Business History Review*, *82*(3), 555–575. doi:10.1017/S0007680500082647

Birch, D. (1987). *Job creation in America: How our smallest companies put the most people to work.* New York: The Free Press.

Brenes, E. R. (2012). *The future of entrepreneurship in Latin America*. Palgrave Macmillan. doi:10.1057/9781137003324

Brown, R., Mason, C., & Mawson, S. (2014). *Increasing the Vital 6%: Designing Effective Public Policy To Support High Growth Firms*. National Endowment for Science Technology & Arts (NESTA). Retrieved from: http://www.nesta.org.uk/sites/default/files/working_paper_increasing_the_vital_6_percent.pdf

Burgoa, T., Herrera, E., & Treviño, J. (2013). Study on Family Business Administration in Mexico: Main Problems and Challenges Confronted. *International Journal of Good Conscience*, *8*(2), 1–22.

Chua, J. H., Chrisman, J. J., & Steier, L. P. (2003). Extending the theoretical horizons of family business research. *Entrepreneurship Theory and Practice*, *27*(4), 331–338. doi:10.1111/1540-8520.00012

CIPI. (2003). *Comisión Intersecretarial de Política Industrial. Observatorio PYME. Primer Reporte de Resultados*. Secretaría de Economía. Retrieved from: http://www.protlcuem.gob.mx/swb/work/models/economia/Resource/965/1/images/ESTUDIOPYMESCIPI.pdf

Coduras, A., Cuervo-Arango, C., & Saiz-Alvarez, J. M. (2016). Interregional Integration of Entrepreneurial Activity in Spain. *Strategic Change-Briefings in Entrepreneurial Finance*, *25*(2), 205–220.

Collins, L., Swart, R., & Zhang, B. (2013). *The Rise of Future Finance: the UK's Alternative Finance Benchmarking Report*. London: NESTA.

Dobrzynski, J. H. (1993). Relationship Investing. *Business Week*, *42*(2), 93-105.

Domínguez, M. C. (2016). Strategy Innovation, Collaboration and Competition in Mexico.*Global Conference and Finance Proceedings*.

Durán-Encalada, J.A., & San Martín-Reyna, J.M. (2013). Estudio comparativo de la empresa familiar mexicana en el contexto mundial. *Revista de Contaduría, Administración e Informática Año*, *2*(5), 21–54.

Durán-Encalada, J. A., San Martin-Reyna, J. M., & Montiel-Campos, H. (2012). Research Proposal to Examine Entrepreneurship in Family Business. *Journal of Entrepreneurship, Management and Innovation*, *8*(3), 58–77.

Dutz, M. (2007). *Unleashing India's Innovation: Towards Sustainable and Inclusive Growth*. Washington, DC: World Bank. doi:10.1596/978-0-8213-7197-8

Dyer, W. G. Jr. (1986). *Cultural Change in Family Firms: Anticipating and Managing Business and Family Transitions*. San Francisco: Jossey-Bass.

Eddleston, K., Kellermanns, F. W., & Sarathy, R. (2008). Resource configuration in family firms: Linking resources, strategic planning and environmental dynamism to performance. *Journal of Management Studies*, *45*(1), 26–50.

Ernst & Young. (2014). *Family Business Yearbook*. Retrieved on July 13th, 2015 from: https://family-business.ey-vx.com/insights/family-business-yearbook-2014.aspx

Expansion. (2015). *Las 500 empresas más importantes de México en 2015*. Retrieved on May 1st, 2016 from: http://expansion.mx/rankings/interactivo-las-500/2015

Ferreira, F. H., Messina, J., Rigolini, J., López Calva, L. F., Lugo, M. A., & Vakis, R. (2013). *Economic Mobility and the Rise of the Latin American Middle Class*. Washington, DC: The World Bank.

Gámez-Gutiérrez, J., & Saiz-Alvarez, J. M. (2012). Microcredits for Women Entrepreneurship: Are They and Effective Tool to Avoid Family Impoverishment? In M. A. Galindo & D. Ribeiro (Eds.), *Women's Entrepreneurship and Economics: New Perspectives, Practices, and Policies* (pp. 167–177). Springer. doi:10.1007/978-1-4614-1293-9_11

García Fuentes, M. (2015). Negocios en familia. Herencia Multiplicada. *Entrepreneur, emprende, crece y triunfa*. Septiembre. Available at: https://mx.zinio.com/www/browse/product.jsp?rf=sch&product Id=500662334#/

García-Tenorio, R. J. & Sánchez, Q. J. (2009). Amenazas y oportunidades: El reto de emprender en la empresa familiar. *Gestión de empresa*, *212*, 66-77.

Grossman, G. M., & Helpman, E. (1991). Quality Ladders in the Theory of Growth. *The Review of Economic Studies*, *58*(1), 43–61. doi:10.2307/2298044

Gupta, A. K., & Govindarajan, V. (1984). Business unit strategy, managerial characteristics, and business unit effectiveness at strategy implementation. *Academy of Management Journal*, *27*(1), 25–41. doi:10.2307/255955

Habbershon, T. G., & Williams, M. (1999). A resource-based framework for assessing the strategic advantages of family firms. *Family Business Review*, *9*(2), 157–170.

Hall, A., Melin, L., & Nordqvist, M. (2001). Entrepreneurship as radical change in the family business: Exploring the role of culture patterns. *Family Business Review*, *14*(3), 193–208. doi:10.1111/j.1741-6248.2001.00193.x

INEGI. (2010). *Economic Census*. Mexican National Institute of Statistics and Geography. Retrieved from: http://www.inegi.org.mx/saladeprensa/boletines/2015/especiales/especiales2015_07_5.pdf

Kantis, H. (2005). The emergence of dynamic ventures in Latin America, Southern Europe and East Asia: An international comparison. *International Journal of Entrepreneurship and Small Business*, *2*(1), 34–56. doi:10.1504/IJESB.2005.006069

Kantis, H. D., & Federico, J. S. (2012). *Entrepreneurial Ecosystems in Latin America: The role of policies*. International Research and Policy Roundtable. Retrieved from http://www.innovacion.gob.cl/wp-content/uploads/2012/06/Entrepreneurial-Ecosystems-in-Latin-America_the-role-of-policies.pdf

Kellermanns, F. W., & Eddleston, K. A. (2006). Corporate entrepreneurship in family firms: A family perspective. *Entrepreneurship Theory and Practice*, *30*(6), 809–830. doi:10.1111/j.1540-6520.2006.00153.x

Kellermanns, F. W., Eddleston, K. A., Barnett, T., & Pearson, A. (2008). An exploratory study of family member characteristics and involvement: Effects on entrepreneurial behavior in the family firm. *Family Business Review*, *21*(1), 1–14. doi:10.1111/j.1741-6248.2007.00107.x

Kelley, D. J., Singer, S., & Herrington, M. (2015). *The Global Entrepreneurship Monitor 2015*. Global Report. Retrieved from: http://www.gemconsortium.org/

Kirkman, A. (1999). The Americas. *Forbes Magazine*. Retrieved on May 13th, 2015 from: http://www.forbes.com/global/1999/0705/0213097a.html

La Porta, R., Lopez-de-Silanes, F., & Shleifer, A. (1999). Corporate ownership around the world. *The Journal of Finance*, *54*(2), 471–517. doi:10.1111/0022-1082.00115

Le Breton-Miller, I., & Miller, D. (2006). Why Do Some Family Businesses Out-Compete? Governance, Long-Term Orientations, and Sustainable Capability. *Entrepreneurship: Theory and Practice*, *30*(6), 731–746.

Lederman, D., Messina, J., Pienknagua, S., & Rigolini, J. (2014). *Latin American Entrepreneurs. Many firms but little Innovation*. Washington, DC: International Bank of Reconstruction and Development/ The World Bank.

Lenartowicz, T., & Johnson, J. (2003). A Cross-National Assessment of the Values of Latin America Managers. *Journal of International Business Studies*, *34*(3), 266–281. doi:10.1057/palgrave.jibs.8400032

Lora, E. & Castellani, F. (2014). *Entrepreneurship in Latin America: A Step Up the Social Ladder?*. World Bank Publications.

Lumpkin, G. T., & Dess, G. G. (1996). Clarifying the entrepreneurial orientation construct and linking it to performance. *Academy of Management Review*, *21*(1), 135–172.

Maloney, W. F., & Sarrias, M. (2012). *What Makes Latin American Managers So Bad?* World Bank.

Maloney, W. F., & Valencia-Caicedo, F. (2012). *Part II: Engineers, Innovative Capacity and Development in the Americas. Background paper for this report*. World Bank.

Mason, C., & Brown, R. (2013). Creating good public policy to support high growth firms. *Small Business Economics*, *40*(2), 211–225. doi:10.1007/s11187-011-9369-9

Miller, D. (1983). The correlates of entrepreneurship in three types of firms. *Management Science*, *29*(7), 770–791. doi:10.1287/mnsc.29.7.770

Naldi, L., Nordqvist, M., Sjoberg, K., & Wiklund, J. (2007). Entrepreneurial orientation, risk taking, and performance in family firms. *Family Business Review*, *20*(1), 33–47. doi:10.1111/j.1741-6248.2007.00082.x

Nordqvist, M., & Merlin, L. (2010). Entrepreneurial families and family firms. *Entrepreneurship and Regional Development*, *22*(3–4), 211–239. doi:10.1080/08985621003726119

OECD. (1997). *Proposed Guidelines for Collecting and Interpreting Technological Innovation Data*. Paris: Oslo Manual.

OECD. (2013). *An international benchmarking analysis of public programmes for high-growth firms*. Paris: OECD LEED Programme.

OECD. (2014). *Latin American Economic Outlook 2015*. OECD iLibrary. Available at: http://www.oecd-ilibrary.org/development/latin-american-economic-outlook-2015_leo-2015-en

Pagés-Serra, C. (2010). *The Age of Productivity: Transforming Economies from the Bottom Up*. Inter-American Development Bank. doi:10.1057/9780230107618

Pistrui, D., Welsch, H. P., & Roberts, J. S. (1997). The re-emergence of family businesses in the Transforming former Soviet-Bloc: Family Contributions to entrepreneurship development in Romania. *Family Business Review*, *10*(3), 221–237. doi:10.1111/j.1741-6248.1997.00221.x

Roper, S., & Hart, M. (2013). *Supporting Sustained Growth Among SMEs – Policy Models and Guidelines*. University of Warwick, Enterprise Research Centre, White Paper No. 7.

Saiz-Alvarez, J. M., & Colvin-Diez, J. (2015). Purple Oceans and the M&A Virtuous Circle: New Concepts for the Theory of Mergers and Acquisitions. In A. Tavidze (Ed.), *Progress in Economics Research* (pp. 139–154). Nova Press.

Saiz-Alvarez, J. M. (Ed.). (2016). Handbook of Research on Social Entrepreneurship and Solidarity Economics. Hershey, PA: Information Science Reference (IGI Global).

Schumpeter, J. A. (1994). *Teoría del desenvolvimiento económico*. México: FCE.

Sharma, P., Chrisman, J. J., & Chua, J. H. (1997). Strategic Management of the Family Business: Past Research and Future Challenges. *Family Business Review*, *10*(1), 1–34. doi:10.1111/j.1741-6248.1997.00001.x

Teece, D. J. (1992). Foreign Investment and Technological Development in Silicon Valley. *California Management Review*, *34*(2), 88–106. doi:10.2307/41166695

Van Stel, A., Carree, M., & Thurik, R. (2005). The effect of entrepreneurial activity on national economic growth. *Small Business Economics*, *24*(3), 311–321. doi:10.1007/s11187-005-1996-6

Ward, J. L., & Aronoff, C. E. (1991). The Power of Patient Capital. *Nation's Business*, *79*(9), 48–49.

Wiklund, J., & Shepherd, D. (2003). Knowledge-Based Resources, Entrepreneurial Orientation, and the Performance of Small and Medium-Sized Businesses. *Strategic Management Journal*, *24*(13), 1307–1314. doi:10.1002/smj.360

World Bank. (2016). *World Development Indicators*. World Bank. Available at: http://data.worldbank.org/data- catalog/world-development-indicators

Young, E. C., & Welsch, H. P. (1993). Major elements in entrepreneurial development in central Mexico. *Journal of Small Business Management*, *31*(4), 80.

Zahra, S. A. (2005). Entrepreneurial risk-taking in family firms. *Family Business Review*, *18*(1), 23–40. doi:10.1111/j.1741-6248.2005.00028.x

Zahra, S. A., Hayton, J. C., & Salvato, C. (2004). Entrepreneurial in Family vs. Non-Family Firms: A Resource-Based Analysis of the Effect of Organizational Culture. *Entrepreneurship: Theory and Practice*, *28*(4), 363–382.

Zahra, S. A., Neubaum, D. O., & Huse, M. (2000). Entrepreneurship in medium-size companies: Exploring the effects of ownership and governance systems. *Journal of Management*, *26*(5), 947–976. doi:10.1177/014920630002600509

KEY TERMS AND DEFINITIONS

Entrepreneurial Activity: Any kind of effort and action aiming to starting a new business or a new venture.

Entrepreneurship: The capacity and willingness to develop, organize and manage a business venture. Its main purpose can be to make a profit or to solve a societal issue.

Established Businesses: Businesses that have been operating for more than 42 months.

Family Businesses and Family Firms: Any business in which two or more family members are involved and the majority of ownership or control lies within a family.

Global Entrepreneurship Monitor: The world's foremost study of entrepreneurship aiming to understand individual's perceptions, attitudes and values regarding entrepreneurship.

Innovation: A process built on different activities that result in a new product, process or service.

Total Early-Stage Entrepreneurial Activity: Businesses that have been operating more than 3 months and less than 42. It includes nascent and new entrepreneurs.

Chapter 7
Pitching and the Other International Practices of Innovation Competitions:
Channel for Youth Entrepreneurship

Rauno Rusko
University of Lapland, Finland

Katja Härkönen
University of Lapland, Finland

Sofia Petäjäniemi
University of Lapland, Finland

ABSTRACT

This chapter focuses on the international practices of innovation competitions. Pitching is a truly global phenomenon in the education and youth entrepreneurship, which has similar structures and contents all over the world. Thus, this international practice follows the ideas of Ohmae (1990) about the borderless world. In other words, the micro-activities and the practices of innovation events are following the joint global macro-level frameworks and practices. However, there seems to be slight differences of these events basing on how these events are named or marketed in the Web. Some of these events are web-based while others involve face-to-face happenings. All these forms of pitching events provide important channel for young entrepreneurship. This chapter focuses on the international practices of innovation competitions. Pitching is a truly global phenomenon in the education and youth entrepreneurship, which has similar structures and contents all over the world. Thus, this international practice follows the ideas of Ohmae about the borderless world. In other words, the micro-activities and the practices of innovation events are following the joint global macro-level frameworks and practices. However, there seems to be slight differences of these events basing on how these events are named or marketed in the Web. Some of these events are web-based while others involve face-to-face happenings. All these forms of pitching events provide important channel for young entrepreneurship.

DOI: 10.4018/978-1-5225-0953-0.ch007

INTRODUCTION

This chapter focuses on the international practices of innovation competitions. Pitching is a truly global phenomenon in the education and youth entrepreneurship, which has similar structures and contents all over the world. Thus, this international practice follows the ideas of Ohmae (1990) about the borderless world. Some of these events are web-based, while others involve face-to-face happenings or both of them (see, e.g. Mccarthy, 2010).

In the 1990s there were two competing popular and influential books about global business: The Competitive Advantage of Nations by Michael Porter (1990a; 1990b) and The Borderless World by Kenichi Ohmae (1990). Porter assumes that internationally successful firms have a home base and that, importantly, multinational companies also have a national background. Strong national competition will challenge the firm to develop its competitiveness in a way that also facilitates international competitiveness (Porter, 1990a; 1990b). However, Ohmae (1990) suggests that national borders and nation states will gradually lose their importance because of consistent universal (young) cultural groups. These groups are based on income levels and education, which of course affect the behavior of individual consumers. These effects are even stronger than national, religious or ethnic borders (Ohmae, 1990).

This chapter focuses on international borderless practices and, particularly, on a specific form of entrepreneurial practice: innovative (mass) events and competitions, such as elevator pitch contests. The entrepreneurial innovations and cultural practices are a combination, which is rarely studied in the management literature (Rossberger, 2014). These kinds of innovation events are a very popular means by which to introduce a business idea, business plan or business model to potential financiers, the so-called "business angels" (Adamczyk, *et al.*, 2012). For example, using the keywords "elevator pitch contest" in the Google search engine results in 1,920,000 hits (as of 18th January, 2014). However, business model innovations, for instance, are still poorly understand in the research literature (see, e.g. Bucherer *et al.*, 2012).

The authors of this chapter recently acted as organizers and planners for one entrepreneurial elevator pitch event and, furthermore, have previously been involved in several other innovation events and contests in northern Finland. This background provided the encouragement for this chapter. However, the case studies chosen for this chapter are mainly international and our aim is to explain the international popularity of innovation events. We seek to discover how and, if so, why the implemented practices, structures and contents of these events are so similar in different parts of the world. In order to answer this research question, we consider it essential to introduce the perspectives of practice theory at the beginning of this chapter.

Thus, this chapter is organized as follows: First we conduct a literature review focused on two themes, the international and macro level practices of innovation competitions, as well as the general features of innovations competitions and events. Then we introduce the research design of the chapter and some international case studies concerning innovation competitions and events. After that, we offer discussion regarding international practices and conventions in terms of innovation events and competitions. Finally, the conclusions of the chapter are presented.

THE GENERAL ROLE OF PRACTICES

In order to understand the importance and role of the contemporary practices of business, it is essential to consider the general role of practices in social sciences and management. Considering common practices is the typical research perspective in contemporary studies of management. The main focus of these studies on practices is on the micro-activities. However, this chapter takes into account - and even emphasizes - the importance of macro level and international level practices. In addition to micro versus macro level activities, this section divides the discussions about practices into three categories: discussions based on social sciences (e.g. sociology), business and management (especially strategy), and on educational studies.

Perhaps the oldest forms of practice theory in the social sciences are to be found in social work. According to Goldstein (1986), early 20th century social work was based on idealism, pragmatism, the moral and spiritual well-being of the clients, financial, health, housing and other material conditions. Gradually, however, "sociology has moved from the study of social structures and normative systems to that of 'practice'" (Olick & Robbins, 1998, p. 108; see also Bourdieu, 1984).

The most famous of the early practice theorists (mainly found in sociology) are Bourdieu, Giddens, Foucault, Garfinkel, Latour, Schatzki and Barnes (cf. Reckwitz, 2002). Bourdieu (1977) developed a theory of practice connected with the concept of habitus, which is based on the interaction between objective structures and practices. Giddens (1979; 1984a; 1984b) developed his version of practice theory within the framework of structuration and he also gave consideration to ancient ethics as a framework for analyzing the relations between bodies, agency, knowledge and understanding.

According to Reckwitz (2002, 245), practice theory - in the sense of authors such as Bourdieu, Giddens, Foucault, Garfinkel, Latour, Taylor or Schatzki - is a type of cultural theory. However, Reckwitz (2002) claims that not every form of cultural theory is actually covered by practice theory. Reckwitz (2002) identifies four different branches of cultural theory: mentalism, textualism, intersubjectivism and practice theory. According to him, practice theory does not place the social aspect in mental qualities (mentalism), nor in discourse (textualism), nor in interaction (intersubjectivism) (Reckwitz, 2002). Reckwitz (2002) defines practice in the following way: "A 'practice' (Praktik) is a routinized type of behavior which consists of several elements, interconnected to one other: forms of bodily activities, forms of mental activities, 'things' and their use, a background knowledge in the form of understanding, know-how, the states of emotion and motivational knowledge. A practice – a way of cooking, of consuming, of working, of investigating, of taking care of oneself or of others, etc. – forms so to speak a 'block' whose existence necessarily depends on the existence and the specific inter- the connectedness of these elements, and which cannot be reduced to any one of these single elements" (p. 249).

There are dozens of definitions of practices in the field of business and management. Røpke (2009), for example, defines practices in the context of consumption: "Practices are what make sense to people in everyday life, and their reputations, decency, self-respect etc. depend on being recognized as competent practitioners. … Although most practices involve the appropriation and use of goods, services and ambience, people first of all think of themselves as being involved in meaningful practices rather than being involved in consumption" (p. 2495).

In the field of management and, particularly, in strategic management discussions, the practice turn of science has been considered. Jarzabkowski *et al.* (2007) note that it is necessary to re-focus research on the actions and interactions of the strategy practitioner. This reinstatement of agency in strategic action is located within the wider practice turn or linguistic turn. Linguistics and discourses are essential parts of strategy in practice (SAP) research (Laine & Vaara, 2007). Further, in the context of strategy research, Neumann (2002) states the importance of the interaction and differences between practice and discourse, which is connected with the issue of the differences and similarities between linguistic turn and practice turn. Furthermore, some researchers have actually questioned the very concept of turn. For example, Simpson (2009) considers that practices have not yet achieved a full breakthrough in the field of management: "a practice turn is indeed underway in organization studies, but there is still some distance to travel before the full potential of this turn is realized" (p. 1329).

Strategy as a practice perspective has several different manifestations. Typical for all these branches of SAP discussions is the aim of bringing human actors and their actions and interactions to the center stage of strategy research (Jarzabkowski & Spee, 2009). It is therefore possible to see an element of the practice turn of social sciences. According to Whittington (2006), there are two traditional perspectives in social theory: those who emphasize the importance of individuals and those who emphasize the role of social forces. Whittington (2006) claims that practice turn overcomes the dualism between individualism and societism. In addition, according to Whittington (2006), management research is increasingly engaging with the practice turn, drawing on it in the fields of technology learning at work, institutional change, marketing and accounting. Thus, the practice turn (and strategy as practice) is associated with learning and change, the important features of both learning school (emergent strategy) and theory-in-use perspective, which emphasizes the importance of learning (Argyris & Schön, 1974; 1987).

Typically, the perspective of strategy as practice emphasizes three themes: praxis, practitioners and practices. In other words, the practices of both organizations and their wider social fields; actual activity: praxis, and actors, which are in this context strategy's practitioners. Whittington (2006) notices, that "social practices are followed in rough and ready ways, according to the exigencies of the situation" (p. 615). Jarzabkowski *et al.* (2007) define practices differently: they are "cognitive, behavioral, procedural, discursive, motivational and physical practices that are combined, coordinated and adapted to construct practice" (p. 11). Although these three themes have different nuances, they are widely accepted as the essential elements of strategizing. In fact, strategizing can even be said to consist of praxis, practitioners and practices (Jarzabkowski *et al.*, 2007).

Both theory-in-use and SAP emphasize the role of practices. However, there are differences in these themes: in theory-in-use the practices and routines of individuals and organizations are restraints in learning and readiness during strategic change, while in SAP they are important sources for emerging strategy-making. Therefore, according to theory-in-use, there is a need for double-loop learning, which also has effects on the theory-in-use of organizations (Argyris, 1976). According to strategy as practice, the strategy consists of practices, while changes in strategies come from the practitioners of strategizing, including the workers within an organization. Single-loop learning only considers the ways by which an organization can achieve current targets, and so will not question the relevance of these targets. On the other hand, double-loop learning also requires critical thinking about the relevance of prevailing targets and so allows for constructive development (Rowe, 1996). Similarly, strategy as practice questions the prevailing mainstream standpoint by sense-making strategizing and is based on critical social constructionism (Laine, 2010) while focusing on implementing the strategy. Although there are some differences in the details, when focusing on strategizing both the strategy as practice and theory-in-use themes are focused on the same practice turn and the doing attitude towards strategic management.

MACRO-LEVEL AND INTERNATIONAL-LEVEL PRACTICES

General Macro-Level and International-Level Perspectives

Although traditional strategy research has focused on micro-economic traditions, due to, for example, the models of Michael Porter, who mainly considers markets, networks and business environments, this has meant that research has focused on the macro level perspectives of firms and markets. These kinds of perspectives are located at the micro level in economics, but in the meso or macro level in the field of management and business studies. The new contemporary branches of strategy research, especially those focused on practices, such as the strategy-as-practice theme, are mainly focused on micro-activities, those being activities between the actors in an organization. However, there are some exceptions in strategy research, where the emphasis is also on meso level and macro level practices (Jarzabkowski & Spee, 2009). This section of the article therefore also takes into account meso level and macro level practices and perspectives in social sciences and management. Figure 1 depicts the general levels of social communities. Typically, the micro level perspective involves the individual, family and other near communities, while the meso level involves some networks of work and leisure, and parts of the local community, but also networks of firms and work (e.g. industry). The macro level consists of state and multinational institutions, such as the EU and multinational enterprises (MNEs).

Tienari and Meriläinen (2010) introduce the features of the global economy using five drivers:

1. Technologizing,
2. Multiculturalism,
3. Media-centrism,
4. Finance-driving,
5. Ecologism.

Figure 1. The levels of society

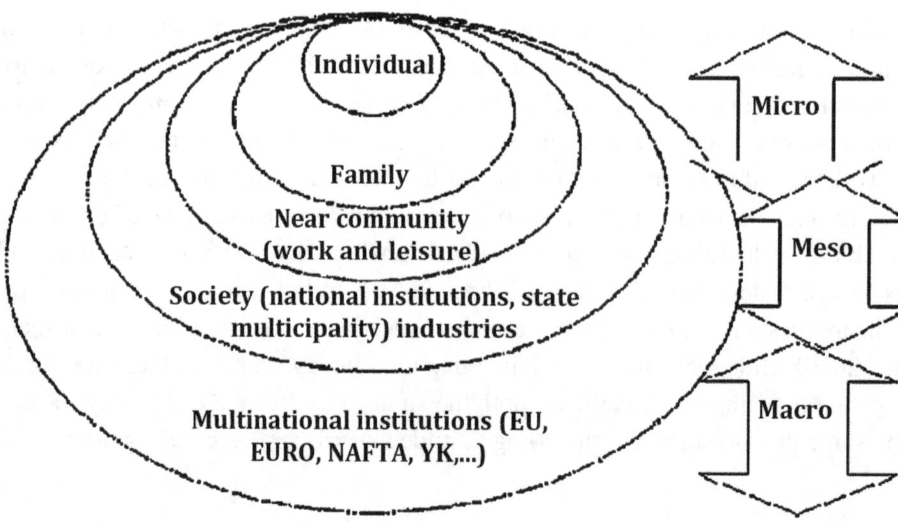

These five drivers are the underlying dimensions in the contemporary global economy. All of these features have changed international business routines and practices. Technologizing takes several forms associated with global trade and information drains. Digitalization also affects the form of technologizing. Due to rapid technologizing, the world is "smaller" and more multicultural in the context of everyday business activities. Also, media-centrism affects the contemporary global practices of economy and business. The actors in business and economy have to take into account the reactions of professional media and social media when deciding on their actions. Finance-driving is another underlying feature of the contemporary global economy. The role of multinational enterprises has grown: they occasionally play even greater roles than the government, multinational associations and organizations. Although the ownership of enterprises has often diversified, the decisions of the enterprises are more often based on the perspectives of the owners and reports of quarterly economic and other financial reports. However, because of drivers such as media-centrism, multiculturalism and technologizing, ecologism has also increased in importance: individuals, enterprises and associations are more aware of sustainable development and ecological issues. Furthermore, the role of citizen-related organizations, such as the campaigning organization Greenpeace, has grown because of information technology developments and social media (Tienari and Meriläinen, 2009).

PRACTICE THEORY AND MACRO-LEVEL AND INTERNATIONAL-LEVEL PRACTICES

According to Giddens (1984), social practices have become the mediating concept between action and structure. Society is seen as being constituted by social practices that are produced and reproduced across time and space. "The basic domain of study of the social sciences, according to the theory of structuration, is neither the experience of the individual actor, nor the existence of any form of societal totality, but social practices ordered across space and time" (Giddens, 1984, p. 2). Although the focus of practice theory and studies seem to be generally on the micro-activities, this will not exclude meso level, macro level and even international level perspectives on practices. One suitable perspective for this is the transnational community (Djelic & Quack, 2010), while another is the "community of practices" (Pietrobelli & Rabellotti, 2010).

The community of practices has the form of organizational or relational proximity, which according to Pietrobelli and Rabellotti (2010) might be even "more important than geographical proximity in supporting the production, the identification, appropriation and flow of tacit knowledge" (p. 25). Thus, multinational firms as well as global value chains may well overcome the absence of geographical proximity. Community of practice, in the sense of Pietrobelli and Rabellotti, is therefore located at the meso level, macro level or even at the international level of practices.

Zucchermaglio and Talamo (2003) emphasize the role of written practices. Generally, each community develops specific communicative practices, both oral and written. According to Zucchermaglio and Talamo (2003), the three main dimensions of community of practice are mutual engagement, joint enterprise and shared repertoire. The mutual engagement dimension emphasizes the centrality of the relations and interpersonal links established between members. Joint enterprise includes negotiations about common goals through interaction and it emphasizes the importance of the project in which community members are mutually engaged. Shared repertoire involves routines, words, instruments, gestures, genres, actions and concepts that have become part of established practices (Zucchermaglio & Talamo, 2003; see also Wenger, 1998).

Taken together, communication channels and platforms as well as face-to-face interactions construct the features of the community of practice. Especially important are the communicative goal and form, such as structure, medium, and language (Zucchermaglio & Talamo, 2003). There have been numerous community of practice discussions concerning the meaning of online communities. According to Hung and Chen (2002), online communities (which are based on the internet) are not actually communities of practices, but are rather "quasi-communities" or the "sub-set of real communities of practices" (see, also Hung & Nichani, 2002). Hung and Nichani (2002) argue that communities of practice have intrinsic dimensions that technology cannot easily bridge (if at all). They suggest that technology has a complementary role in that it can be used to support the communities of practices, but not to replace them (Hung & Nichani, 2002).

Conversely, Zucchermaglio and Talamo (2003) state that the interorganizational community of practice can actually even be based on emails. Their longitudinal analysis of emails revealed several genres in the community of practice and the phases of community development. The distribution of the messages showed how the community made different use of genres over time (Zucchermaglio &Talamo, 2003).

Although the strategy as practice perspective is mainly focused on micro-activities (Chia & MacKay, 2007), there are some exceptions. Jarzabkowski and Spee (2009) connect the strategy as practice perspective with all three levels: micro, meso and macro. According to definitions provided by Jarzabkowski and Spee (2009), strategy as practice as a research topic is concerned with the actual "doing" of strategy: who does it, what they do, how they do it, what they use, and what implications this has for shaping strategy. They identify three types of practitioners:

1. Individual,
2. Aggregate actors within an organization, and
3. Extra-organizational and aggregate actors.

When these three types are considered with micro level, meso level and macro level typology, there are nine alternative positions (Figure 2) (Jarzabkowski & Spee, 2009).

In this context, the domains of "C", "F" and "I" are particularly interesting. According to Jarzabkowski and Spee (2009), "Domain C examines the relationship between individuals and macro-praxis, where macro refers to institutions, markets or industries" (p. 76), "Domain F examines the relationships between aggregate practitioners within organizations and macro-praxis in terms of institutions, industries or sectors" (p. 78), and "Domain I examines the association between extra-organizational actors (or other aggregate actors) and macro-praxis" (p. 80). Extra-organizational actors in this context are, for example, researchers and government policies.

Additionally, Jarzabkowski and Spee (2009) note the lack of macro level (or institutional level) studies in strategy as practice discussions. They mention two articles in this context: the study of Vaara, Kleymann and Seristö (2004) and the study of Lounsbury and Crumley (2007). The first of these studies explains the rise of alliance activity as the dominant mode of strategy praxis within the airline industry, while the second explains the institutionalization of active money management as the dominant practice in the US mutual fund industry. Schatzki (1996) emphasizes that doings and sayings are linked, and identifies three major avenues of linkage. He defines practice-as-entity as:

Figure 2. Typology strategy as practive framework
Suggested by Jarzablowski& Spee (2009).

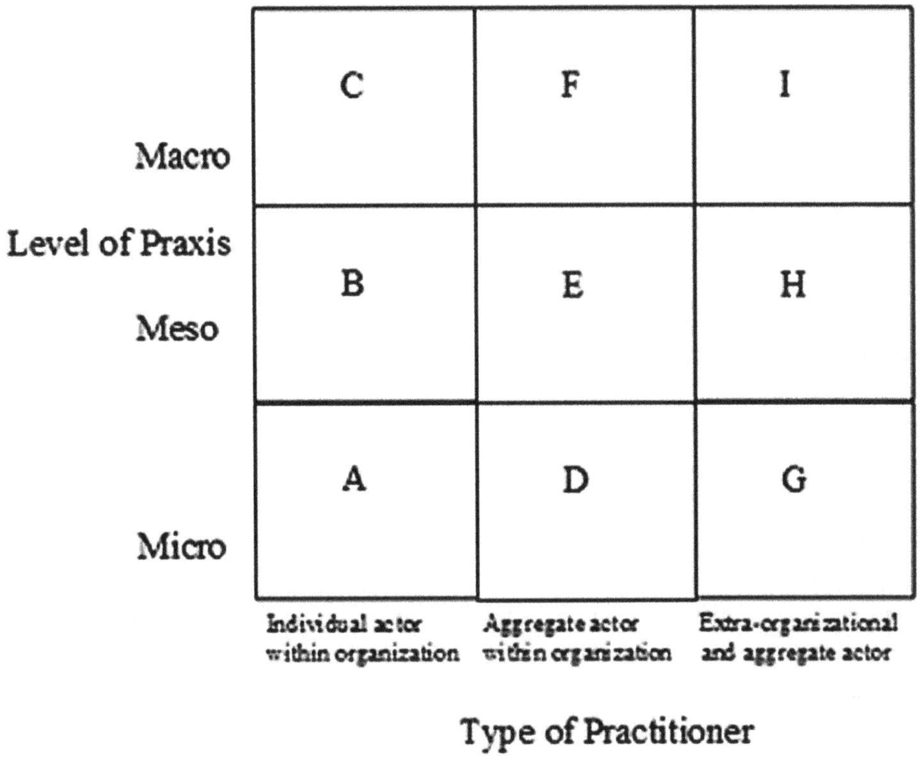

... a temporally unfolding and spatially dispersed nexus of doings and sayings.... To say that the doings and sayings forming a practice constitute a nexus is to say that they are linked in certain ways. Three major linkages are involved: through understandings, for example, of what to say and do; through explicit rules, principles, precepts and instructions; and (3) through what I will call "teleoaffective" structures embracing ends, projects, tasks, purposes, beliefs, emotions and moods. (Schatzki, 1996, p. 89)

The Creative and Innovative Practices

The contemporary literature concerning management has focused on the study of innovative practices, for example on the context of innovation technology and project management (Reich, Sauer and Wee, 2008). However, the research perspectives mostly emphasize the creative and innovative practices of cities and other public communities (Bradford, 2004; Borch, 2008; Arthurs, 2012).

Surprisingly, studies focusing on the creative and/or innovative practices of start-ups are not common in management literature. Neither innovation events and nor design events have regularly been introduced from the perspective of practices (cf. Leong and Clark, 2003). This study focuses on the general practices of innovation events and, especially, elevator pitch events in order to fill this gap in the literature. The focus is on the international practices of these events.

However, the management literature does contain some themes, such as crowdsourcing, which are practical tools for improving innovation and creativity. Crowdsourcing emphasizes the role of customers as a part of the production process. Howe (2006), introduces crowdsourcing as an alternative to outsourcing and defines it in the following way: "... crowdsourcing represents the act of a company or institution taking a function once performed by employees and outsourcing it to the undefined (and generally large) network of people in the form of an open call. This can take the form of peer-production (when the job is performed collaboratively), but is also often undertaken by sole individuals. The crucial prerequisite is the use of the open call format and the large network of potential laborers" (p. 5).

THE GENERAL FEATURES OF INNOVATION COMPETITIONS AND EVENTS

In the management literature, in addition to crowdsourcing, there are theoretical concepts that have direct and indirect connections to innovation competitions, such as open innovation strategy and value co-creation (Ebner *et al.*, 2009; Rusko, 2013; 2014). In this section, we introduce these concepts and, in particular, their connections to innovation competitions and innovation events.

One form of crowdsourcing is based on open competition, where active customers are developing the products and services based on the needs of the firms (Rusko, 2013). In this form of open competition, crowdsourcing has a relatively long history. According to Lampel *et al.* (2012): "As far back as the 18th century, organizations and governments staged 'design competitions' that challenged innovators to submit prototypes to competitive evaluation in nonmarket settings" (p. 71). Thus, crowdsourcing has both physical (traditional) forms and virtual forms. That is, business practice either means literally to outsource an activity to the crowd or a form of outsourcing not directed to other companies but to the crowd by means of an open tender or open call via an Internet platform (Howe, 2006; Burger-Helmchen & Penin, 2010; Schenk & Guittard, 2009; Rusko, 2013; 2014). The literature on crowdsourcing is relatively new, but already the ideas of Wind and Rangaswamy (2001) about consumers as co-creators in the digital marketplace fit well within the category of crowdsourcing.

Ebner *et al.* (2009) studied crowdsourcing in virtual communities, which have IT-supported idea competitions. Following the ideas of Chesbrough (2003), there are two paradigms: the closed innovation paradigm and the new paradigm (open innovation paradigm). The closed innovation paradigm has obsolete features because of erosion factors, such as the availability and mobility of skilled technology workers, the expansion of the venture capital market, unused technologies and the increased supply of highly capable external suppliers (Ebner *et al.*, 2009). The "new" paradigm has two main features: a) crowdsourcing opens the company's innovation funnel, enabling the development of new ideas, and b) the underlying philosophy of open innovation that "We is smarter than me", leading to better selection of ideas and better development of innovations (Ebner *et al.*, 2009, 343). It should be noted that these perspectives are based on the underlying assumption that the activator of the innovations is a company. However, the main focus of the article of Ebner *et al.* (2009) is wider and also involves idea competitions organized by public organizations, non-profit organizations and individuals (Table 1).

According to Ebner and colleagues, idea competitions might be arranged by online, offline or both of them. Especially crowdsourcing activities are often based on online platform (Rusko, 2013), but traditional pitching events has often face-to-face elements (see e.g. Konczal & Motoyama, 2013).

There might be at least three main motives for starting innovation competitions and events:

1. The need to speed up and intensify the research and development process of a company or strategic alliance by constructing an open innovation competition (cf. Rusko, 2013).

Table 1. Criterions and features of idea competitions

Criteria	Features (Examples)
Organizer	Companies, public organizations, non-profit organizations, individuals
Timeline	Very short (s/min), short(days), middle (weeks), long term (semester)
Evaluation	Performance-oriented, participation-oriented
Incentives	Monetary prizes, non-cash prizes
Context	Products, processes
Problem specification	High (searching for a solution of a specific problem), low (general)
Elaborateness	Complexity, quality, condition
Target group	Qualified (by age, interests), not qualified
Composition of groups	Single, team
Idea reviewers committee	Specialists, non-professionals (VIPs)
Idea review	Person, process, context, product
Nature of competition	Online, offline, mixed

Ebner et al., 2009, p. 346.

2. The need to advertise and develop the brand of a company or alliance by establishing a startling happening in the form of open innovation company (e.g. a competition concerning the new name of the company, shop or other business unit).
3. The general need to develop and speed up the innovation processes of start-ups in society or in a specific branch of business or geographic area (Konczal & Motoyama, 2013).

The list above is clearly organizer-oriented. The aims of the innovation competitions vary according to the organizers and their aims. These features have also been observed by Ebner, Leimeister and Krcmar (2009) and Lampel, Jha and Bhalla (2010).

Ebner *et al.* (2009) note that the idea of integrating competitions into the HR processes of a company seems to be very fruitful. In their study, a mixture of cash and non-monetary prizes was a successful means by which to motivate various participant groups (Ebner *et al.*, 2009). Further, Lampel *et al.* (2010) mention several of the factors included in Table 1. They notice the growing popularity of design competitions as well as their changed form: "Whereas past design competitions were set up a single competitive episodes, current design competitions are more likely to be multi-episode or even permanent process using the Internet" (Lampel *et al.*, 2010, p. 83). It is noteworthy that both of these papers (Ebner *et al.*, 2009; Lampel *et al.*, 2010) have identified the importance of virtual platforms and crowdsourcing. Lampel *et al.* (2010) see connections between the popularity of innovation competitions and open innovation tendency. They do note, however, that breakthrough innovations are less likely to emerge from open innovation (See also Laursen and Salter, 2006; Dunlap-Hinkler *et al.*, 2010).

RESEARCH DESIGN

Methodologically, this article is based on the case study strategy, which includes several cases exploiting the most appropriate research methods in the analysis. Thus, the case study strategy enables several research techniques and methods to be used in the same study (Yin, 1983; Eriksson and Kovalainen, 2008). In this study, the cases are based on innovations competitions and events. It is typical to study such events using case study methods. In event research, the case study is often used due to its instrumental value – to gain an understanding about the directorship in a diversity dynamic event and to elaborate on, and extend, the relevant theory (Denzin & Lincoln, 2005; Cilliers & May, 2012).

The cases used in this study are innovation competitions and the research phenomenon is the joint or separate features and practices of these competitions. There are thousands of innovation or (elevator) pitch competitions. We did not consider it necessary to study all of these competitions. At first, we decided to choose the first ten hits in a Google search using the keywords "elevator pitch event" (where the search line included quotation marks). The search was made on 26 February, 2014. The results of this search were confusing (see the next section) and therefore we decided to try another set of keywords, "startup pitch event", using the Google search. Again, the first ten hits (as of 28 February 2014) were used in the analysis. Since both of these searches mainly failed to identify the most famous international pitch competitions, we also included three well-known international events in the case study sample. Furthermore, we have familiarized the contents of dozens of other pitching events and also participated in several pitching events as organizers. All these material and experiences have effects on the content and outcomes of the analysis. Thus, the research process is emergent, meaning the methods and research techniques are flexible based on the preliminary results of the research. This kind of flexibility is typical for qualitative studies (Moisander & Valtonen, 2006).

In addition to the sample of cases about innovation competitions, the perspective of this study is also based on the experiences of the authors as organizers and participants in the organizing units of the local innovation competitions or pitch events. The chosen cases do not cover these pitch events, but the experiences of the authors will strengthen the perspectives of this study. We decided in this article to focus on innovation events that were not organized by the authors. Thus, this article is not an ethnographic study, but instead is a case study concerning innovation competitions and their (cross-national) practices and features.

CASE STUDIES OF INNOVATION COMPETITION AND EVENTS

In this section, we introduce some international competitions and events in order to compare their different features. One interesting question is whether it is possible to identify new joint characteristics among these cases or whether they follow the features already found in contemporary management studies. The underlying aim is also to uncover international conventions and practices among these innovation events as well as possible processes and structures, which maintain and reproduce these international practices of innovation events.

In the management and business literature, it seems that innovation competitions and crowdsourcing are considered as nearly a synonym (Lampel *et al.*, 2010; Ebner *et al.*, 2009). Crowdsourcing has several definitions. While following the popular definition, "the practice of obtaining participants, services, ideas, or content by soliciting contributions from a large group of people, especially via the Internet" (Swan, 2012, p.1), it is possible to draw the conclusion that the one of the earliest innovation competitions (e.g. Longitude Prize, an award organized by British government in 1714 for the invention of a simple way of determining a ship's longitude (Birch & Heffernan, 2014; Sobel, 2004)) is also based on the construction, which is today known as crowdsourcing.

Furthermore, some contemporary discussions use the elevator pitch or pitching concepts in the context of the crowdsourcing (and crowdfunding) activities (Marom & Sade, 2013). Typically, innovation competitions include activities, which are also known as "pitching" (Lorah, 2013). Thus, contemporary innovation competitions are often known as pitching or elevator pitching competitions.

Due to the mass of different elevator pitching events, we decided to restrict the number of the main cases included in the present study. The initial aim was for this section to introduce only ten elevator pitch events, which were picked from the first results of the "elevator pitch event" Google search on February 26, 2014. However, finally the number of the studied cases was significantly larger. This methodology, which is mainly based on the outcomes of search engine might decrease the role of subjectivity in the results of this article, but emphasizes those pages that have succeeded in their Google search engine optimization. Since we do not consider videos in the analysis or the pages of the same organizer twice, the case study examples where actually among the first fifteen hits in the Google search. The analysis in this section is mainly based on direct citing from the homepages of each event. The first page details the main features of these pitching events or competitions, and for that reason we will make particular use of the first pages of the elevator pitch events in our analysis.

As a result of the Google search, we identified the following events as the case study targets:

1. **UC San Diego Entrepreneur Challenge:**http://challenge.ucsd.edu/elevator-pitch
2. **The Annual MIT GSW Elevator Pitch Training and Competition:** http://mitgsw.org/2013/workshop/elevator-pitch-contest-page/
3. **MIT $100K Entrepreneurship Competition:** http://www.mit100k.org/
4. **Network for Teaching Entrepreneurship Elevator Pitch 2013:** http://www.elevatorpitch2012.com/
5. **The Elevator Pitch Competition organized by Entrepreneurship, Rensselaer Polytechnic Institute:**http://www.eship.rpi.edu/elevator.php
6. **The Elevator Pitch Competition Organized by Purdue University:** http://www.purdue.edu/discoverypark/entr/competitions/elevator_pitch_contest.php
7. **CEO National Elevator Pitch Contest:** https://www.facebook.com/CEONationalElevatorPitch
8. **Annual Elevator Pitch Competition with Rice Alliance, the OEDK and the Brown School of Engineering:** http://rcel.rice.edu/elevatorpitch
9. **Annual Elevator Pitch Plus Contest for Chicago Businesses:** http://www.chicagocitytreasurer.com/small-business-entrepreneurs/elevator-pitch-contest/
10. **Elevator Pitch Competition (University of Idaho):** http://www.uidaho.edu/cbe/experientiallearning/view/elevatorpitch

Most of these pages define the concept of an elevator pitch or pitching competition. Below are some of the definitions obtained from the case study sample:

An elevator pitch is a brief overview of an idea for a product, service, or project that can be delivered in the space of an elevator ride (say, thirty seconds or 100-150 words). The goal is to interest your target enough in your idea that he or she wants to meet with you again. Potential investors often judge the quality of an idea and team on the basis of its elevator pitch, using them to quickly weed out bad ideas. (MIT GSW Elevator Pitch, 2013)

An "elevator pitch" is a carefully crafted message designed to explain a business to a potential investor. The catch — it must be given in the time that it takes to ride up an elevator. (NFTE, 2013)

The pitches will describe the designs, the problem or need being solved, the competitive advantages, and the commercial potential of the devices and other products. (Rice Center, 2014)

An elevator pitch is named because sometimes your most important pitch is not the formal business plan presentation you make to investors but short, opportunistic pitches when you share an elevator ride, a cab, or bump into someone at a reception. Your pitch is an overview of an idea for a product, service, or project. The name reflects the fact that an elevator pitch should be possible to deliver in the time span of an elevator ride. While a typical elevator pitch is at least 1 minute, a reception pitch will likely be a bit longer, and might be more interactive. The term is typically used in the context of an entrepreneur pitching an idea to a venture capitalist or angel investor to receive funding. Venture capitalists often judge the quality of an idea and team on the basis of the quality of its elevator pitch, and will ask entrepreneurs for their elevator pitches to weed out bad ideas and weak teams. If interested they will exchange contact information and ask for more information or give the team an opportunity for a longer, more formal presentation. In addition to entrepreneurs a variety of other individuals such as project managers, salespeople, policy-makers, and job seekers use an elevator pitch to get their point across quickly. (University of Idaho, 2014)

According to the definitions above, typical elevator pitch events include, for example, the following features:

1. Business idea or plan.
2. Public presentation.
3. Fast presentation (even 30 seconds is enough).
4. Jury or potential financiers.
5. Possibility to have longer discussions afterwards with potential financiers.
6. Prize for the best pitches (This is generally included in these pages, but not those cited above).

Although these features were expected, the geographical distribution of this case study sample, which covers ten innovation competitions identified through a Google search using the keywords "elevator pitch event", was confusing (Table 2). Some 90% of the identified innovation competitions were implemented in the USA, with Estonia being to only other country to feature. Even in the Estonian case, the competition was partly organized by MIT, USA. Another interesting result was that most of the innovation competitions seemed to be organized by North-American Universities (80%) while the rest of the competitions were organized by associations of educational institutions.

Generally, our research team was confused by the geographical distribution of the case study sample. Keywords in a web search play an essential role to the outcome. Therefore, we made a *second web search*, again using the Google search (February 28, 2014), but this time using the keywords "startup pitch event". Since startup is a very popular contemporary term referring to starting enterprises, we assumed

Table 2. Innovation competitions identified by a Google search using the keywords "elevator pitch event"

Organizer	Country, Place	Time
1. UC San Diego Entrepreneur Challenge	San Diego, USA (national)	Annual
2. MIT Global Startup Workshop, Cambridge	Tallinn, Estonia (International)	Annual
3. MIT, Massachusetts Institute of Technology	Massachusetts, USA (national)	Annual, three different competitions
4. Network for Teaching Entrepreneurship	New York, USA (national)	Several (even daily)
5. Rensselaer Polytechnic Institute	New York, USA (national)	Annual
6. Purdue University	Indiana, USA (national)	Annual
7. Collegiate Entrepreneurs Organization	USA (mainly national, but also international elements)	Annual
8. Rice Center for Engineering Leadership	Texas, USA (national)	Annual
9. City Treasurer's Office, City of Chicago	Chicago, USA (national)	Annual
10. University of Idaho	Idaho, USA (national)	Twice a year

Reviewed in February 26, 2014.

that this search would also take into account the latest innovation competitions. In this second search, the following events were identified:

1. **Startup Pitch Night:** http://www.startuppitchnight.com/startuppitchnight/
2. **Pitch San Francisco'13:** http://www.pitchevent.com/
3. **Startupbootcamp Istanbul:** http://www.startupbootcamp.org/events/istanbul-pitch-day-3.html
4. **Startupbootcamp Pitch Day Athens:** http://www.startupbootcamp.org/events/startupbootcamp-pitch-day-athens.html
5. **Startup Pitch Night:** http://www.meetup.com/New-York-Startup-Pitch-Night
6. **PitchForce:** http://www.meetup.com/IGNITE-Fundraisers-monthly-startup-pitch-events/
7. **Ultra Light Startups:** http://ultralightstartups.com/
8. **DukeGEN Startup Showcase:** http://www.dukegen.com/angel-pitch-event
9. **SwitchPitch NYC:** http://switchpitch.com/switchpitch-nyc/
10. **Hong Kong Young Entrepreneur Club:** https://www.facebook.com/pages/Hong-Kong-Young-Entrepreneur-Club/154156511301909

In this second case study sample, the distribution of the features of the innovation competitions was more scattered. Now 60% of the events in the case sample were organized in the USA. The keywords "startup pitch event" resulted in several types of organizers. Now only in one case out of ten was the innovation event organized solely by a university (Duke University) in Britain. Mostly, these startup pitch events were organized by other (entrepreneurship) organizations or associations. In two cases out of ten, the organizers were, according to the websites of the events, individuals. Furthermore, there were disparities among the cases in terms of how many times per year the event is organized. A couple of these innovation events were so new that it was impossible to identify the regularity of the event. Four out of ten events were actually arranged monthly or even more frequently (Table 3).

Table 3. Innovation competitions provided by a Google search using the keywords "startup pitch event"

Organizer	Country, Place	Time
1. Startup Pitch Night Meetup™ Groups	Several places, mainly in USA (13 cities), also in Canada (2), Belgium, Great Britain and Israel.	Several events per month.
2. Pitch San Francisco'13 Pitchevent	San Francisco, USA (national)	Annual (biennial)
3. Startupbootcamp, Istanbul (Bahcesehir University.	Istanbul, Turkey (national, part of international chain)	Annual? (New event)
4. Startupbootcamp Pitch Day, Athens (Microsoft Innovation Center.	Athens, Greece (national, part of international chain)	Annual? (New event)
5. Startup Pitch Night	New York, USA (national)	Monthly
6. PitchForce/Max Shapiro	San Francisco, USA (national)	Monthly
7. Ultra Light Startups/Graham Lawlor	New York, Boston, USA (national)	Monthly
8. DukeGEN Startup Showcase, DUKE University	Durham, Great Britain (national)	Twice a year (smaller events monthly)
9. SwitchPitch NYC	New York, USA (national)	Annual? (new event)
10. Hong Kong Young Entrepreneur Club	Hong Kong, China (national)	Less than annual

Reviewed in February 28, 2014.

The comparison between these two Google searches, for "elevator pitch event" and "startup pitch event" respectively, reveals that the elevator pitch concept is particularly popular with North American educational organizations, such as universities. The elevator pitch concept is almost a national expression in the USA. However, according to the results, the startup pitch concept is more international although even with those keywords the majority of events originated in the USA. Furthermore, the variety of organizers is higher in the case of the startup pitch expression when compared with the elevator pitch expression. For the latter, nearly all of the events were organized by universities, whereas innovation competitions that emphasize the startup pitch are mainly organized by individual business actors, entrepreneurship organizations or associations.

The analysis, which covered these 20 samples, was based on separate innovation events, which were mainly national events (in the USA). Already the analysis has shown that the innovation competitions have international aspects although most of the cases were arranged in the USA. However, because innovation competitions and pitch events are undeniably international phenomena, we introduce more carefully three well-known innovation events in order to identify the most essential international perspectives for pitch event practices. We decided to introduce the following three pitching events, all of which have well-known international status: NEXT Berlin, MIT Global Startup Workshop (Tallinn) and Slush. The MIT Global Startup Workshop in Tallinn was also involved in the case study sample based on the keywords "elevator pitch event" and was briefly mentioned above. This event is, however, similar to the NEXT Berlin and Slush events, but is, importantly, a typical international (European) innovation competition.

NEXT Berlin

NEXT Berlin is the European conference for digital business and is hosted by the digital agency SinnerSchrader. This two-day conference and once a year organized event is held in Berlin, Germany. The European web industry consider the NEXT Berlin events to be the most important networking and trend events. NEXT Berlin will provide debates on tomorrow´s topics of the digital economy in Europe. Even though these events are arranged in Berlin, the speakers are very international. The speakers, of whom there are more than 100 every year, come from all over Europe and the US. In these events, for example, business developers, executives and service providers can get in touch with marketing experts and international pioneers in business, technology and science.

NEXT Berlin events also arrange pitching arenas for start-ups and their innovation ideas or products. You could get testing for your ideas or products in front of international entrepreneurs and many potential investors. In 2013, almost 100 start-ups applied for the NEXT Berlin start-up pitch from where a jury chose the top 30. After that, the distinguished jury of the event voted the lucky 12 finalists from the community´s top 30. Finally, 12 international finalists had the chance to present their visions and business models live on stage at NEXT13. The 12 finalists came from Germany, Austria, Portugal, United Kingdom, Sweden, France and Italy. The winner for 2013 came from Austria and received a prize package with a total value of 20,000 euro and a lot of media-attention.

The ninth event, NEXT14, will take place on May 5-6, 2014 in centrally located in Berlin and the event´s the main idea or agenda is 'This in the new normal' (NextBerlin, 2014).

MIT Global Startup Workshop

The Annual MIT Global Startup Workshop (MIT GSW) will be held every year in a different location. The MIT GSW was founded in 1997 and the first MIT GSW was held in Cambridge, Massachusetts in March 1998. At first, the MIT GSW´s goal was to stimulate and nurture entrepreneurship worldwide, but in 2007 the MIT GSW expanded its mission to include all aspects of the entrepreneurial ecosystem. Today MIT GSW wants to build stronger, more productive entrepreneurial ecosystems around the world and a global support network for entrepreneurs.

The MIT GSW has been organized on six different continents, except Antarctica, and the recent years´ locations have been Tallinn, Estonia (2013), Istanbul, Turkey (2012), Seoul, South Korea (2011) and Reykjavik, Iceland (2010). The MIT GSW 2014 will be held in Marrakech, Morocco. Every country has the opportunity to host the MIT GSW conference. Organizers of the event will choose the next host of the conference from a group of applicants and they encourage all interested parties to apply. MIT GSW will offer a great deal of benefits for hosting parties, for example, to showcase host´s organization and country to entrepreneurs, investors, executives, government leaders and other supporters of entrepreneurship.

Included in the MIT GSW event is the Elevator Pitch Training and Competition, which gives innovators from around the globe the opportunity to refine presentation skills, increase confidence and present their ideas in 60 seconds or less. The Elevator Pitch Competition also gives the opportunity to win cash prizes. First place will bring to winner 1500$, second place 500$ and third place 250$. Also, the crowd favorite pitch will be awarded 250$. Pitches will be evaluated based on four categories, which are presentation, opportunity, solution and the question of whether the investor wants to meet the competitor again. During the presentation, competitors will not be allowed to use props of any kind or visual aids. The MIT GSW Team will choose the winner and two runners-up and their decisions are final (MIT Global, 2013).

Slush

Slush is the leading startup conference in northern Europe and Russia. Situated in Helsinki, Finland, it brings together the early-stage startup scene to the region along with top-tier angel investors and venture capitalists from Europe and the US. Slush is a nonprofit, two-day technology and startup event where the most promising companies from northern Europe, Russia and beyond gather to meet with top-tier investors and media from all around the world. It is organized to allow the next generation of great companies to have the ecosystem they deserve to succeed on a global level. It brings the best investors and companies to the same place and makes sure that the right companies meet the right investors in one-on-one meetings.

At the core of Slush is the Slush Pitching Competition for early-stage startups and the Slush Demo Showcase for companies that have already developed their products. In the first round 100 top startups get to pitch their ideas on one of the event stages to a jury including experienced entrepreneurs and investors. After the first day, the 20 best startups will be selected for round two on the next day. Finally, the four best startups will move on to finals. In the Demo Showcases, only two finalists from among the 20 competitors will be selected to pitch on day two. Both winners will be awarded €10 000 and a unique Finnish glass statue. One of the pitching competition finalists will also receive the AppCampus award of €10 000.

In 2008 to 2012, Slush transformed from a 300-person event to become one of the top startup events in the world, attracting more than $40 billion worth of venture capital, some of the world's top media, plus 3,500 attendees in 2012. In 2013, Slush became one of the largest events of its kind in the world, bringing together $60 billion venture capital and over 5000 attendees to Helsinki. The latest event also attracted investors from the Asian market for the first time. Slush is organized yearly by Startup Sauna, the Helsinki-based seed accelerator which has recently rewarded as the top young university accelerator in the world by UBI Index (Slush, 2014).

Table 4. summarizes and compares the features of these three innovation events. Although they are all located in Europe, they all have connections to Northern America, particularly in the case of MIT GSW Tallinn, which is part of a global chain of innovation competitions. Participation in all of these events requires the payment of a fee, but the underlying financial potential is enormous. This is a key element of the event and will of course significantly increase the attraction of these events.

THE INTERNATIONAL PRACTICES AND CONVENTIONS IN INNOVATION EVENTS AND COMPETITION

The analysis above proved that elevator pitch events have several common features. Furthermore, we noticed that the historical background to pitching competitions stretches back as far as the eighteenth century, to the Longitude Prize, an award organized by British government in 1714. This competition had several features similar to contemporary pitching events: the public presentation, jury, prize and an organizer from the public sector.

Our analysis revealed that elevator pitching competitions are typically organized by educational institutions, such as universities, and that the competitors and participants in these events are mainly relatively young university students. These pitching competitions seem to be especially popular in the USA. Nearly all of our case study examples came from the USA. However, this is only a half-truth: there are several international mass events in different parts of the world, which also follow the typical features of pitching events. This is particularly true when considering startup pitching competitions. What is common within both the North American and international pitching events is the same language: English. In Finland, for example, even in national or local pitching events, which are organized using the

Table 4. The features of three famous international innovation competitions

	Slush	NEXT Berlin	MIT GSW
Organizer	Non-profit organization	Company	University
Since	2008	2006	1997
Target Market	• Northern-Europe • Russia • Asia	• Europe • US	Global
Target group(s)	• Investors • Startups • Companies • Students • Industries	• Developers • Designers • Business developers • Startups • Companies	• Companies • Education • Financiers • Government agents
Special focus of business fields	• Gaming • Health • CleanTech • Information and Cybersecurity	• Web/digital industry • Service design	• Mobile technology • Health technology • Energy • Fashion
Location: Permanent/Floating	• Helsinki, Finland • Permanent	• Berlin, Germany • Permanent	• Tallinn, Estonia • Floating
Duration	2 days	2 days	3 days
Fee in 2013	• Normal €200 • Early stage startups €100 • Angel ticket €500 • Students €50 • Media free	• Normal €200 • Others n/a	• Normal $850 • Early bird $750
Attendees	5000	1700	300
VC money	~ $60 billion	n/a	n/a
Other	Online live stream	Spin-off NEXT service design conference	• Event is organized by MIT students • Conferences has been held on 6 continents
Pre-qualification/open	Pre-qualification	Pre-qualification	n/a
No. of competitors	100	30	n/a
Rounds	3	2	2
Prize	• €10,000 • Glass statue	• €20,000 • Professional coaching	$1,500
Jury	• Entrepreneurs • Investors	• Entrepreneurs • Investors	n/a
Time per pitch	n/a	n/a	60 sec

local Finnish language, the innovation competitions are announced using words like “pitching”, “pitch” or “pitsaus”, (Arkko-Saukkonen and Merivirta, no date; Nuortila, 2013; Tervo, 2013). Clearly, only the last of these is a Finnish expression. However, in the Finnish language this expression does not have any particular meaning although it does resemble the word “pizza”.

Innovation and design competitions, which have various titles, such as “elevator pitching event” or “startup pitch event”, are international social practices. They contain several levels: the individual, micro level, meso level, macro level and, finally, the international level. These competitions have different meanings and practices at each of these levels. The general established structure is the driver of these practices at each level.

At the individual level, the practices, practitioners and praxis (c.f. Whittington, 2006; Jarzabkowski *et al.*, 2007) are similar in each innovation competition. Among the practitioners, the participants in the competition are typically innovative students at universities/colleges or young (potential) entrepreneurs, who have invented a potential business idea or business model. Normally, there are 1 to 5 (or 7) team members. The jury consists of the financiers and/or teachers at the colleges or universities. The audience is based on students and teachers, other potential entrepreneurs, potential financiers and media presenters.

The conventional practices at the individual level involve a short public presentation (typically 0.5 to 5 minutes), stage, speaker(s), comments and questions from the jury, potential for further private discussions between participants and members of jury, and often prizes for the best teams.

The praxis in this context is based on the several different kinds of practices: the same organizers annually organize similar innovative competitions. The same jury members, especially if they are financiers, might participate in several similar events during the same year. The speaker(s) at the events might be professional speakers or innovators, or members of the organizing firm or institution. Typically, they are familiar with these kinds of situations.

More generally, at the micro level (excluding individuals), the practitioners are organizations and firms, which are organizers or participants in the competition or financiers. In this context, the practices are based on the conventions of these firms and organizations. Typically, the organizer organizes the same competition annually. Another possibility is that the same organizer organizes a chain of similar competitions in different parts of country or even globally. The financing and competing firms might participate in several innovation competitions a year or annually in the same event. These drains of conventions and practices generate the praxis.

The meso and/or macro level includes networks of firms and organizations participating in the same innovation competition, whether in the role of organizer, financier or competitor. For some industries, these innovation competitions are more important than for others. This is the case in the technical industries. These kinds of competitions will maintain the level innovation in the business and make sure that there are fresh ideas being generated and so guaranteeing investments in the future of the industry.

Finally, the international level contains the whole international institution of the innovation competition or elevator pitch competition, where the individuals, firms and organizations, networks and industries have their own roles to play. Actually, at the international level, the innovation competitions, such as pitch competitions, have several cross-sectional features, which are based on individual, micro, meso and macro levels. Those practices and features of the individual, micro, meso and macro levels in innovation events are common internationally. That is, they are also international practices. Figure 3 describes the multi-level international practices in innovation competitions, such as in pitch events.

At the beginning of this article, we introduced, in accordance with the suggestions by Tienari and Meriläinen (2010), five drivers of the global economy: technologizing, multiculturalism, media-centrism, finance-driving and ecologism. Digitalization is an element of technologizing. The contemporary innovation competitions use intensive digital networks and social media as a platform to advertise and inform the potential audience about these competitions. Thus, technologizing and media-centrism are important drivers of these innovation competition practices. Also, multiculturalism is an important feature in innovation competitions although the English language seems to be a key part of the competitions. These competitions are arranged all over the world and so are implemented in national languages although concepts such as pitch and startup are often mentioned in the names of these events. Further, finance-

Figure 3. Individual, micro, meso, and international level practices in innovation competitions

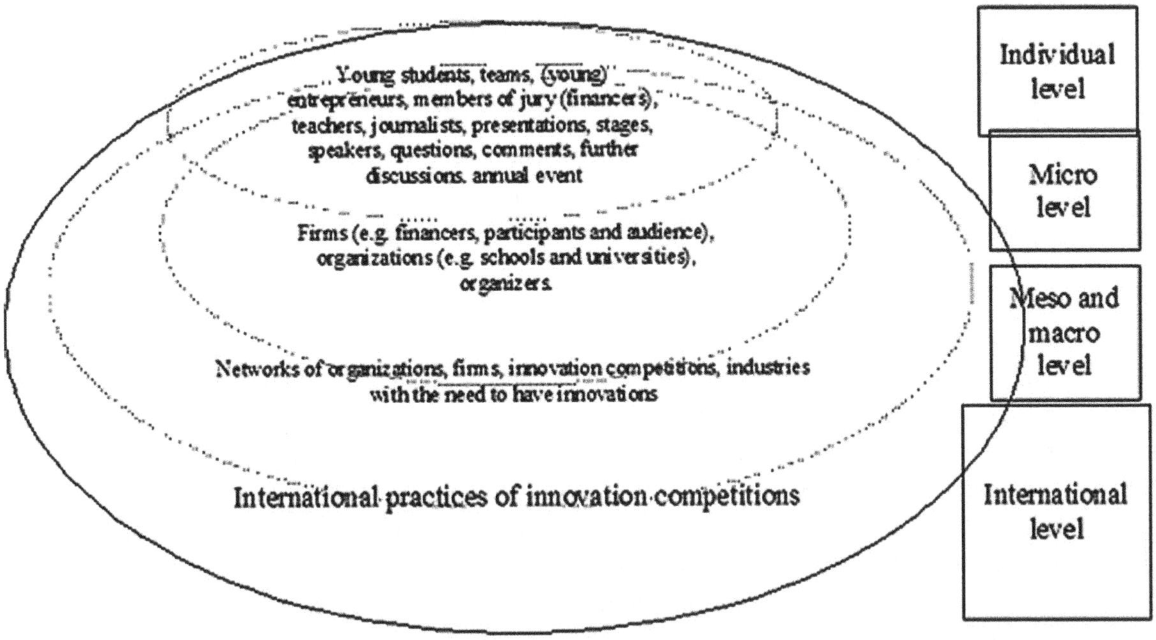

driving is an important feature in these competitions. The jury members are often potential financers of the presented ideas. The main motive for pitching is to obtain financial support via the innovation competition. However, ecologism is not the main driver in typical innovation competitions. Of course, there are also innovation competitions that are totally focused on green-thinking, such as the Pitch Me Green program (Pitch Me Green, 2014).

CONCLUSION

Ohmae (1990) emphasizes that we live in a borderless world. By this, he suggests that the meaning of national borders is decreasing because of consistent universal cultural groups. According to Ohmae (1990), these groups are based on income levels and education, which in turn affect the behavior of individual consumers. These effects are even stronger than national, religious or ethnic borders. Porter (1990a; 1990b) also considers international markets. He maintains that, despite the increasing internationalization of business, nations still have importance in the form of the "competitive advantage of nations". According to Porter (1990a; 1990b), even multinational enterprises have a (national) home base, which of course affects their business practices.

The analysis in this article focused on the international practices involved in innovation competitions, such as pitch events. The case study analysis showed that the roots of these pitch events seem to be found in the USA since most of the pitch competitions are launched in that country. Even some important international pitch competitions follow the format of North American (e.g. MIT) universities or organizers. Language is an important channel and tool in the context of practices (Laine and Vaara, 2007). In the case of innovation competitions, the most important language is English. Even those innovation competitions that use national languages have been named according to English words, such as pitch, pitching, startup or a combination of pitch and startup. This suggests that there is a home base in the branch of innovation competitions. Such a base was initially found in Great Britain, where perhaps the earliest innovation competition was organized, and is currently in Northern America, where most of the contemporary innovation competitions are organized. In the Longitude Prize innovation competition held in 1714 in Great Britain, the organizer was a public sector body, namely the British government. Still today, in contemporary innovation competitions, the role of the public sector is often important since organizers are typically universities while the competitors and audience are also drawn from the university.

The case study analysis, which is based on the contemporary pitching events, shows that the roots of contemporary entrepreneurial pitch competitions are found in the USA although a similar format is in use globally. The innovation competitions that emphasize expressions such as "elevator pitch event" are typically organized in the USA and that the organizer is a university or another educational unit. Conversely, in innovation competitions with expressions "startup pitch event" the variety of organizing institutions is wider and includes individual actors, entrepreneurship organizations and associations. Further, these competitions are mostly outside of the USA. Pitching events principally involve new firms or business students with new innovations that are systematically founded by older enterprises and entrepreneurs. Thus they are relevant channel for youth entrepreneurship. This systematic structure is perhaps the most important international practice in the context of contemporary innovation competitions. This study shows the multi-level characteristics of innovation competition, which contains several practices about startups, their financing and development work, which do not differ across the countries.

These findings show that there is some kind of community of practice surrounding the innovation competitions, which transmits the features of such competitions all over the world. In this case, these communities of practices are possibly online communities, which can be called "quasi-communities" (Hung & Chen, 2002; Hung & Nichani, 2002). Furthermore, these results reveal the importance of geographical proximity. In some parts of Northern America, the number of pitch events is huge, which suggests a real local community of practices in certain areas.

Although contemporary innovation competitions have many multinational practices, such as the short presentation of ideas, a jury consisting of potential financiers, and the common naming of competitions, there are still some variations between them. Some of these competitions are organized biennially, others annually, while some are held monthly or even more frequently. Furthermore, there are variations in the locations of the events since some organizers organize the event in the same place every time, some organizers use the same format all over the world or country, while several events can even be organized at the same time. However, the audience and participants of the events are mainly young people, who are often students of the universities or other educational organizations.

Lampel *et al.* (2010) noted that contemporary design (or innovation) competitions are often multi-episodic or permanent processes, which use the internet as platform. We noticed these same features of innovation competitions. Although many competitions are face-to-face events, the role of the Web e.g. as information and advertising channel is important.

Ultimately, these practices and their variations are perhaps not the main point. Arguably, the main point is instead the entrepreneurship and the progress of both national and global economies. For example, according to Schumpeter (1939; 1942), progress in the economy requires new firms or startups, which maintain the high level of economic activity in society through the process of creative destruction. The contemporary pitching events provide an arena where new firms are systematically founded with the aid of current enterprises and entrepreneurs. This systematic structure is perhaps the most important international practice in the context of contemporary innovation competitions.

FUTURE LINES OF RESEARCH

This chapter shows several themes for further studies. Pitching events are innovation events. Some of these events are web-based while others involve face-to-face happenings. This distinction provides the possibilities for the careful comparison between the features of the face-to-face and the web-based pitching events. Furthermore, innovation practices are global. One important perspective for further study is to generally study global innovation practices, that is the other practices of innovation - in addition to innovation competitions and pitching events.

REFERENCES

Adamczyk, S., Bullinger, A. C., & Möslein, K. M. (2012). Innovation contests: A review, classification and outlook. *Creativity and Innovation Management*, *21*(4), 335–360. doi:10.1111/caim.12003

Argyris, C. (1976). Single-loop and double-loop models in research on decision making. *Administrative Science Quarterly*, *21*(3), 363–375. doi:10.2307/2391848

Argyris, C., & Schôn, D. (1987). Reasoning, action strategies, and defensive routines: The case of OD practitioners. *Research in Organizational Change and Development*, *1*, 89–128.

Argyris, C., & Schon, D. A. (1974). Theory in action. *The Journal of Transformative Studies.*

Arkko-Saukkonen, A., & Merivirta, M. (n.d.). *Creative Steps-Työpajan Eteneminen Askelittain*. Creative Steps. Kemi-Tornio University of Applied Sciences.

Arthurs, K. E. (2012). *Creative cities, creative spaces, and urban policy: the impacts of regulations on artist-run centres and independent arts spaces in Vancouver, British Columbia*. Academic Press.

Birch, K. E., & Heffernan, K. J. (2014). Crowdsourcing for Clinical Research–An Evaluation of Maturity. *Proceedings of the Seventh Australasian Workshop on Health Informatics and Knowledge Management* (HIKM 2014).

Borch, O. J., Førde, A., Rønning, L., Vestrum, I. K., & Alsos, G. A. (2008). Resource configuration and creative practices of community entrepreneurs. *Journal of Enterprising Communities, People and Places in the Global Economy*, *2*(2), 100–123.

Bourdieu, P. (1977). *Outline of a Theory of Practice* (R. Nice, Trans.). doi:10.1017/CBO9780511812507

Bourdieu, P. (1984). *Distinction: A social critique of the judgement of taste*. Harvard University Press.

Bradford, N. J. (2004). *Creative cities: structured policy dialogue report*. Canadian Policy Research Network, Family Network.

Bucherer, E., Eisert, U., & Gassmann, O. (2012). Towards systematic business model innovation: Lessons from product innovation management. *Creativity and Innovation Management*, *21*(2), 183–198. doi:10.1111/j.1467-8691.2012.00637.x

Burger-Helmchen, T., & Pénin, J. (2010). *The limits of crowdsourcing inventive activities: What do transaction cost theory and the evolutionary theories of the firm teach us*. In *Workshop on Open Source Innovation*, Strasbourg, France.

Chesbrough, H. W. (2003). *Open Innovation: The New Imperative for Creating and Profiting from Technology*. Boston: Harvard Business School Press Books.

Chia, R., & MacKay, B. (2007). Post-processual challenges for the emerging strategy-as-practice perspective: Discovering strategy in the logic of practice. *Human Relations*, *60*(1), 217–242. doi:10.1177/0018726707075291

Chicago City Treasurer. (n.d.). *Elevator Pitch Contest*. Retrieved from: http://www.chicagocitytreasurer.com/small-business-entrepreneurs/elevator-pitch-contest/

Cilliers, F., & May, M. (2012). The directors' roles in containing the Robben Island Diversity Experience (RIDE). *SA Journal of Industrial Psychology*, *38*(2), 11–21. doi:10.4102/sajip.v38i2.986

Denzin, N. K., & Lincoln, Y. S. (2005). *The Sage handbook of qualitative research*. London: Sage.

Djelic, M. L., & Quack, S. (Eds.). (2010). *Transnational communities: Shaping global economic governance*. Cambridge University Press. doi:10.1017/CBO9780511778100

Duke. (n.d.). Retrieved from: http://www.dukegen.com/angel-pitch-event

Ebner, W., Leimeister, J. M., & Krcmar, H. (2009). Community engineering for innovations: The ideas competition as a method to nurture a virtual community for innovations. *R & D Management*, *39*(4), 342–356. doi:10.1111/j.1467-9310.2009.00564.x

Elevator Pitch. (2012). Retrieved from: http://www.elevatorpitch2012.com/

Eriksson, P., & Kovalainen, A. (2008). Qualitative methods in business research. *Sage (Atlanta, Ga.)*.

Facebook. (n.d.). Retrieved from: https://www.facebook.com/CEONationalElevatorPitch

Facebook. (n.d.). Retrieved from: https://www.facebook.com/pages/Hong-Kong-Young-Entrepreneur-Club/1

Giddens, A. (1979). *Central problems in social theory: action, structure and contradictions in social analysis* (Vol. 241). University of California Press. doi:10.1007/978-1-349-16161-4

Giddens, A. (1984). *The constitution of society: introduction of the theory of structuration*. University of California Press.

Goldstein, H. (1986). Toward the integration of theory and practice: A humanistic approach. *Social Work*, *31*(5), 352–357.

Howe, J. (2006). The rise of crowdsourcing. *Wired Magazine, 14*(6), 1-4.

Hung, D., & Nichani, M. (2002). Differentiating between Communities of Practices (CoPs) and Quasi-Communities: Can CoPs exist Online?. *International Journal on E-Learning*, *1*(3), 23–29.

Hung, D. W. L., & Chen, V. D. T. (2002). Understanding how thriving Internet quasi-communities work: Distinguishing between learning about and learning to be. *Educational Technology*, *42*(1), 23–27.

Jarzabkowski, P., Balogun, J., & Seidl, D. (2007). Strategizing: The challenges of a practice perspective. *Human Relations*, *60*(1), 5–27. doi:10.1177/0018726707075703

Laine, P. M. (2010). *Toimijuus strategiakäytännöissä: Diskurssi-ja käytäntöteoreettisia avauksia*. Väitöskirja, Sarja A-1.

Laine, P. M., & Vaara, E. (2007). Struggling over subjectivity: A discursive analysis of strategic development in an engineering group. *Human Relations*, *60*(1), 29–58. doi:10.1177/0018726707075279

Lampel, J., Jha, P. P., & Bhalla, A. (2012). Test-driving the future: How design competitions are changing innovation. *The Academy of Management Perspectives*, *26*(2), 71–85. doi:10.5465/amp.2010.0068

Laursen, K., & Salter, A. (2006). Open for innovation: The role of openness in explaining innovation performance among UK manufacturing firms. *Strategic Management Journal*, *27*(2), 131–150. doi:10.1002/smj.507

Leong, B. D., & Clark, H. (2003). Culture-based knowledge towards new design thinking and practice—A dialogue. *Design Issues*, *19*(3), 48–58. doi:10.1162/074793603768290838

Lorah, K. P. (2013). *Disruptive by design: the promise of the tech-enabled design competition* (Doctoral dissertation). Massachusetts Institute of Technology.

Lounsbury, M., & Crumley, E. (2007). New practice creation: An institutional perspective on innovation. *Organization Studies*, *28*(7), 993–1012. doi:10.1177/0170840607078111

Marom, D., & Sade, O. (2013). *Are the life and death of a young start-up indeed in the power of the tongue?* Lessons from Online Crowdfunding Pitches.

Meet Up. (n.d.). *PitchForce*. Retrieved from: http://www.meetup.com/IGNITE-Fundraisers-monthly-startup-pitch-events/

MeetUp. (n.d.). Retrieved from: http://www.meetup.com/New-York-Startup-Pitch-Night

MIT. (n.d.). Retrieved from: http://www.mit100k.org/

MIT Global. (2013). *MIT Global Startup Workshop*. Reviewed March 4, 2014 from http://mitgsw.org/2013/

MIT Global Start Up. (2013). *Elevator Pitch Contest*. Retrieved from: http://mitgsw.org/2013/workshop/elevator-pitch-contest-page/

Moisander, J., & Valtonen, A. (2006). Qualitative marketing research: A cultural approach. *Sage (Atlanta, Ga.)*.

Neumann, I. B. (2002). Returning practice to the linguistic turn: The case of diplomacy. *Millennium*, *31*(3), 627–651. doi:10.1177/03058298020310031201

NextBerlin. (2014). *This is the New Normal*. Review March 4, 2014 from http://nextberlin.eu/

Nuortila, E. (2013). *Crowdfunding as the forms of financing in independent movie*. Helsinki: Metropolia University of Applied Sciences.

Ohmae, K. (1990). *The borderless world*. New York.

Olick, J. K., & Robbins, J. (1998). Social memory studies: From "collective memory" to the historical sociology of mnemonic practices. *Annual Review of Sociology*, *24*(1), 105–140. doi:10.1146/annurev.soc.24.1.105

Pietrobelli, C., & Rabellotti, R. (2010). *The global dimension of innovations systems and enterprise upgrading–Linking innovation systems and global value chains* (No. 025). SLPTMD Working Paper Series.

Pitch Event. (n.d.). Retrieved from: http://www.pitchevent.com/

Pitch Me Green. (2014). *Pitch Me Green. Mentoring program for early stage green and clean technology ventures*. Reviewed March 4, 2014 from http://www.pitchmegreen.com/

Porter, M. E. (1990a). The Competitive Advantage of Nations. *Harvard Business Review*, *68*(2), 73–93.

Porter, M. E. (1990b). *The Competitive Advantage of Nations*. Free Press. doi:10.1007/978-1-349-11336-1

Purdue University. (n.d.). *Elevator Pitch Contest*. Retrieved from: http://www.purdue.edu/discoverypark/entr/competitions/elevator_pitch_contest.php

Reckwitz, A. (2002). Toward a Theory of Social Practices A development in culturalist theorizing. *European Journal of Social Theory*, *5*(2), 243–263. doi:10.1177/13684310222225432

Reich, B. H., Sauer, C., & Wee, S. Y. (2008). Innovative practices for IT projects. *Information Systems Management*, *25*(3), 266–272. doi:10.1080/10580530802151210

Rensselaer. (n.d.). Retrieved from: http://www.eship.rpi.edu/elevator.php

Røpke, I. (2009). Theories of practice — New inspiration for ecological economic studies on consumption. *Ecological Economics*, *68*(10), 2490–2497. doi:10.1016/j.ecolecon.2009.05.015

Rossberger, R. J. (2014). National Personality Profiles and Innovation: The Role of Cultural Practices. *Creativity and Innovation Management*, *23*(3), 331–348. doi:10.1111/caim.12075

Rowe, C. (1996). Evaluating management training and development: Revisiting the basic issues. *Industrial and commercial. Training (New York, N.Y.)*, *28*(4), 17–23.

Rusko, R. (2013). The Redefined Role of Consumer as a Prosumer: Value Co-Creation, Coopetition, and Crowdsourcing of Information Goods. In P. Renna (Ed.), *Production and Manufacturing System Management: Coordination Approaches and Multi-Site Planning*. IGI Global. doi:10.4018/978-1-4666-2098-8.ch009

Rusko, R. (2014). Crowdsourcing for Transcultural Marketing and Innovation. In B. Christiansen, S. Yıldız, & E. Yıldız (Eds.), *Transcultural Marketing for Incremental and Radical Innovation* (pp. 382–394). Hershey, PA: Business Science Reference; doi:10.4018/978-1-4666-4749-7.ch018

Schenk, E., & Guittard, C. (2009, December). *Crowdsourcing: What can be Outsourced to the Crowd, and Why*. In *Workshop on Open Source Innovation*, Strasbourg, France.

Schumpeter, J. (1942). *Creative destruction. Capitalism, socialism and democracy*. Harper Perennial Modern Classics.

Schumpeter, J. A. (1939). *Business cycles* (Vol. 1). New York: McGraw-Hill.

Simpson, B. (2009). Pragmatism, Mead and the practice turn. *Organization Studies*, *30*(12), 1329–1347. doi:10.1177/0170840609349861

Slush. (2014). *SLUSH*. Reviewed March 4, 2014 from http://www.slush.org/

Sobel, D. (2004). Longitude: The True Story of a Lone Genius who Solved the Greatest Scientific Problem of His Time. *Academy of Management Learning & Education*, *3*(2), 220–220. doi:10.5465/AMLE.2004.13500465

StartUp Boot Camp. (n.d.a). *Istanbul*. Retrieved from: http://www.startupbootcamp.org/events/istanbul-pitch-day-3.html

StartUp Boot Camp. (n.d.b) Retrieved from: http://www.startupbootcamp.org/events/startupbootcamp-pitch-day-athens.html

StartUp Pitch Night. (n.d.). Retrieved from: http://www.startuppitchnight.com/startuppitchnight/

Swan, M. (2012). Crowdsourced Health Research Studies: An Important Emerging Complement to Clinical Trials in the Public Health Research Ecosystem. *Journal of Medical Internet Research*, *14*(2), e46. doi:10.2196/jmir.1988 PMID:22397809

SwitchPitch. (n.d.). *SwitchPitch is coming to New York*. Retrieved from: http://switchpitch.com/switch-pitch-nyc/

Tervo, T. (2013). *The Content is King! How to Sell the Documentary Idea for the Production Company?*. Helsinki: Metropolia University of Applied Sciences.

The Entrpreneur Challenge. (2016). *Elevator Pitch Competition*. Retrieved from: http://challenge.ucsd.edu/elevator-pitch

Tienari, J., & Meriläinen, S. (2009). *Johtaminen ja organisointi globaalissa taloudessa*. Helsinki: WSOYpro Oy.

Ultra Light Start Ups. (n.d.). Retrieved from: http://ultralightstartups.com/

University of Idaho. (n.d.). Retrieved from: http://www.uidaho.edu/cbe/experientiallearning/view/elevatorpitch

Vaara, E., Kleymann, B., & Seristö, H. (2004). Strategies as discursive constructions: The case of airline alliances. *Journal of Management Studies*, *41*(1), 1–35. doi:10.1111/j.1467-6486.2004.00419.x

Wenger, E. (1998). *Communities of Practice: Learning, Meaning and Identity*. Cambridge, MA: Cambridge University Press. doi:10.1017/CBO9780511803932

Whittington, R. (2006). Completing the practice turn in strategy research. *Organization Studies*, *27*(5), 613–634. doi:10.1177/0170840606064101

Wind, J., & Rangaswamy, A. (2001). Customerization: The next revolution in mass customization. *Journal of Interactive Marketing*, *15*(1), 13–32. doi:10.1002/1520-6653(200124)15:1<13::AID-DIR1001>3.0.CO;2-#

Yin, R. K. (1983). *Case Study Research: Design and Methods* (3rd ed.). Sage Publications.

Zucchermaglio, C., & Talamo, A. (2003). The development of a virtual community of practices using electronic mail and communicative genres. *Journal of Business and Technical Communication*, *17*(3), 259–284. doi:10.1177/1050651903017003001

KEY TERMS AND DEFINITIONS

Community of Practice: The joint conventions among groups of practitioners without organizational borders.

Elevator Pitching Event: An innovation competition or event, which are normally organized by (North American) universities.

Finance-Driving: A phenomenon, where the role of multinational enterprises has grown: Although the ownership of enterprises has often diversified, the decisions of the enterprises are more often based on the perspectives of the owners and reports of quarterly economic and other financial reports.

Media-Centrism: The actors in business and economy have to take into account the reactions of professional media and social media when deciding on their actions because of media-centrism.

Practice Turn: A tendency in social science, which is associated with the importance of practices learning and change.

Startup Pitching Event: An innovation competition or event, which is mainly organized by (entrepreneurship) organization or association besides educational institution.

Technologizing: Due to rapid technologizing, the world is "smaller" and more multicultural in the context of everyday business activities.

Chapter 8
A Neuromarketing Perspective on Measuring Marketing Influence at the Unconsciousness Level

Ioana Iancu
Babeș-Bolyai University, Romania

ABSTRACT

In a context characterized by an inflation of marketing messages, it is imperious to understand how consumers succeed in making the buying decision. Starting by briefly describing the structure and the role of the brain and the differences between consciousness and unconsciousness, the paper aims to investigate the way neuromarketing can help in comprehending the feelings of the consumers, the way products or services match the consumers' needs, and the way companies can discover the insights of decision-making process. This paper can be perceived either as a guide for the companies that aim to find more on the way people manage information and make decisions or as a comprehensive description on human being marketing behavior that can serve both business, academic environments and consumers.

DOI: 10.4018/978-1-5225-0953-0.ch008

INTRODUCTION

In a context characterized by a dynamic market, by increasingly sophisticated consumers and by an immense amount of information, the way companies work in order to accomplish individuals' expectations become a complex issue. Therefore, the main questions that can be asked are what consumers really feel and think on a certain product or service, how that product or service should match the consumers' needs, and to what extend companies can discover the insights of decision-making process.

Considering the above background and the need to deeply enter the secrets of consumers, the present paper aims to theoretically investigate the role of unconsciousness on the individuals' decisions at the marketing level. In this respect, issues like how the brain works and neuromarketing are some of the main themes assessed.

This paper can be perceived either as a guide for the companies that aim to find more on the way people manage information and make decisions or as a comprehensive description on human being marketing behavior that can serve both business and academic environments and consumers. First, using the insights of this paper companies can develop instruments in order to find more valuable information on what exactly consumers' want and need. Second, academics can use the information as an inspiration source for their further experiments and in order to improve the already used instruments. Finally, consumers can find information about themselves and about the way brain works and, thus, uses it either in their shopping behavior or on other social decisions.

Starting by analyzing insights of the brain and the role of consciousness and unconsciousness, the paper emphasizes the way the modern marketing can complete traditional marketing by shifting the focus from the product to the complexity of the consumer. Thus, the core concept of this paper is neuromarketing. Based on this concept, the main discussed areas are related to the instruments of neuromarketing, the way it offers extra information comparing with the traditional research methods, its entrepreneurship applicability and its ethical perspectives.

BACKGROUND

An Insight of the Brain

Although representing only 2% of the total body mass, the brain is responsible for all the marketing behavior and it consumes approximately 20% of the body energy (Morin, 2011, p.134). Renvoisé and Morin (2007) consider that the simplified structure of the brain is the following one: the reptilian brain or R-complex (responsible with instinct), the middle brain or the limbic system (responsible with emotions), and the new brain or neo-cortex (responsible with rational thinking). Being entities for which the survival process is the most important one, the human behavior is mostly controlled by the reptilian brain (Morin, 2011, p.134). In comparison with the new brain that is slow in reaction, it effort itself in order to answer to a certain stimulus, it is intelligent, conscious and partially controllable, the reptilian brain is 500 million years old, is the first that reacts to incentives, it is always on, it is unconscious and incontrollable (Renvoisé and Morin, 2007). In the same respect, emphasizing that there are a lot of aspects related to impressions, intuitions and decisions that are far to be deeply understood, Kahneman (2011) splits the brain in two parts: System 1 (reptilian brain) and System 2 (new brain). The two concepts have been initially used by Keith Stanovich and Richard West. System 1 is described as being automatic, permanent, quick, effortless, and involuntary. In contrast, System 2 is the part of the brain that pays more attention to the mental effort, it can realize complex calculus, and it is associated with decision and with concentration (Kahneman, 2011).

The description of the parts of the brain and the way they work and react is important in the marketing context, mainly considering that almost 80% of the information received by a consumer is internalized at the unconsciousness level (Morin, 2011, p.134). Put it differently, although the cognitive functions of the brain are treasured, in a context in which individuals receive more that 10.000 messages per day, a large part of this information is lost unless it addresses the reptilian brain (Morin, 2011, p.135). Therefore, one can conclude that a message in general and a marketing message in particular are efficient mainly if it contains stimuli that can activate mainly the instinctual and unconscious part of the brain, namely the reptilian brain. Based on this issue, this chapter is going to underline the domain of neuromarketing and its several techniques already used by companies or academics, or techniques and experiments that can be used in order to gain business efficiency and knowledge.

Consciousness and Unconsciousness

The evolution of the techniques studying the marketing influence at the brain level (neuromarketing) is mainly due to the fact that researchers have realized that the process of decision making is not a rational one and that it is impossible to make an optimal rational decision in which all the alternatives are taken into consideration (Ramsøy, 2014). Although there is a part of the literature emphasizing that consumers' decisions are conscious and deliberated processes, another part of the literature debates the role of unconsciousness within the decision making process (Tom et al., 2007). For instance, in a study, thirsty consumers that have been exposed to smiley individuals have consumed higher quantities of juice and have been willing to pay more for the consumption in comparison with the same type of persons exposed to sad individuals (Winkielman, Berridge, & Wilbarger, 2001 in Tom et al., 2007). In the same respect, Tom et al. (2007) claims that consumers pay more on brands they like than on alternative brands. They also talk about two types of effects: *mere exposure effect* implying that a repeated or simple exposure to a certain stimulus results in developing an affective positive reaction on the stimulus; the *endowment effect* implies that a consumer perceives an object as being more valuable if he/she possesses that object. (Tom et al., 2007, pp.118-119). While the first effect increases the preference for a certain object (affection without cognitive mediation), the second effect increases the value of the object (cognition without affective mediation) (Tom et al., 2007, p.119).

Thus, the decisions are made based on two processes: a quick unconscious process and a rapid decision experience. Ramsøy (2014) considers that it is necessarily to understand both conscious and unconscious brain. Conscious events are those brain activities individuals can report with accuracy, in optimal conditions and with a minimum distraction. In contrast, unconscious events are those one knows that exist but there is a lack of ability to develop an accurate report on them (Baars, Ramsøy, & Laureys, 2003, p.672). While the unconscious processes imply limited and rigid processing of information, the conscious processes are associated with a much more vast and flexible processing of information, the effect on thinking and behaving being deeper (Shanahan & Baars, 2005 in Ramsøy & Skov, 2014, p.1).

Shiv and Fedorikhin (1999, pp.288-289) talk about the affective-cognitive model of behavior and consider that, in the context of exposure to certain alternatives during a task, there are two processes that can take place: an affective one and an cognitive one. While it is likely that fist one happens automatically, the second one is likely to be controllable focusing on the consequences of the chosen alternative.

Moreover, while the consumer decision is likely to be based on affective aspects when the resources processing is constraint, the lack of constraints implies rather cognitive reactions (Shiv & Fedorikhin, 1999, pp.288-289).

Cognitively speaking, the objectives are mentally represented as schemas, similar with stereotypes or attitudes. Once having these constructs activated, the actions associated with them will be realized (e.g. the objective of saving money will lead to the behavior associated with it) (Bargh, 1990 in Chartrand et al., 2008, p.190). Another part of the literature sustains that in order to accomplish an objective, this must be associated with positive feelings (Custers & Aarts 2005 in Chartrand et al., 2008, p.190). Within this context, the activation of the objectives can be done either consciously, when the individual develops an objective and act in order to accomplish it, or unconsciously, when an objective has been frequently associated with a certain situation, that objective being automatically activated when the circumstances of that situation are recognized (Bargh, 1990 in Chartrand et al., 2008, p.190).

Dijksterhuis, Smith, van Baaren, and Wigboldus (2005 in Simonson, 2005) claim that there are a large number of decisions taken at the unconsciousness level and they are strongly affected by the environment factors. For instance, an adult in a supermarket that sees a child running is likely to unconsciously buy something that reminds him/her about childhood (Dijksterhuis, Smith, van Baaren, & Wigboldus, 2005 in Simonson, 2005, p.212). However, Simonson (2005) defend the theory that stresses that individuals are rather conscious when they make buying decisions, the above mentioned example being rather an accident than a behavioral model.

Bargh (2002, p.280) is interested in finding to what degree the individual is conscious about what influences him/her during buying behavior and to what degree he/she can control it. He stresses that there are two ways in which unconscious elements can be activated: either subliminal, when the stimuli are not accessible at the conscious level, or supraliminal, the individual being conscious about the stimuli but not about their effect (Bargh, 2002, p.282). For instance, there are a large amount of individuals that buy tickets to Super Bowl only in order to see the commercials. Thus, they know they are influenced, but they do not know which the effect of this influence is (Bargh, 2002, p.283).

Chartrand et al. (2008) aim to analyze whether random exposure to certain incentives (for instance, certain prints) can activate different buying objectives and whether it can influence the final decision. In the buying behavior context, there are a lot of studies showing that the individual is directed on accomplishing the objectives, these ones being the key motivational element within the decision making process (Chartrand et al., 2008, p.189). However, there is a smaller amount of literature talking about the factors that determine the choice of those particular objectives and their accomplishment. Thus, in contrast with the traditional research that stresses that individuals are perfectly conscious about their objectives regarding the choices, the newer literature talks about the possibility that the objectives can be unconsciously chosen (Chartrand et al., 2008, p. 189). Chartrand et al. (2008), based on several experiments, draw the conclusion that a subliminal exposure to certain brands (either associated with luxury products or cheaper products) activates objectives associated with these brands (e.g. the objective of buying a luxury product, or the objective of buying a cheaper product).

In the context of exposure to a certain stimuli, Berkowitz (1993 in Shiv & Fedorikhin, 1999, pp.279-280) talks about three types of processes. First, the information regarding the stimulus is perceived in an automatic associative process, affective sometime, that happens before the cognitive processes of evaluation, interpretation, schemas development, attributions and strategy. Second, the information on the incentive is the subject of a deliberative and cognitive process that can improve or aggravate action tendencies. Third, the affective significance of the results following a cognitive process can result in more complex affective reactions and in slower action tendencies in comparison with preliminary affective reactions, the deliberative process making the decision slower.

Considering this context, a study presents the following experiment. The individuals have been asked to memories either a number of 7 digits or a number of 2 digits, this task being described as very important. While the subjects' attention has been focused on learning the numbers, the subjects have been asked to choose between a very appealing but unhealthy chocolate cake and a very healthy bun non-appealing fruit salad. The results shows that the subjects are inclined to choose the unhealthy cake when they have a difficult task, as memorizing a 7 digits number (Shiv and Fedorikhin, 1999, p.279). Put it differently, in the case in which the rational part of the brain is occupied, the intuitive and impulsive brain has a stronger influence on behavior (Kahneman, 2011). Thus, in the context of a binary choice where one of the alternatives (chocolate cake) is much more associated with affection and less with cognition in comparison with the second alternative (fruit salad), the decision depends very much both on the level of processing the resources of the task and on the impulsive behavior (Shiv & Fedorikhin, 1999, p.288).

Another experiment was conducted in the same room in which the tea is usually served at the British University. Every individual had the freedom to put in a box a certain amount of money they considered to be enough for their consumption. Although there was a list with suggested prices, it was not mandatory to respect it. The experiment implied as well the existence of banner above the list with prices for 10 weeks time period, every week the image on the banner being changed. There were used images either with flowers, or with eyes part of the head of men and women. There was no explanation given for these pictures. The results showed that in the days in which there was shown eyes, the contribution for the consumed quantity increased almost three times than in the days in which the image represented flowers. Thus, it can be said that images with people watching the consumers, either they are happy, aggressive or neutral, can influence the level of obedience and the quantity of the given money (Kahneman, 2011).

The above-described literature emphasizes the role of the brain and of unconsciousness within the consumer behavior. At the same time, it underlines the need to deeply asses this behavior with other instruments than traditional ones. Therefore, neuromarketing might be a domain that helps in better understanding the influences of different incentives on the consumers' decision process.

A GENERAL PERSPECTIVE ON NEUROMARKETING

Victoria Phan (2010, p.14) claims that advertising becomes extremely advanced from a scientifically point of view. In this context and in a strong relation with the subliminal messages and the role of emotions within the decision making process, neuromarketing is a relatively new domain that describes clinical information on the way the brain functions and explains the consumer's behavior based on neurological activity (Fugate, 2007, p.385). Being a combination between neuroscience and marketing, neuromarketing has appeared in 2002 and its instruments seems to be more and more used in the business environment (Morin, 2011, p.131). In 2004, The Economist has stressed that Jerry Zaltman is the first one that proposed, in the mid '90, the use of imagistic technology in marketing (Wilson et al., 2008, p.390). However, another source stresses that the concept of neuromarketing has been invented in 2002 by the professor Ale Smidts from the Erasmus University and it has been recognized two years later at the first conference on neuromarketing (Phan, 2014, p.15). American companies as Brighthouse or SalesBrain are the first ones that have offered neuromarketing services (Morin, 2011, p.132).

Thomas Ramsøy (2014) claims that there is difference between *neuromarketing* and *consumer neuroscience*. While the first uses neuroscience having the commercial goal to improve the consumer's knowledge and the marketing effects, the second one is rather an academic exercise meant to understand consumer's psychology and behavior (Ramsøy, 2014).

Neuromarketing is a research domain that helps gather relevant information in the responses of the brain to different stimuli from marketing (Renvoisé´ and Morin, 2007 in Murphy et al., 2008, pp.293-294). The goal of neuromarketing is to offer objective information on the consumer's perceptions without registering his or her subjective answers (Murphy et al. 2008, pp.293-294). Lee, Broderick, and Chamberlain (2007 in Wilson et al., 2008, p.390) consider that neuromarketing implies the application of neuroscience methods and instruments in order to analyze and understand human behavior in its relation with the market and the changes within it. At the same time, neuromarketing uses neuro-imaging techniques in order to efficiently sell products (Lee et al., 2007, p. 200).

Thomson (2003 in Wilson et al. 2008, p.390) describes neuromarketing as a domain in which the methods specific for the neurological domain are used in advertising. In the same respect, Lee, Broderick, and Chamberlain (2007 in Wilson et al. 2008, p. 390) consider that neuromarketing implies the application of neuroscience methods in analyzing and understanding the human behavior in its relationship with the market and the changes within it. Thus, one can claim that neuro-imaging technology will contribute to the understanding of the way individuals take decisions and of the methods used by specialists to influence those behaviors (Wilson et al. 2008, p.390).

In addition, neuromarketing is perceived as being useful within the process of developing a new model of consumerism behavior. Thus, based on the neuromarketing instruments, the practitioners have access to valid data on consumers and the consumers can learn to know themselves better and to better understand their decision making process (Butler, 2008, pp.416-417). In the '50, two researchers from the McGill University conclude that within our brain there is an area that can be named the "center of pleasure". Based on this discovery, it can be said that humans are motivated by things that make them feel better, even when it comes to shopping (Nobel, 2013). However, although individuals can say to a certain degree what they want, what they like or how mush they would pay for a certain product, they cannot clearly say why. This is the context in which neuromarketing brings the necessary instruments that can help in better understanding the human behavior (Nobel, 2013).

While the instruments of neuronal imaging are widely used in economics (neuroeconomics), they are relatively slowly adopted in marketing. Lee *et al.* (2007) offers several explanations for that. First, neuroscience can be an intimidating domain in an academic context. Second, considering the costs implied for purchasing the necessary technology, universities cannot afford to empirically study this domain (Lee *et al.*, 2007, p. 199). Finally, neuromarketing raises a large amount of discussions regarding the search for the buy button in the brain of the consumers (Lee *et al.*, 2007, p.199).

The actors that can be involved in the neuromarketing research are advertising agencies and marketing departments (that aim to present their products, design, campaigns, to measure the individuals' reactions regarding the efficiency of promoted messages), media departments (that aim to determine the best media space for a campaign and the mix of efficient strategies), companies (that aim to understand the consumers' preferences), strategists from the political domain (that aim to evaluate the communication and debate styles), or film producers (that aim to take decisions regarding the best actors to use, the best end of a movie, the best scenes of a trailer in order to get the highest level of efficiency) (Bercea 2015). Thus, neuromarketing instruments can help in gathering relevant information mainly before a product is being launched on the market.

The first research in the neuromarketing field has been published by McClure, Li, Tomlin, Cypert, L. M. Montague and P. R. Montague in *Neuron* in 2004 (Morin, 2011, p.132). The research is conducted by using a set of blind and anonymous tests and passive experiments using Functional Magnetic Resonance Imaging (fMRI). The authors aim to investigate both the neuronal response that can be correlated with the preference for Coca-Cola or Pepsi, and the influence of brand image on the decision making process (McClure *et al.*, 2004, p.379). The research starts from three main research questions. The first one refers to the cognitive and neurological responses in the context in which the two soft drinks are tasted without knowing the brand. The second analyzed issue refers to the cognitive and behavioral influence in the context in which the subject knows the brand of the tasted soft drink. Based on these two aspects, the third one refers to a possible correlation between the behavioral preference and the neurological response measured with fMRI (McClure et al. 2004, p.380).

The researchers claim that the two soft drinks included in the experiment, Coca-Cola and Pepsi, are very similar from the point of view of chemical composition, the second one being only a little bit sweeter. However, the study concludes that the cultural messages, related to the brand, correlated with the content can influence the way individuals perceive a product (McClure et al. 2004, p.379). Thus, knowing the brand of the soft drink, especially when it comes to Coca-Cola, biases the consumers' preferences. When the subjects do not know which soft drink they are consuming, a large majority declares they prefer, at the taste level, Pepsi. In this particular context, limbic system of the brain, the one responsible with emotions and instinctual behavior is activated. In contrast, when the subjects know which of the two soft drinks they are consuming, a large majority prefers Coca-Cola. In this case, a different part of the brain, namely the frontal cortex, responsible with executive functions, with thinking and planning, is activated (McClure et al. 2004). Put it different, although they like better the Pepsi taste, the subjects claim they prefer Coca-Cola, mainly developing this response on the brand image rather than on taste (Phan, 2010, p.15). The insight of this conclusion is that a commercial or a set of commercials have the capacity to make an individual act in contradiction with what their bodies are actually asking for (Phan, 2010, p.16). Thus, there is the capacity of certain external factors, as brand, brand reputation, social desirability, to influence the consumers' preferences and their buying behavior.

The Instruments of Neuromarketing

The instruments used in neuromarketing can be grouped in at least three categories (Zurawicki, 2010 in Bercea, 2012). First, there are the instruments used to register the metabolic activity of the brain (functional Magnetic Resonance Imaging - fMRI, Positron emission tomography - PET). Second, there are the instruments used in order to measure the electrical activity of the brain (electro-encephalography – EEG, magneto-encephalography (MEG), Transcranial magnetic stimulation - TMS, Steady State Topography - SST). Finally, there are the instruments that do no require the registration of brain activity. They refer either to facial reactions (Eye Tracking, Facial coding, Facial electromyography), or to body reactions (Physiological response measurement, Galvanic skin). The most used techniques are fMRI, EEG, and magneto-encephalography (Fugate, 2007, p.386) at the brain level, and Eye Tracking, Facial coding and galvanic skin at the body level (Bercea, 2012). Usually, there is a combination of techniques used within an experiment. These instruments are non-invasive (Morin, 2011, pp.132-133). The following part of the chapter will comprehensively describe the most important of them.

Electroencephalography (EEG), although being an older method, it is still considered to be efficient regarding the measurement of brain activity (Morin 2011, p.132). In the 17th century scientists have discovered that individuals are controlled by electrical impulses, the first electroencephalograph being built at the beginning of the 20th century and the first demo being conducted by Hans Berger (Ohme & Matukin, 2012).

Neurons, the cells responsible for the cognitive responses, are more than 100 billions. At the same time, there are trillions of synaptic connections that represent the base of the neuronal circuit. When an individual is exposed to a stimulus (e.g. a commercial) the neurons are activated and they produce small electricity. This electricity can be associated with different emotional states. In marketing, EEG data are obtained by using a helmet with electrodes that is placed on the head of the subject and that is able to make up to 10.000 registrations per second (Morin, 2011, pp.132-133). However, by using EEG the precise location in the brain in which the neurons have been activated cannot be detected, mainly when it is about deeper areas in the brain. EEG devices can measure electrical signals only at the surface of the brain. Moreover, although data obtained by using EEG are valuable in the case of measuring the effect of an ad, they are not sufficient to understand the entire cognitive process that determines a certain activity (Morin, 2011, p.133). Studies that use EEG are based on inputs (e.g. images) presented on the screen of a computer, and an experiment can take approximately 30 minutes (Ohme & Matukin, 2012).

Although it is not sensitive at the deepest brain structures, EEG has a high temporal resolution (milliseconds), which means that it can detect very succinct brain events. At the same time, EEG has a small spatial resolution, the obtained information depending on the number of electrodes on the helmet. The more electrodes, the better the spatial resolution is and the more information is obtained (Ariely & Berns, 2010, p.288). One of the most common measurements regards the left-right asymmetry of the frontal EEG, meaning the measurement of alpha coefficient (alpha band 8-13 Hz) (Ariely & Berns, 2010, p.288). There are several types of electrical signals measured with EEG. For instance, while delta signals are register in lethargy state, alpha signals are registered in relaxation state, and beta signals are registered in conscious activity state (Ohme & Matukin, 2012).

The first studies in the psychology using EEG have been conducted around 1979. Davidson has been one of the first cognitive specialists that have studied the relationship between emotions and electrical signals in the brain. There are studies that have proved that the electrical models are located in the frontal part of the brain. For instance, while alpha signals in the left frontal lobe indicate positive emotions, electrical activity in the right part of the frontal lobe is usually correlated with negative emotions, with the desire to leave an action (Morin 2011, pp.132-13). By using EEG, there can be detected the scenes of a commercial that have the highest emotional intensity, or there can be evaluated the reaction of a subject to different elements of an ad (e.g. image, sound, words, special effects etc.). Moreover, there can be predicted to what degree a song will be liked by the public, a strategy will be efficient, a packaging will be preferred, or a story will be powerful (Ohme & Matukin, 2012).

Ohme and Matukin (2012), in one of their studies, about the analysis of *Balls* commercial for Sony Bravia LCD. The ad has been launched in November 2005 and won Gold Lion at Cannes Lions International Advertising Festival in June 2006 and Grand Prix at Midsummer Awards in London in June 2006 (the original commercial can be seen at https://www.youtube.com/watch?v=R8WkIC7bMnE). The special element of the ad is a number 250.000 colored real balls that are launched to freely jump on San Francisco streets (Ohme, & Matukin, 2012). Based on this commercial, the authors are interested in understanding which is the elements that have made this commercial so appreciated and which are the scenes with the highest emotional appeal. Thus, completing a traditional research the authors have conducted a neuromarketing study using EEG on a number of 45 subjects. One of the most intense emotional scenes is the moment in which a frog comes into play (17th second of the ad), this scene being filmed by chance.

Moreover, the last part of the commercial in which informational aspects on the product are presented is considered to have a stronger emotional effect than the first part of the ad in which Sand Francisco city is visualized (Ohme & Matukin, 2012). Interestingly, in comparison with the neuronal results, at the declarative level (focus groups), the respondents claim that the most pleasant images are that of the balls falling. The research also tries to determine the importance of the music in this commercial (the song "Heartbeats", singer - José González, composer - The Knife) and its impact on the entire ad. Thus, when the commercial runs without music, the impact is lower (less positive emotions) than in the case of the ad with music. The same thing happens in the case in which the song is presented without images. Although in this case the image and the sound create a necessary unity, there are neuromarketing studies that claim that it is not a rule that this mix is efficient (Ohme & Matukin, 2012).

Magnetoencephalography (MEG) exists from the middle of the '60s and has the role of measuring the magnetic field of the brain and of developing the magnetic map of it (Morin 2011, p.134). MEG has a temporal and spatial resolution better than EEG. It is happening due to the fact that the magnetic field is less biased by the cranium and its artifacts in comparison with the electrical field (Ariely and Berns 2010, p.288). However, MEG is limited to measuring brain activity only on its surface. Thus, it is not a proper method for measuring brain activity in the sub-cortical areas (Morin 2011, p.134). The literature talks about studies that use MEG and that have succeeded in finding a correlation between certain frequency bands and certain controllable cognitive tasks (e.g. recognizing objects, accessing verbal memory, remembering specific events etc.). Thus, MEG is rather recommended in measuring activities that imply specific tasks than for exploratory experiments that imply investigating cognitive (cortical) or emotional (sub-cortical) functions (Morin 2011, p.134).

Functional Magnetic Resonance Imaging (fMRI) is used in measuring the blood flow in the brain. When the neurons are activated by a stimulus (e.g. commercial), they need energy that is transported in blood. In marketing studies using fMRI, the acronym BOLD (Blood Oxygen Level Dependant) is used. It refers to the oxygen quantity in blood. All the studies that use fMRI consider that a change in BOLD signal represent a valid measurement of neuronal activity (Morin 2011, p.134). Although the resolution in the care of fMRI is better than in the case of EEG and the deepness of activity measurement is higher, the first method is slower. There is a lateness of several seconds between the moment in which the neuron is activated and the moment in which a change in BOLD is remarked (Morin 2011, p.134). Moreover, although fMRI has the advantage of offering information from the profound areas of the brain, certain regions (e.g. frontal orbital cortex) are affected by artifact signals that can bias the data (Ariely and Berns 2010, p.288).

In the case of fMRI, the researchers can isolate systems of neurons associated with different functions of the brain. An example used in the literature is that of an individual that watches a print commercial. In this particular case, a part of the 125 millions of visual neuronal receptors in each eye are activated. The neuronal signals are transmitted to that part of the brain responsible with the pupils and that initiate the eye movement on the print. Other signals are issued, through the optical nerves, to the opposed part of the brain in such a manner that the left side of the print is perceived by the right hemisphere and the right part of the print is perceived by the left hemisphere (Dubuc 2007 in Wilson 2008, p.391). Thus, by using fMRI, it is possible to detect neuronal activity associated with the sight and with the cognitive and affective responses related to the print (Wilson 2008, p.391).

In the context in which the isolation of the neuronal systems formed of 100 billions of neurons in the human brain is a complex task, fMRI is the tool capable of locating the active systems by comparing the images in the moment the brain is active with the images in the moment in which the brain is not active (Wilson 2008, p. 391). Put it different, first the brain is scanned when it is not running the task and then the brain is scanned while developing the task. Thus, the activity in the brain is detected and it is interpreted by taking into account the lateness of the data, the movements of the head, the breathing or the heart beating (Wilson 2008, p.392). There is the need of involving many subjects in order to compare the activated locations in the brain (Wilson 2008, p.392).

Simply put, the brain activity follows the following pattern: the signals are transmitted from a neuron to the other through neurotransmitters, through synapse to the receptors (Carey 2005 in Wilson 2008, p.391); the synaptic activity of the neurons' network generates blood flow in that particular region (Raichle & Mintun 2006 in Wilson 2008, p.391); the blood brings a supplementary quantity of oxygen and hydrogen that increase the magnetic field during scanning, field that is detected by fMRI device (Matthews & Jezzard 2004 in Wilson 2008, p.391).

Although fMRI can be used in gathering data based on visual, aural or gustative stimuli, the scanning process can cost even 1000$ a day and there are needed at least 20-30 subjects (Karmarkar in Nobel, 2013). However, there is research that can predict the future popularity of a certain product. Such an example is a study conducted by Emory University in which the specialists succeed in correlating certain neuronal activities with the musical success. For instance, the song Apologize from OneRepublic was very well perceived both in scanning and in tops (Berns & Moore in Nobel, 2013).

Besides the neuro-imaging techniques, neuromarketing uses instruments that measure body reactions. Thus, *eye tracker* measures the way the subject looks at a certain stimulus, the level of pupil dilatation, where he/she is looking and for how long (Bercea 2013). The aim of the analysis is that of assessing the visual perception, the cognitive intention, the interest in and the importance given to an observed object (Zurawicki 2010 in Bercea 2013). Moreover, the measurement of the physiological response to certain incentives is a method through which there can be obtained data related to emotional effects of a certain stimulus at the heart level, blood pressure, skin conductivity (e.g. sweating), stress hormones in saliva, or muscles contractions of the face (Bercea 2013).

Although the potential of neuromarketing is vast, Lee *et al.* (2007) consider that the main aspect the neuronal imaging is focusing on is branding advertising, and consumer behavior. For instance, while EEG can be used in order to asses the impact of certain visual inputs, MEG can be used in order to understand the way cognitive and affective commercials generate different reactions in the brain. Another marketing specific issue that can be analyzed using neuronal imaging is the decision behavior of the consumer (for instance, the difference between predictable choices and unpredictable ones based on gender) (Lee *et al.*, 2007, p.201). Murphy et al. (2008, p.295) claims that by combining these methods, the value of the data increases.

NEUROMARKETING AS A COMPLETION FOR THE TRADITIONAL RESEARCH METHODS

Although marketing is perceived as a deleterious domain, it helps matching the products with the consumer through guiding the design and presentation of the product in such a way in order to correspond with individuals' preferences (Ariely & Berns, 2010, p.284). Traditional research methods used in marketing (opinion survey, focus group, interview etc.) tries for a long time now to predict the efficiency of marketing campaigns. However, considering that emotions are a strong and complex mediator of the way consumers process and understand messages, these traditional methods usually fail to develop valid results (Morin, 2011, p.132). Some researchers wrongly assume that individuals are capable to describe their cognitive processes, although they have a large amount of subconscious elements. Moreover, there are reasons for which consumers prefer to bias the offered information, as the incentives, time constraints, interviewees' pressure etc. (Morin, 2011, p.132).

As Plato claims, an individual is formed of two major elements: reason and emotion. However, most of the research has been based the human behavior's explanations solely on rational judgment (The Economist, 2005 in Fugate, 2007, p.386). In contradiction, nowadays, based on social, technological and economic evolution, the buying decision is perceived as the result of the combination between rationality and emotions. Using neuromarketing instruments, Fugate (2007, p.386) claims that rational and emotional thinking are co-dependent. In the same perspective, Damasio stresses that emotions are responsible with attracting consumer's attention and with their motivation in order to focus the rationality on certain elements (Damasio in Coy, 2005 in Fugate, 2007, p.386). In contrast with the traditional research methods that focus on self-evaluation and on the ability and desire of the individual to give exact and correct answers, instruments used by neuromarketing can detect data that are not necessarily visible at the consciousness level (Petty and Cacioppo, 1983 in Lee et al., 2007, p.200). Schafer emphasizes that neuromarketing, in contrast with the traditional methods, has the ability to focus on specific elements belonging to unconsciousness within the buying behavior (Schafer, 2005 in Fugate, 2007, p.386).

The tradition research methods used in analyzing the consumer behavior do not efficiently succeed in describing the emotions, these descriptions being dependent on the desire and on the competency of the individuals to emit real data (Morin, 2011, p.131). Thus, neuromarketing completes the traditional methodology through collecting data especially at the brain level (Morin, 2011, p.132). Thus, neuro-imaging techniques offer a fascinating alternative to traditional research. These techniques offer the possibility to find valuable information on the way a message in perceived and processed at the unconsciousness level and on the future success of that message (Morin 2011, pp.132-133).

Usually, specialists use research methods as focus group, opinion survey or market tests. On one hand, while some of these methods (e.g. focus group or opinion survey) are relatively simple to apply and cheap, they imply a large amount of possible errors. On the other hand, while using these techniques one can obtain precise data (e.g. market tests), the costs of application are very high and the product must exist already on the market in order to develop the tests (Ariely and Berns 2010).

While the tradition research methods can analyze the reactions of the consumers only after the product or the service is released on the market, neuromarketing techniques are used both before and after releasing it. Thus, concept-products can be tested and eliminated if they seems un-attractive, a large amount of resources being saved or reused for more efficient products (Ariely and Berns, 2010). For instance, the willingness to pay has started to be measured using fRMI. A specific example is an experiment that implies snacks' consume. Thus, the willingness to pay is associated with activity in the part of the brain named frontal orbital cortex and pre-frontal cortex. The authors of the study claim that similar parts of the brain are activated when the subjects anticipate a pleasant taste, hear a nice music, receive money or experienced a social prize (Ariely and Berns, 2010, p.285). In conclusion, one can say that neuromarketing completes the traditional research methods by adding information regarding the reaction at the brain or body level (Morin, 2011, p.132).

Ariely and Berns (2010) consider that, beside that neuromarketing brings information that is not detectable by using conversational research methods, it is likely that using neuromatketing instruments might be a quicker and low cost alternative comparing to other methods including in the phases before the moment in which the product is distributed on the market. Specialists claim that individuals do not know to comprehensively articulate their preferences and that the human brain hide information related to the real preferences of the consumer. Neuromarketing, as a domain that has the capacity to detect this hidden information could be used both for stressing what individuals want and what they would buy (Ariely & Berns 2010, p.284).

However, there are common features both traditional research methods and neuromarketing have. First, regarding the similarities with the quantitative research methods, the most important ones are the following: the measurement of beliefs, knowledge, attitudes, attention, memory or memories; the development and testing of the hypothesis and causal effects; the use, analysis, interpretation and triangulation of statistical data; the objective observation role of the researcher; the possibility to generalize data, although the sample is small in the case of neuromarketing studies (the brain mechanisms are similar from person to person), predictions development regarding the consume behavior (Bercea 2013). Second, regarding the similarities with the qualitative methods, neuromarketing has the following characteristics: the objective is that of gathering deep information through studying certain phenomena; a small number of needed subjects (15-20 individuals are enough in order to have internal validity) (Bercea 2013). Third, there are some similarities of neuromarketing with both types of research methods: the resulted data can be images, numbers, statistics; the main objectives are the description, the explanation, the prediction (similar with quantitative methods), and the exploration, the discovery and the construction (similar with qualitative methods); the subjects' behavior is analyzed in controlled conditions and due to technical evolution, there is the possibility to develop experiments with a natural context (e.g. supermarkets) (Bercea 2013).

Although using neuroscience instruments as a completion of the traditional methods is useful regarding the decisional process of the consumer, the literature reflects on several concerns regarding this domain (Yoon et al., 2012, p.484). Being a domain that implies a different way of thinking and new theoretical perspectives, one of the concerns regards the high costs implied for educating especially the students. At the same time, the needed infrastructure is costly. Another concern states the necessity to calibrate the practitioners' expectations regarding the way they perceive the efficiency of the neuromarketing results, this domain not being able to make perfect predictions on consumer choice (Yoon et al., 2012, p.484).

NEUROMARKETING AND ITS ENTREPRENEURSHIP APPLICABILITY

Due to market globalization, to technological evolution and to changes in the purchase and consumption behavior, the contemporary marketing is constantly challenged in adapting itself (Pop, Iorga, 2012, p.632). Thus, the need for interdisciplinary and the need of clarifying the way people make decisions are increasing concerns for both marketing theorists and practitioners. In this context, the neuromarketing with its specific tools is considered a viable solution for the companies in order to better understand costumers in respect to different domains, as corporate economy, environmental economy, human recourses, or customer relations management (Pop, Iorga, 2012, pp.633-634).

While the rational hypothesis of classical economy that states that people are rational and follow their own interests is not longer valid, more recent theories emphasize that people are intuitive, impulsive, their decisions being rather filled with emotional attachment (Pop, Iorga, 2012, p.634). Therefore, practitioners are challenged to understand the irrational behavior in order to correlate "processes, such decision-making, judgment, encoding memory or emotions with marketing concepts, such as positioning, brand loyalty and consumer reaction to marketing messages" (Perrachione, 2008 in Pop, Iorga, 2012, p.634).

Neuromarketing is contradictory perceived by the researchers. While some are consider neuromarketing as a science, as a valid field of research that goes beyond the simple application of neuroimaging tools in order to sell products (Lee et al., 2007 in Pop, Iorga, 2012, p.636), others consider it as only a simple business tool and label it as "consumer neuroscience" (Hubert and Kenning, 2008 in Pop, Iorga, 2012, p.636). Consumer neuroscience is defined as an interdisciplinary field (psychology, neuroscience and economics) that analyzes the way the brain is affected by advertising and marketing strategies (Lee et al. 2007 in Agarwal and Dutta, 2015, p.458). Interestingly, although consumer neuroscience and neuromarketing are often understood as being the same thing, consumer neuroscience is associated with academic research and neuromarketing with the practitioners (Plassmann et al., 2011 in Agarwal and Dutta, 2015, p.458). Other specialists stress the importance of neuromarketing as a mechanism that helps entrepreneurs to deliver the right message to the right people (Georges and Badoc, 2010 in Pop, Iorga, 2012, p.636). While, Constensen (2011 in Pop, Dabija and Iorga, 2014, p.28) claims that neuromarketing is a component of marketing by means of which the customer behavior can be understood by interpreting psychological and neuropsychological information, neuromarketing is believed to be helpful for entrepreneurs when making decisions in both their communication with consumers and in their communication with partners (Pop, Dabija and Iorga, 2014, p.28).

Marketing research is considered to be a very important part of the marketing efforts, although it is perceived as being costly and time consuming. In the same respect, accurate results are needed for having the most efficient marketing decisions (Egrie and Bietsch, 2013). Consumer satisfaction is one of the most significant factors that determine the success of a business mainly in the context of relationship marketing. Satisfaction implies both buying the same product/service again and recommending the product to other potential consumers. At the same time, satisfaction relates to both customers' judgment of the quality of the product and costumer's evaluation of the interaction experience with the brand provider (Dapkevicius and Melnikas, 2009, p.17). However, there are large amounts of unknown variables that can determine a decision-making. In this context, neuromarketing is considered a solution (Egrie and Bietsch, 2013). Google is one of the companies that use neuromarketing research. Google specialists are using neuroscience tools in order to learn new information about the way the consumers perceive different web pages and to help companies to have their webpage as the top pick. The main variables that can be tested with neuromarketing tools are mainly visual appeal and emotional engagement (Egrie & Bietsch, 2013).

Another interesting example is that belonging to the neuroscientists form the University of California, Los Angeles, that have scanned the brain of individuals while watching Super Bowl commercials (Suomala et al., 2012, pp.12-13). This type of expertise is very important mainly taken into consideration that there are enormous amount of money invested in launching these commercials during this American football event. The main results of the research emphasize that while a very few commercials stimulate empathy, a large number of other ads provoke anxiety, fear, and insecurity (Suomala et al., 2012, pp.12-13). These results can be considered as being valuable for any entrepreneur in his or her decision to further use or not similar messages to the target group of the company.

When talking about using neuromarketing tools indoor of a company, a problem that is frequently mentioned in the literature is the need for training. Neuromarketing implies interdisciplinary. Thus, a combination of backgrounds in statistics, psychology, computer science might be an advantage (Smidts et al., 2014, p.260). In the same respect, Smidts et al. (2014, p.263) discuss three technical advances in consumer neuroscience: continuing innovation in scanner strength, speed and stability in order to provide improvements in spatial and temporal resolution; advances in modeling of behavior in order to elucidate processes that generate choice; and advances in analysis of algorithms that can identify multivariate patterns of brain activity correlated with future behavior.

There is an increasing number of companies that use neuromarketing research. Some of the main domains are media, advertising, media planning, automotive, technology, online auction or food. In the same respect, several exact examples are Yahoo, Hyundai, Microsoft, EBay, Daimler. The main analyzed issues are the following: how the brain react in the morning hours, the reaction to advertising, the reaction to viewing a product, understanding human (emotional) interaction with computers, adjusting products or commercials etc. (Flores, Baruca and Saldivar, 2014, p.78).

The marketing entrepreneurs are increasingly adopting neuroscience techniques in understanding consumer behavior. Levalloise et al. (2013 in Smidts et al, 2014, p.260) is talking about an increase from 13 neuromarketing companies in 2008 to more than 60 in 2012. Nowadays, on the website of the Neuromarketing Science & Business Association (NMSBAc) are listed more than one hundred companies that provide neuromarketing services. In addition, there are research companies that acquire leading neuromarketing companies proving the value of this neuroscience approach. Such an example is Nielsen that has bought in 2011 Neurofocus, a very appreciated neuromarketing company (Smidts et al, 2014, p.260).

A business model is composed of three main stages: the development of *value proposition* (product design and development), the development of *value delivery* (the channels used in order to address the message to consumers), and the development of *value capture* (the pricing formulas needed in order to generate income) (Teece, 2010 in Kamp, 2014, p.426). A market failure can intervene in each of these stages. While the literature defines a market failure as a mismatch between supply and demand (Kamp, 2014, 425), neuromarketing is considered to be a domain that can eliminate a part of the problems. For instance, by using neuroscience tools, practitioners can better understand the demand characteristics and influence the investments in new products and services, in innovation or in entrepreneurial initiatives.

One of the reasons the new products introduced in the market fail is the mismatch between the attitude and actual purchase behavior of the consumer (Agarwal and Dutta, 2015, p.457). Neuroscience approaches have the ability to predict choices and thus, have large implications for marketers (Plassmann, 2015 in Agarwal & Dutta, 2015, p.458). In this respect, neuroscience tools can be applied for testing the branding and marketing strategies before being implemented and can choose the most emotional strategies by using a celebrity endorsement or an association with a social or environmental cause. All these perspectives can help both in designing an efficient campaign and in avoiding the waste of money on implementing inefficient strategies (Agarwal & Dutta, 2015, p.458).

Desiring to know how consumers select their products, companies invest significant amount of recourses in order to better understand the purchasing behavior and what people really like (Pop, Iorga, 2012, p.640). Ariely and Berns (2010 in Pop, Iorga, 2012, p.640) identify several areas in which neuromarketing finds relevant application for the entrepreneurs. The first area is related to *designing food and beverage products*. In this respect, variables as flavor, texture, looks are important features. For instance, Plassmann et al. (2008 in Smidts et al., 2014, p.259) used fMRI to assess the role of information about product quality, as the price, on the perception of the product in terms of post-consumption rationalizing or in terms of changes in actual taste perception. The results show that changing the price of otherwise identical wines can affect the interpretation of taste pleasantness during sampling sessions (Plassmann et al., 2008 in Smidts et al., 2014, p.259). The subjects rate the expensive wine as being much more pleasant (the part of the brain that experiences pleasure being activated) (Dapkevicius and Melnikas, 2009, 19). Booth and Freeman (2014) name several situations in which neuromarketing can be useful.

For instance, while it is considered that slogans and headlines in ads for food can increase sensory appeal (Elder and Krishna, 2010 in Booth and Freeman, 2014, p.178), the names of foods can affect the neural response to aroma (de Araujo *et al.*, 2005 in Booth and Freeman, 2014, p.178).

The literature consider that ad testing is the most popular area of neuromarketing practical application, the brain reactions implying a valuable added information to the results from the traditional research, especially at the prediction level (Ariely and Berns, 2010 in Smidts et al, 2014, p.260). By using neuromarketing instruments, advertisers can collect and assimilate information on consumers at the personal level in their endeavor to create intimate and lasting relationships with a brand (Rapp et al., 2009, p.51).

The second area refers to *architecture for designing buildings*. Neuromarketing can be used in order to assess the perceptions on the building design by using virtual reality or augmented reality (Pop, & Iorga, 2012, p.640).

The third area discusses the *movie industry* and the role neuromarketing have in evaluating the efficiency of certain scenes or of using certain characters. The same idea fits very well for music industry or for commercial industry (the way certain ads are designed, the music they use, the way websites are organized etc.). Thus, aspects as emotions, interest, trust, loyalty can be measured (Pop, & Iorga, 2012, p.640). In this respect, Stallen et al. (2010 in Smidts et al., 2014, p.259) talks about the need of companies to evaluate the brain reaction on celebrity endorsement.

Moreover, Shiv and Yoon (2012 in Agarwal & Dutta, 2015, p.459) emphasize ways in which neuroscience can be efficient: "providing opportunities and guidelines to facilitate theoretical development; facilitating new empirical tests of standard theoretical claims; explanations for observed heterogeneity within and across consumer groups; and novel mechanisms for considering the physiological context and the role of numerous biological factors, including hormones and genes, on consumer preferences and decisions".

In the same respect, Fugate (2007) underlines entrepreneurial application of neuromarketing by developing several perspectives. The first one talks about testing advertising effectiveness. Although there is no direct connection between arousal and behavior or purchase intent, the advertising specialists can use brain scan to predict the strength of an ad recall or the emotional attachment with the ad (Fugate, 2007, p.387). The second perspective refers to testing product appeal, meaning the efficiency of the design o the product. The third perspective emphasizes celebrity endorsement. In this respect, studies have shown that dopamine and phenylethylamine flood the brains in the moments in which consumers see familiar faces. It is believed to have the same reaction when world of mouth is used in the case of recommendations from close friends or family (Fugate, 2007, p.388). The fourth perspective refers to logo and brand selection, higher level of brain activation signifying a higher level of involvement, of relevance and importance to the person (Fugate, 2007, p.388). The fifth perspective deals with media selection. Besides analyzing which channel is the most appropriate for releasing a message, the research can be enlarged to the way consumers scan and read text, ads, images of even web pages (Reynolds, 2006 in Fugate, 2007, p.389). In this case, while an efficient message for the right part of the brain would be short and frequent television messages, an efficient message for the left side should be print and complete with factual information (Fugate, 2007, p.389). Not recommending a special media, Renvoise and Morin (2005 in Fugate, 2007, p.389) claim that the delivered message should be more visual and less verbal.

Although neuromarketing is generally perceived as a promising domain, there are authors claiming that that there are biases that can be encountered and that can make the data redundant. Several examples are the following: when observed, individuals act and react differently than in a usual context, there are no strict rules about the relationship between brain activity and behavior, the variables of the life stage are much more influential on behavior than brain arousal (James, 2004 in Fugate, 2007, p.391). Thus, neuromarketing is considered very unlikely to solve the problems marketing is confronting.

Moreover, although neuromarketing can provide valuable insights, it is criticized for being conducted in artificial environment and has limitations regarding the applicability in the real world (Barkin, 2013, p.50).

In the same respect, Medina (2004 in Fugate, 2007, p.391) considers that the technology is still imprecise and that neuroscientific understanding is not yet sufficient to understand a complex issue as consumer behavior. Another important concern states that there is the inability to generalized data from a small sample to a wide and diverse population (Nature Neuroscience, 2004 in Fugate, 2007, p.391), an example here being the diverse understandings of humor in different cultures (Harris, 2006 in Fugate, 2007, p.391).

Neuromarketing and Ethics

Being perceived as a domain that aims to find the buy button in the brain, neuromarketing is the subject of numerous criticisms and ethical discussions (Lee *et al.*, 2007, p.199). Although the majority of the instruments used by neuromarketing are not invasive, the scientific community has suggested and already works on a neuroethical code that aims to protect the involved entities (Murphy et al., 2008, p.295). There are at least two aspects that can create the connection between neuromarketing and non-ethical behavior (Ramsøy, 2014). First, generally speaking, the companies are aware of the ethical issues, social responsibility actions (Corporate Social Responsibility) getting advantage of the neuromarketing results as much as possible. Second, there is a false impression that neuromarketing can discover the buy button in the consumer brain (Ramsøy, 2014).

Thus, the literature talks about the need to protect the vulnerable groups, as individuals with neurological diseases, with psychological disorders, children or other protected groups (Murphy et al., 2008, p.295). In the same respect, the emphasis is on the following aspects: the need to provide complete information regarding the privacy protection of the subjects, the experimental protocol, the risks implied, the use of data, the possibility of retreat from the experiment anytime and regardless of the reason, the correct presentation of information in mass-media, and the control over the offered incentives for participants (Murphy et al., 2008, p.298).

Bercea (2015) underlines three types of responsibilities when it comes for the ethics in neuroscience. First, it is the responsibility on the subject. The subjects needs to be informed regarding the entire research process, to have the benefit of finding out examples of similar research already developed and without being used technical language. Second, there is the responsibility on the consumer. It implies using the results of the experiments in such a manner that these to be useful for the consumer and emphasizing honesty. The consumer must have the final word in the process of marketing decision. Finally, the literature stresses the responsibility on the researcher. Thus, being a growing and developing domain, it is necessary that neuromarketing to be accompanied by an open ethical guide (Bercea 2015).

One of the issues that preoccupy the neuromarketing specialists is the need to eliminate de problem of reverse inference. Usually, based on a study that involves neuroscience instruments, the statement that is designed is the following one: if the cognitive process X takes place (for instance, computing the desire to pay), then the Z zone from the brain is activated. The reverse inference implies changing the two components of the statement. Empirically, it is impossible to say that activation in certain zone of the brain implies a certain unique cognitive process. A single area of the brain can be activated by a large number of distinct cognitive processes (Plassmann, Ramsøy, Milosavljevic, 2012, p.29). Considering this context, one way of controlling the bias is that of finding the measure in which the interest region is selectively activated by a specific mental process. Thus, if a region is activated by a large number of cognitive processes, the activation of the region is not significant in relationship with the cognitive process chosen. However, if a region is selectively activated nu a mental process, then it can be validly infer the relationship between the cognitive process and the activation of brain area, a certain selectivity factor being required to be computed (Poldrack, 2006 in Plassmann, Ramsøy, Milosavljevic, 2012, p.30).

An exploratory research conducted among neuromarketing specialists around the world investigates the level of interest the neuromarketing companies have in the ethical issues (Pop, Dabija and Iorga, 2014). The survey includes the participation of 52 companies from Europe, North America, South America and Asia. The surveyed subjects are policy makers (chief executive officers, administrative officers, chief neuromarketing officers and chief research officers). Based on this survey, the authors develop a ranking with the most important areas in which neuromarketing tool are used: video ads creation, creation of communication campaign, printed ads/banners creation, product design, packaging, online media campaigns, sales promotion, entertainment (Pop, Dabija and Iorga, 2014, p. 34). Regarding the ethical issues mentioned by the specialists, the research mentions the following: the need to transparently and precisely inform the subjects, the need to guarantee data privacy and security, the fear that a third party might have access to disturbing information about subjects, and the security of data processing methods (Pop, Dabija & Iorga, 2014, p. 34).

The Neuromarketing Science & Business Association (NMSBA) has an important role regarding the issue of ethics in neuromarketing. NMSBA is a global association composed of researchers interested in neuromarketing domain and that aims to discuss the latest research in the domain, to develop an international network of knowledge in the domain, and to improve the quality of the domain (NMSBAa). Among the most important aspects mentioned in the ethical code this organization is proposing are the following: the importance of respecting a high degree of research professionalism, of respecting the subjects implied, of providing correct, complete and honest information, of offering the subjects the possibility to leave the experiment anytime, of respecting the privacy of the data, and of cautiously using the data only with the subjects' agreement (NMSBAb). This, based on these aspects and on what Racine et al. (in Bercea, 2015), is stating, confidentiality and protecting the privacy of the subjects are the most important aspects related to the way neuromarketing as a domain is entitled to use the neuroscience instruments.

Although there are organizations that militate against neuromarketing by underlying that the use of medical instruments in marketing purpose can deteriorate the collective mental, the optimists are emphasizing that the results in this domain can be useful for the society. For instance, it can be mentioned the contribution of neuromarketing for the public services industry, this domain being capable of leading to efficient anti-drug or anti-smoking campaigns (Phan, 2010, p.16).

In addition to all the above mentioned shortcomings of neruomarketing, in the majority of the studies using neuroscience instruments, one of the selection criteria of the subjects is related to the hand they are using for writing. Thus, the individuals that are left-handed are excluded form the experiments (Willems et al., 2014). In the context in which the left-handed people differently perceive the stimuli, rather using the right part of the brain, it is consider that their inclusion into experiments bring too much variety of the data. However, Willems et al. (2014) advocate for the introduction of left-handed people into experiments buy invoking that these individuals represent a significant part of the population (approximately 10%) and that losing them implies losing the representativeness of the results. Moreover, it is probable that by including them, the scientists could find new and relevant information on the way the brain works (Willems et al., 2014).

CONCLUSION

Concluding, although being a theoretical paper, the present study intents to treat the issue of unconsciousness reactions in business and marketing environments from a practical perspective. The brain and the way it works become important variables in the study of decision making process. Therefore, the measuring task of these processes implies a deep and careful design. In this respect, neuromarketing, as a new domain, is proposed as a solution in order to improve the knowledge specialists have.

Finding relevant information directly from the human brain and using neuromarketing instruments in order to gather and evaluate influences to market incentives are not totally well perceive processes. However, as this paper clearly states, there is an increasing need to better understand human behavior in general and in marketing in particular. Moreover, by using neuro-imaging instruments in marketing, there is a win-win situation. While entrepreneurs can find relevant information in order to develop their marketing strategies, consumers can both find relevant information about their own behavior and learn how companies are trying to persuade them. In addition, this paper emphasizes that there is a large number of experiments that can be replicated and that can complete the traditional methods in order to increase the amount of knowledge on human behavior.

REFERENCES

Agarwal, S., & Dutta, T. (2015, December). Tanusree Dutta. Neuromarketing and consumer neuroscience: Current understanding and the way forward. *Decision*, *42*(4), 457–462. doi:10.1007/s40622-015-0113-1

Ariely, D., & Berns, G. S. (2010, April). Neuromarketing: The hope and hype of neuroimaging in business. *Nature Reviews. Neuroscience*, *11*(4), 284–292. doi:10.1038/nrn2795 PMID:20197790

Baars, B. J., Ramsøy, T. Z., & Laureys, S. (2003, December). Brain, conscious experience and the observing self. *Trends in Neurosciences*, *26*(12), 671–675. doi:10.1016/j.tins.2003.09.015 PMID:14624851

Bargh, J. A. (2002, September). Losing Consciousness: Automatic Influences on Consumer Judgment, Behavior, and Motivation. *The Journal of Consumer Research*, *29*(2), 280–285. doi:10.1086/341577

Barkin, E. (2013, July). The prospects and Limitations of Neuromarketing. *Customer Relationship Management.*

Bercea, M. D. (2012). *Anatomy of methodologies for measuring consumer behavior in neuromarketing research.* Retrieved January 2014, from http://www.lcbronline.com/index_files/proceedingsemc12/12emc023.pdf

Bercea, M. D. (2013). *Quantitative versus qualitative in neuromarketing research.* Munich Personal RePEc Archive. Retrieved June 2015, from http://mpra.ub.uni-muenchen.de/44134/

Booth, D. A., & Freeman, R. P. J. (2014). Mind-reading versus neuromarketing: How does a product make an impact on the consumer? *Journal of Consumer Marketing*, *31*(3), 177–189. doi:10.1108/JCM-08-2013-0674

Butler, M. J. R. (2008). Neuromarketing and the perception of knowledge. *Journal of Consumer Behaviour*, *7*(4-5), 415–419. doi:10.1002/cb.260

Chartrand, T. L., Huber, Shiv, & Tanner. (2008). Nonconscious Goals and Consumer Choice. *Journal of Consumer Research, 35.*

Dapkevicius, A., & Melnikas, B. (2009). Influence of Price and Quality to Customer Satisfaction: Neuromarketing Approach. Science – Future of Lithuania, 1(3).

Egrie & Bietsch. (2013). *Marketing Research Ethics: How Consumers feel about Neuromarketing*. Academic Press.

Flores, Baruca, & Saldivar. (2014). Is Neuromarketing Ethical? Consumers Say Yes. Consumers Say No. *Journal of Legal, Ethical and Regulatory Issues, 17*(2).

Fugate, D. L. (2007). Neuromarketing: A layman's look at neuroscience and its potential application to marketing practice. *Journal of Consumer Marketing*, *24*(7), 385–394. doi:10.1108/07363760710834807

Kahneman, D. (2011). *Think fast and slow*. New York: Farrar, Straus and Giroux.

Kamp, B. (2014). Overcoming Market and Network Failures by Means of Embedded Business Model Innovation. *Boletin de estudios economicos*, *69*(213), 423-442.

Lee, N., Broderick, A. J., & Chamberlain, L. (2007). What is 'neuromarketing'? A discussion and agenda for future research. *International Journal of Psychophysiology*, *63*(2), 199–204. doi:10.1016/j.ijpsycho.2006.03.007 PMID:16769143

McClure, S. M., Li, J., Tomlin, D., Cypert, K. S., Montague, L. M., & Read Montague, P. (2004). Neural correlates of behavioral preference for culturally familiar drinks. *Neuron*, *44*(2), 379–387. doi:10.1016/j.neuron.2004.09.019 PMID:15473974

Morin, C. (2011). Neuromarketing: The New Science of Consumer Behavior. *Soc*, *48*(2), 131–135. doi:10.1007/s12115-010-9408-1

Murphy, E. R., Illes, J., & Reiner, P. B. (2008). Neuroethics of neuromarketing. *Journal of Consumer Behaviour*, *7*(4-5), 293–302. doi:10.1002/cb.252

NMSBA. (n.d.a). *The Neuromarketing Science & Business Association*. Retrieved June 2015, from http://www.nmsba.com/

NMSBA. (n.d.b). *The Neuromarketing Science & Business Association*. Retrieved June 2015, from http://www.nmsba.com/Ethics

NMSBA. (n.d.c). *The Neuromarketing Science & Business Association*. Retrieved April 2016, from http://www.nmsba.com/neuromarketing-companies

Nobel, C. (2014). HBS Working Knowledge, Neuromarketing: Tapping Into the 'Pleasure Center' of Consumers. In Consumer Neuroscience and Neuromarketing (2nd ed.). Academic Press.

Ohme & Matukin. (2012). *Small Frog That Makes a Big Difference*. DOI: .10.1109/MPUL.2012.2189169

Olteanu, B. (2015, January). Neuroethics and Responsibility in Conducting Neuromarketing Research. *Neuroethics*. doi:10.1007/s12152-014-9227-y

Phan, V. (2010). *Neuromarketing. Who Decides to Buy?* The Triple Helix.

Plassmann, H., Ramsøy, T. Z., & Milosavljevic, M. (2012). Branding the brain: A critical review and outlook. *Journal of Consumer Psychology*, *22*(1), 18–36. doi:10.1016/j.jcps.2011.11.010

Pop, Dabija, & Iorga. (2014). Ethical Responsibility Of Neuromarketing Companies In Harnessing The Market Research – A Global Exploratory Approach. *Ethical Responsibility of Neuromarketing Companies in Harnessing the Market Research – a Global Exploratory Approach. Organizations' responsibility to Consumers, 16*(35).

Pop, N. A. (2012). Ana Maria Iorga. A New Challenge For Contemporary Marketing – Neuromarketing. *Management & Marketing Challenges for the Knowledge Society*, *7*(4), 631–644.

Ramsøy, T. (2014). Consumer Neuroscience and Neuromarketing (2nd ed.). Academic Press.

Ramsøy, T. Z., & Skov, M. (2014). Brand preference affects the threshold for perceptual awareness. *Journal of Consumer Behaviour*, *13*(1), 1–8. doi:10.1002/cb.1451

Rapp, J., Hill, R. P., Gaines, J., & Mark Wilson, R. (2009, Winter). Advertising and Consumer Privacy. Old Practices and New Challenges. *Journal of Advertising*, *38*(4), 51–61. doi:10.2753/JOA0091-3367380404

Renvoisé, P., & Morin, C. (2007). *Neuromarketing: Understanding the "Buy Button" in Your Customer's Brain*. Thomas Nelson.

Shiv, B., & Fedorikhin, A. (1999, December). Heart and Mind in Conflict: The Interplay of Affect and Cognition in Consumer Decision Making. *The Journal of Consumer Research*, *26*(3), 278–292. doi:10.1086/209563

Simonson, I. (2005). In Defense of Consciousness: The Role of Conscious and Unconscious Inputs in Consumer Choice. *Journal of Consumer Psychology*, *15*(3), 211–217. doi:10.1207/s15327663jcp1503_5

Smidts, A., Hsu, M., Sanfey, A. G., Boksem, M. A. S., Ebstein, R. B., Huettel, S. A., & Yoon, C. et al. (2014). Advancing consumer neuroscience. *Marketing Letters*, *25*(3), 257–267. doi:10.1007/s11002-014-9306-1

Suomala, J., Palokangas, L., Leminen, S., Westerlund, M., Heinonen, J., & Numminen, J. (2012, December). Technology Innovation. *Management Review*.

Tom, G., Nelson, C., Srzentic, T., & King, R. (2007). Mere Exposure and the Endowment Effect on Consumer Decision Making. *The Journal of Psychology*, *141*(2), 117–125. doi:10.3200/JRLP.141.2.117-126 PMID:17479582

Willems, R. M., Van der Haegen, L., Fisher, S. E., & Francks, C. (2014). On the other hand: Including left-handers in cognitive neuroscience and neurogenetics. *Nature Reviews. Neuroscience*, *15*(3), 193–201. doi:10.1038/nrn3679 PMID:24518415

Wilson, R. M., Gaines, J., & Hill, R. P. (2008). Neuromarketing and Consumer Free Will. *The Journal of Consumer Affairs*, *42*(3), 389–410. doi:10.1111/j.1745-6606.2008.00114.x

Yoon, C., Gonzalez, R., Bechara, A., Berns, G. S., Dagher, A. A., Dubé, L., & Spence, C. et al. (2012). Decision neuroscience and consumer decision making. *Marketing Letters*, *23*(2), 473–485. doi:10.1007/s11002-012-9188-z

Youtube, R. S. (2015). Retrieved from https://www.youtube.com/watch?v=R8WkIC7bMnE

Youtube. (n.d.). Retrieved September 2015, from https://www.youtube.com/watch?v=IGQmdoK_ZfY

Chapter 9
Digital Entrepreneurial Charity and Solidarity for Social Change:
KIVA and the "Social-on-Linezation" of Poverty

José Manuel Saiz-Alvarez
Tecnologico de Monterrey, Mexico

Jorge Colvin-Díez
Schiller International University, Spain

ABSTRACT

Microcredit has been studied from many perspectives. In this chapter, we analyze KIVA, the most important Person-to-Person microfinance organization from the perspective of social change, and we study how it has impacted on the nascent of a new wave of entrepreneurs known as digital entrepreneurial charity. Applied to KIVA, we analyze the impact of the digital space and its Internet-based Peer-to-Peer Lending to create social change in the poor, while alleviating the poverty thanks to solidarity and charity. This work concludes affirming that banking the poor and education, with the intensive use of Internet-based devices, is the best way to alleviate poverty in our digital and globalized economic world.

DOI: 10.4018/978-1-5225-0953-0.ch009

INTRODUCTION

The nascent field of microcredit is growing rapidly and attracting increased attention from many sectors. There is something inherently interesting and appealing about microfinance and increasingly new stories as KIVA. These extraordinary people, who started helping others after the seminal work carried out by the Nobel Peace Prize laureate Muhammad Yunus, came up with brilliant ideas, and against all the odds, succeed at creating new products and services that dramatically improved people's lives.

KIVA can be defined as a social change based a new entrepreneurial movement. This is why not only we have analyzed as a background the principles of microfinance as a way for alleviating poverty, but also for impulsing social change derived:

1. From a financial approach and the banking of poverty;
2. From an increasingly used Digital Space and its Internet-based Peer-to-Peer Micro-lending;
3. From an Entrepreneurial approach;
4. From Organizational dynamics; and
5. From Virtuous Approach.

As a result, and by means of this in-depth study, it has allowed us to focus on Kiva.org or the *Social-on-Linezation* of poverty.

Does KIVA experience and success can be applied to other NGOs? What is the impact of ICT on microfinance? What are the advantages using Joint Liability Lending? In order to answer these questions, the objective of this chapter is to analyze how the strategy and the good results obtained in KIVA can serve as learning from practice to other NGOs pursuing similar objectives. To do this, the chapter begins describing the theoretical background that exists on this issue. Then the model of KIVA's loan cycle will be described to analyze the impact of the digital space through this case study. Finally, future research directions and conclusions will be added.

THEORETICAL BACKGROUND

Alleviating the Poverty: Main Global Initiatives

Before understanding the deepness of the case of KIVA within its *social-on-linezation* of poverty, it does make important to understand what really is poverty and the principles of microfinance which underlines as a new wave of alleviating the poverty. According to the World Bank Group (n.d.), there are an estimated 2.5 billion financially excluded adults today, with almost 80 percent of those living under $2 per day, and having no accounts at formal financial institutions. This matter holds back the fulfillment of the World Bank Group's goals of eradicating extreme poverty by 2030 while increasing the share of income held by the bottom 40 percent of the population. The World Bank's Global Financial Inclusion Database (Global Findex) reports that three-quarters of the world's poor lacks of a bank account, as poverty, costs, travel distances and the often burdensome requirements involved in opening an account, do not allow them to be current account's holders. Only 25 percent of adults earning less than $2 a day have saved money at a formal financial institution. Being "unbanked" is linked to income inequality, so they are exposed to be victims of loan sharks.

On the other hand, from micro, small and medium-sized enterprises (MSMEs) perspective, there is still a lack of access to credit and savings services that would enable them to grow. Indeed, and according to the World Bank Group (n.d.), about 200 million formal and informal businesses in developing countries lack the financing they need to grow, with a financing gap of $2.1 trillion to $2.6 trillion for MSMEs in developing economies. As a result, business growth is inexistent in many cases, and a survival economy is created, so arise in the poorest countries, populist political movements based on a strong state intervention, even tending toward dictatorship, which further impairs the development and financing of these companies.

In this scenario, over the last few decades and from private sector, we are seeing an increasing growth of different types of financial-services providers for poor people offering new possibilities, such as non-government organizations; cooperatives; community-based development institutions, like self-help groups and credit unions; commercial and state banks; insurance and credit card companies; telecommunications and wire services (World Bank, n.d.); post offices; and other points of sale. But all this type of new services targeted towards future clients is valid for emergent countries only, so the poor continue in the same situation in countries and regions mainly in the African continent.

Other initiatives have been coming up with from developed countries, and in particular:

1. The well-known G-20, which has made financial inclusion a permanent policy priority by establishing the Global Partnership for Financial Inclusion (GPFI), which includes the World Bank Group (WBG), the International Finance Corporation (IFC), and the Consultative Group to Assist the Poor (CGAP)[1] and
2. The Alliance for Financial Inclusion and the Organization for Economic Co-operation and Development (OECD) which have been designated as implementing partners.

Principles of Microfinance

Regarding these new wave of possibilities, it posed the microfinance as one of the most effective means of poverty alleviation whose principles (Ramírez & Anderson, 2007) are based on:

1. Poor people need a variety of financial services, not just loans;
2. It is a powerful tool to fight poverty by poor people earning more, build their assets, and cushion themselves against external shocks;
3. It means building financial systems that serve the poor. Indeed, although microfinance is often seen as a marginal sector—a "development" activity that donors, governments, or social investors might care about, but not as part of the country's mainstream financial system, microfinance can reach the maximum number of poor clients only when it is integrated into the financial sector;
4. Microfinance can pay for itself, and must do so if it is to reach very large numbers of poor people. Achieving sustainability means lowering transaction costs, offering more useful services to the clients, and finding new ways to reach more of the unbanked poor;
5. Microfinance is about building permanent local financial institutions. Which need to attract domestic savings, recycle those savings into loans, and provide other services;
6. Microcredit is not always the answer. In many cases, other tools will alleviate poverty better—for instance, small grants, employment and training programs, or infrastructure improvements;

7. Both interest rate ceilings hurt poor people by making it harder for them to get credit, and the scarce and uncertain supply soft money from donors or governments should be avoided, as they usually set them at levels so low that microcredit cannot cover its costs;
8. The role of government is to enable financial services, improving the environment for "micro-businesses" and access to markets and infrastructure, and not to provide them directly, and finally,
9. Donor funds should complement private capital, not compete with it.

All these principles are summed up in just one phrase: 'the poor know that credit is their only opportunity to break out of poverty' (Yunus, 1999: 59). Credit loans granted individually or in groups (Joint Liability Lending) to bring social and economic changes in the short term, and these effects are stronger when they are directed to women. Poor women are often more conservative in their investment strategies, and are often more easily swayed by peer pressure and the intervention of loans officers, making women more reliable bets for banks worried about repayment (Armendáriz & Morduch, 2007). Besides, women are more concerned about children's health and education than men, and they are usually overrepresented among the poorest of the poor, and are too often oppressed by their husbands and by prevailing social norms. As a result, they are more socially vulnerable.

MICROFINANCE, SOCIAL CHANGE, AND FINANCE: THE "FINANCIALIZATION" OF POVERTY

One effect of conceptualizing poverty in financial terms is that it enables the 'financialization of development' (Roy, 2010: 31). Roy uses this term to refer to the integration of the world's poor into global financial markets, especially by way of the Consultative Group to Assist the Poor's (CGAP) version of microfinance. But still the academic literature on conceptualizing poverty from financial perspective remains inconclusive. For instance, as we have seen in the previous epigraph, the World Bank's global poverty line has an 'essentially mono-dimensional conception of "extreme poverty" (as lack of income or consumption)' (Edward, 2006: 381) by defining the extremely poor those who live under \$1/day and as poor those who live under \$1.25/day, giving, therefore, exclusively a financial indicator. Financially speaking, the need of broader definitions of poverty is quite clear, being necessary to analyze a multidimensional character of poverty. This multidimensional approach we can be seen on microfinance when, not only does it enable the creation of 'social value maximization of the disposable income of the poor' (Chun, 2005: 14) but also microfinance is being one of the effects of poverty's process of banking, too.

Taking into account this multidimensional approach, it makes essential:

1. Understanding of poor people's financial profile through methodologies ranging from financial diaries to ethnographic research and in particular, being the poor people made knowable not just low incomes but also fluctuating and unpredictable income, which forced them to develop sophisticated money management skills centering on borrowing and saving (Collins *et al.*, 2009: 3);
2. New financial values, based on financial literacy education which is the forefront of any poverty alleviation initiative.

According to a World Bank (n.d.), financial literacy has come to play an increasingly prominent role in financial reform in both developed and developing countries[2]. These training initiatives are focused on expanding poor people's knowledge of formal financial services, budgeting and asset building. Indeed, promoting savings is a central component of financial literacy strategies, working to inculcate practices of planning for the future rather than succumbing to day-to-day struggle (Ilcan & Phillips, 2010).

To sum up, the social change of microfinance from financial approach is observing the poverty from a multidimensional perspective:

1. Showing that poor fiscal subjects are also rebranded as financial consumers, being targeted by the marketing of formal financial instruments designed specifically for them. —indeed, microfinance has captured the corporate imagination of global financial giants like Citibank and Deutsche Bank, who see the potential for profits to be made at the next financial frontier (Roy, 2010) and
2. Researching the poor people's lives while educating and promoting financial values by making financial schemes which take the social, cultural and religious complexities of poor people's lives into account.

Social Change and Digital Space: Internet Peer-to-Peer Micro-Lending

Moving towards new forms of lending beyond microfinance traditional literature, we found the Internet Peer-to-Peer literature where, instead of specifically being the lending associated with group borrowing, there also exists literature that highlights the new forms of Internet-based Peer-to-Peer micro-lending, both from technology (Chae *et al*, 2016; Lu, Wang & Li, 2016; Cha, Yang & Kim, 2016; Jung *et al*, 2016) or social sciences (Li, Park & Park, 2016; Carr *et al*, 2016). Obviously, there is a lot of controversy, as long as these forms of lending are similar to online shopping (Hartley, 2010). We refuse this consumer-focused conception, as the primary objective of this Internet Peer-to-Peer micro-lending is to help others, so social values prevail over economic ones.

Meanwhile, although we analyze later the Case Study of KIVA on the online peer-to-peer micro-lending landscape, the Internet-based peer-to-peer micro-lending has become an increasingly prominent global means of capital acquisition, and also an increasingly popular form of crowd-sourced microfinance (Hartley, 2010). Thus, differing in its mission and scope, we observe a wide lending institutions which exist alongside Kiva.org such as:

1. Prosper.com, launched in 2006, is the pioneer in Internet-based peer-to-peer micro-lending, allowing individuals to provide capital to borrowers in need while concurrently obtaining interest on their investment. Within Prosper, borrowers can request between loans of between $1,000 and $25,000, and stipulate terms under which they'd be amenable to borrowing, While Prosper provides loans to individuals in need, much of the site is dedicated to lenders;
2. Similar to Prosper.com and touting its ability to provide online matching of savvy investors and smart borrowers, there is other institution named as Lending Club which also offers lenders the opportunity to, while assuming some managed risk, earn a return on their investment. Borrowers with a sufficiently high credit score can seek personal loans for amounts between $1,000 and $25,000 online;
3. Accion.org which is a private non-profit umbrella microfinance institute under which Acción USA, the Acción Network, iLab, EB-Acción Savings and Loans, and Center for Financial Inclusion at Acción International operate. Established in 1961, Acción has been a global leader in microfinance

since 1973. Its business model is sustainable thanks to borrowers pay interest on loans sufficient to cover the expenses of loan creation, though remunerative returns are not provided to the lender; and finally,
4. TrickleUp, Global Giving, Grameen Foundation, UniThrive, among others.

In the end, all these Internet based Peer-to-Peer micro-lending platforms have on common that they have other motivations beyond the "Solidarity" lever and the Social change as neither they do provide zero-interest loans to borrowers in need nor they are uniquely international.

Microfinance, Social Change, and Entrepreneurship

It has long been argued that an organization's entrepreneurial approach is positively associated with a number of organizational outcomes related to growth and success in the microcredit environment (e.g., Moss & Meyskens, 2015; Wiklund & Shepherd, 2005), which consist of (Moss & Meyskens, 2015):

1. **Autonomy:** Where individuals or teams are able to act independently to foster new ideas and bring those to fruition (Lumpkin & Dess, 1996), being free to actively pursue new opportunities without being constrained. The outcomes of autonomy are quite clear such as innovation, the creation of new businesses, and organizational effectiveness;
2. **Competitive Aggressiveness:** As several offensive tactics aimed at gaining a strong competitive position, or aggressive reactions intended to maintain a position in the face of a strong threat from industry rivals (D'Aveni, 1994). According to Prahalad and Hart (2002) and comparing to the developed world's microenterprises, the developing economies' microenterprises are harder to survive due to the scarcity of resources their survival;
3. **Innovativeness:** Whose consequences are very close to the previously developed concept of autonomy, meaning the propensity of companies to introduce new products, processes, and routines as the result of experimentation and creativity (Lumpkin & Dess, 1996) and to be committed to disrupting current market conditions by stimulating new demand or creating novel resources combinations;
4. **Proactiveness:** This which means anticipating changes forthcoming in the environment and position themselves in advance (Lumpkin & Dess, 1996). For instance, organizational initiatives through crowdfunding, such as KIVA, has allowed microfinance to become a globalized phenomenon, placing opportunities within reach of microenterprises (Zahra, Newey, & Li, 2014) and finally;
5. **Risk-Taking:** By reflecting the capacity and comfortability in launching new, bold, and often costly action in the face of considerable uncertainty (Moss & Meyskens, 2015).

According to Moss and Meyskens (2015) and within the microcredit environment, those microenterprises which have these attributes belonging to entrepreneurial attitudes will be perceived presumably as more desirable investments and therefore, in our view, these will be more proactive to promote a social change by building and executing strong business models, which will lead to revenue growth and profitability.

This entrepreneurial spirit among KIVA borrowers can be observed in the importance of microenterprises supported by KIVA (more than 60%), and the strong collaboration with more than 300 field partners located in 84 countries and 450 volunteers around the world. When applied this entrepreneurial spirit to business, two more concepts are born:

1. **Social Entrepreneurship:** When all efforts are focused on ameliorating the standard of living of the population influenced by the entrepreneur, and
2. **Social Business:** When mature companies use Corporate Social Responsibility (CSR) and ethics to benefit population at risk of social exclusion, or even in social exclusion derived from economic poverty and social rejection.

Microfinance, Social Change, and Organizational Dynamics

Since the 1976 founding of Grameen Bank, lending to groups has become an increasingly popular tool for poverty alleviation. It has also become a widely studied phenomenon from Organizational dynamics and in particular, the potential for loan default behind these dynamics. Capital is provided to groups who – through having mutually re-enforcing loans– put internal pressure on those likely to default to maintain their credibility. Therefore, the theory and practice behind group lending is fundamentally both one of peer pressure encouraging repayment and a way of cutting back on transaction costs associated with loan distribution and collection.

"Group lending" within microfinance literature therefore typically refers to a single lender – often a microfinance institution (MFI) –providing capital to groups of borrowers rather than groups of lenders providing capital to one borrower. On the borrower's side, microcredit institutions employ various institutional structures to provide loans to group-based borrowers[3].

The problem is that this literature is predominately focused on *borrower-side* dynamics with the goal of lowering borrower default rates –more traditional forms of microfinance, rather than the nascent field of Internet-based peer-to-peer micro-lending (Hartley, 2010) based on *lending-side* dynamics[4].

Therefore, before looking into the Case Study of KIVA and according to Hartley (2010), we will outline the main characteristics of these novel *lending-side* dynamics which are many-to-many platform named as "crowd-sourced" microfinance, improving the outcome thanks to the homogeneity of the group (Hermes, Lensink, & Habteab, 2006)

Microfinance, Social Change, and the Virtuous Approach

One of the highlights of social change of the microcredit is its revolutionary orientation to organizational virtue. This is defined as "ethical character traits that are learnt from an accumulative perception of a firm's behaviors in everyday business life, that drives internal and external stakeholder satisfaction, and that is aligned with its ethical values used for strategic positioning" (Chun, 2005, p. 272), and is an integrated set of beliefs that inform the organization's ethical traits and virtuous behavior.

According to Payne *et al* (2011), an organization's virtuous approach can impact on:

1. The organization's processes, methods, and decisions (Cameron, Bright, & Caza, 2004) and how they conduct their operations (Moss & Meyskens, 2015), and
2. The manner in which it outlined itself to its stakeholders.

This virtuous approach is strengthened when social values are being incorporated into it, generating a broad social impact, mainly on the poor. As a result, a kind of private-public cooperation is created which causes positive externalities.

The character traits associated with organization's virtuous approach has been analyzed by numerous researchers (e.g., Murphy, 1999; Shanahan & Hyman, 2003; Solomon, 1992, 1999). In this respect, Paine (1991) includes principles such as honesty, fairness, reliability, trustworthiness, and commitment to others. Traits which are deeply developed by Chun (2005), and Moss & Meyskens (2015) by providing an empirically verified list of six dimensions of organization virtue, namely:

1. **Conscientiousness:** The extent to which an organization can be considered dependable and hard-working (Barrick & Mount, 1991);
2. **Courage:** The attainment of success in achieving a desired goal or objective and the level of effort taken to accomplish the outcome (Harris, 2001);
3. **Empathy:** It is portrayed by the ability to perceive or consider the feeling of others (Sawyer, 1975);
4. **Integrity:** The trustworthiness and honesty of an individual or organization (Butler & Cantrell, 1984);
5. **Warmth:** Characterized by open and pleasant expressions of courtesy, friendliness, and generosity to others (Chun, 2005); and finally,
6. **Zeal:** An organization virtue characterized as including elements such as humor, excitement, passion, or fun.

These researches have also been analyzed in order to improve the microenterprises' performance. For instance, in case of *conscientiousness*, conscientious employees are more likely to have a calling orientation in their work, and are thus more productive (Wright & Cropanzano, 2004) as well as microenterprises that display empathy are concerned, reassuring, supportive, and sympathetic, and empathic organizations are thought to have higher stakeholder satisfaction (Chun, 2005) by creating a social capital defined by trust and cooperation. Both factors are vital for baking the poor.

Referred to microcredit environment, an organization's virtuous approach will likely increase lender confidence, which in turn will increase the likelihood of funding (Moss & Meyskens, 2015). The assurance of this lender confidence will invite others to lend, so a virtuous circle is created. In fact, microfinance is one of those small ideas that turn out to have enormous social and economic implications, as this lending movement has grown, especially in Asia and Latin America, through cross-pollination thanks to Internet. As a result, and by recognizing its energy and activity, the United Nations designated 2005 as the International Year of Microcredit.

THE KIVA.ORG ICT-BASED MODEL: THE "SOCIAL-ON-LINEZATION" OF POVERTY

Based on Information and Communication Technologies (ICTs), Kiva.org (KIVA) as the world's first Peer-to-Peer micro-lending website has seen an incredible growth over the past years. Media exposure such as an appearance on the Oprah Winfrey Show and a mention in Bill Clinton's book, *Giving*, resulted in an enormous number of lenders coming to the KIVA website in hopes of lending money to entrepreneurs in the developing world.

In 2005, KIVA introduced microfinance to the Internet, and since then, has created a global community of people connected through lending. By partnering with microfinance institutions (300 field partners) in over 84 different countries, 1,395,969 KIVA lenders, $831,829,700 in loans and a 98.36%

repayment rate (Kiva, n.d.), KIVA has given small businesses in developing countries the opportunity to grow and thrive. The company's microfinance partners post profiles of loan applicants on the website, where anyone with a credit card and an Internet connection in United States, Canada, Japan, Europe and beyond, can provide small loans to the businesses of their choice. Loans conceded depending on local ONGs as they make selection of aid applicants.

KIVA's Social Change Indicators

Before analyzing social change from previous quoted approaches, first of all, we should measure what social change is and how it works from KIVA perspective. Social change is defined by KIVA as a way to "create good in the world, a lot of good", which is measured by bringing financial concepts into the social reality through social performance. According to KIVA, this social performance is "one way" that they "try to maximize the good created [...] by partnering with organizations that go above and beyond to generate positive outcomes for the communities they serve" (Kiva, n.d.).

Using the KIVA social indicators, which are named "KIVA Social Performance Badges", based on their "Field partners", we found out in which social change –and thus, social reality- is focused. From a KIVA perspective, this organization is focused on:

1. **Anti-Poverty Focus:** Working locally with other microfinance institutions to combat poverty;
2. **Client Voice:** By serving and adapting their business practices and product offerings to meet their needs;
3. **Family and Community Empowerment:** Offering support services that address the needs of their clients' families: their health, education, and/or well-being;
4. **Entrepreneurial Support:** By training and support to help people start, manage and grow their businesses;
5. **Facilitation of Savings:** By promoting savings as a practice to the people they serve, and finally,
6. **Innovation:** By embracing technology and innovation to better address the needs of the people they serve.

The final result of this social performance badges is to fight for a better society defined by its social justice, equal opportunities, and reduced poverty degree.

The main tool is the KIVA's Loan Cycle, as we can see on Figure 1.

Intentionally or subconsciously people identify risks, consider their potential impacts, and develop strategies to manage them. In this sense, the poor are retaining risk oriented, as they try to avoid or to share risk through groups, although their quality of living is very low. This is why the Joint Liability Lending (JLL), especially among women, is preferred when lending through KIVA (Domen, 2011).

Most poor people, particularly those working in the informal economy and in rural areas, earn and spend in cash. To handle a cash transaction where there is no bank branch, banks have at least two ICT options. They may use an Automated Teller Machine (ATM) that can accept, store, and dispense cash, or they can use devices placed at an outlet where cash is kept on hand. Both situations, however, are not at disposal in remote rural areas where microfinance has a key role to play.

Complementary to KIVA lending, and derived from microfinance, the poor can also opt for creating basic micro-savings, medium-term commitment savings plans, endowments, and annuities, as shown in Table 1.

Figure 1. Model of KIVA's loan cycle
Source: adapted from Kiva.org web.

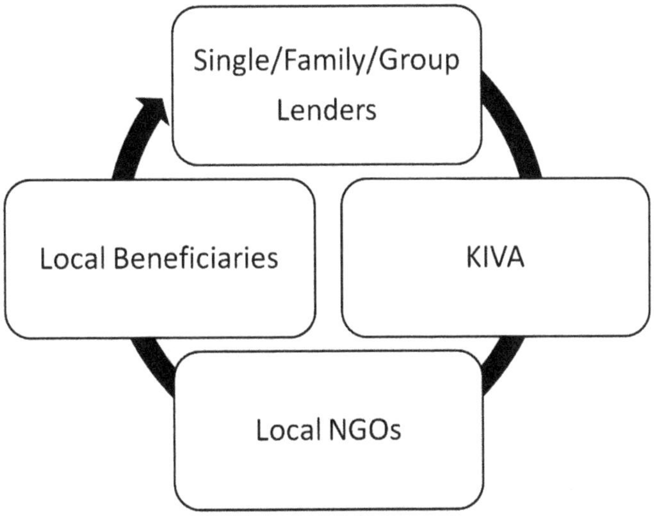

Regarding Table 1, microcredits are increasingly well established and are profitably practiced by specialists and more recently, by formal banks. This fact is fostered by micro-borrowing to complement the traditional systems of inter-generational care that have been used within the poor. Globalization tend to create an emergent medium-class society, and mainly in the BRICS (Brazil, Russia, India, China, and South Africa), the poor are being banked very rapidly. As a result, financially-based activities have emerged as investment options for investors in the Third World, such as insurance brokerages, development of efficiency-enhancing infrastructures, efficiency claims settlement companies, long-term savings custodian, and licenses for joint-ventures in third countries for successful domestic SMEs and NGOs with good microfinance operations.

Table 1. Current and future microfinance products focused on the poor

Products	Opportunities	Challenges
Microcredit	Loans can be stored and converted to larger assets (like land)	---
Basic microsavings	Focused on the reliable poor with current accounts and passbook savings	Safety of deposits
Medium-term commitment savings plans	Clients amass financial assets over time in rhythm with their own savings capacity	• Fund management • Inflation losses • Privacy and client identity
Endowments	Endowments attach life insurance policies to commitment savings plans, and therefore add further security	• Partnership • Risk of overselling and fraudulent selling
Annuities	Directed towards formally-employed low-paid workers	Formal actuarial skills are required

Source: adapted from Rutherford (2008).

The main problem for banking the poor is that the high cost of technology infrastructure is beyond the capacity of individual MFIs, so it is necessary to find a balance between product innovation (Point of Sale (PoS) Devices, smart cards for the poor and Internet-based microfinance), local realities, and business process change, as appropriate technologies must be designed for scale. This is why all microfinance initiatives are born in the First World countries, as Internet-based technologies and the facility for receiving donations anywhere in the world. We think that this technology gap will continue over time, despite the economic growth achieved by emergent countries.

Social Change from a Financial Approach: Banking the Poor

Kiva.org has become one of the world's first Peer-to-Peer micro-lending website (Flannery, 2007, 2009). The social change is measured not only by social indicators but also by financial indicators pointed out at the beginning of this chapter, and by the Boone indicator, a relatively new method of competition measurement able to measure competition on a yearly basis in market segments affected by microfinance without considering the entire market to avoid distorsions (Kar, 2016). According to this indicator, and over the period 2003-2010, India and Nicaragua had the most competitive microfinance loan markets, while competition among the microfinance institutions in Bangladesh and Bolivia declined significantly over time. The organization's website presents its visitors, many of whom learn about microfinance through the site, with the World Bank/CGAP financialized explanations of poverty and the importance of microfinance in its alleviation. From financial approach, this suggests a revolutionary social change in developed countries by giving financial support and contributing to KIVA's success (Black, 2009: 277) and in turn arising a "intimate mediations with poverty' (Roy, 2012: 149). Building up financial connections materialized in millions of microloans, the social impact emerges among thousands of lenders' inboxes and blog posts from KIVA fellows, telling their offline and online stories (Gajjala & Birzescu, 2011). It is important remark that the borrowed money does not directly go to a person, if not it is transferred into a local KIVA partner (MFI), which fill another loan to that end borrower. All being possible thanks to Internet-based technologies, which make these connections possible. Financially, this method allows the aggregation of thousands of microloans and their disbursement to local MFI partners as intermediaries. Flannery, KIVA's co-founder and CEO, called it "connected capital".

Table 2. KIVA's supporters

More than USD 5 million	HP Foundation / Omidyar Network
USD 1 million or more	Pepsico Foundation / Deutsche Bank / Pearson / Metlife Foundation / Grameen-Jameel Microfinance / Google.org / Capital One / Moody's / Research Now / The Diller-von Furstenberg Foundation / Skoll Foundation / Ernst & Young / PayPal
USD 500,000-1,000,000	The Fossil Foundation / The Pershing Square Foundation / The Cisco Foundation / Citibank / TripAdvisor / Dermalogica / Innovations for Poverty Action
USD 250,000-500,000	Maybank Kim Eng / Silatech / Cisco / Yahoo
USD 100,000-250,000	Chevron / Keen / Yelp / Staples / Kabbage / Godadday / The Hall Steps Foundation / Henry L. Hillman Foundation / VitaCoco / The Mullago Foundation / CNF Development Bank / Orrick / Rita Allen Foundation / Richard King Mellon Foundation / Isabel & Alfred Bader Fund
USD 50,000-100,000	Intuit / QBE / PNC / MEDC / NYSE Euronext / Chegg Foundation / The Owlsey Brown II Family Foundation / Stock Yards Bank & Trust / Tillamock / The Lan Martin Group

Source: Adapted from Kiva.org.

According to Schwittay (2013), the "connected capital" have certain characteristics such as:

1. Long-term vision;
2. Catalytic by taking big risks in order to have big impacts;
3. Accountable as KIVA lenders want feedback on their loans; and
4. Democratic by being the KIVA lenders hundreds of thousands.

What results is a relationship between people and money that enables action towards poverty alleviation through microfinance and consequently has a positive social impact. In other words, this "connected capital" allows banking the poor by avoiding usury and by inserting them into the market. The world has sufficient productive resources for all mankind. The main problem is their unfair product and economic resources distribution.

KIVA has changed the interrelationships between money and social impact by:

1. Arising new concepts, such as social capital which sets the capital as force for good;
2. Bridging the worlds of finance and development ("social world") into one world: *the global financial-capitalist marketplace*;
3. Helping to its partner MFIs to enter into *this* market (Gajjala & Birzescu, 2011);
4. Emulating the same feeling of well-known *philanthro-capitalists* through microfinance, leaving out the charity and celebrating a new equality based on entrepreneurial partnerships (Flannery, 2009). As Premal Shah, KIVA's President observes, 'we let the average person be like a Bill Gates or a Rockefeller' (cited in Schwittay, 2013); and finally,
5. Widening the vocabulary of finance by including other forms of return not thought before such as being KIVA lenders paid with stock success stories[5].

Social Change from Entrepreneurial and Virtuous Approach: Digital Entrepreneurial Charity

The charitable sector has witnessed considerable experimentation with new ideas and practices, in particularly those adopted from the entrepreneurial sphere (Wagner, 2002; Moody, 2008). For instance, KIVA is a charitable nonprofit that combines microfinancing and internet technology to enable individuals to lend (without interest) small amounts of money to entrepreneurs of the developing world, representing a fertile ground for charitable giving. Charitable giving based on imaginary conceptions and how individuals envision the world and their role in it (Godelier, 1986), that make lending through KIVA meaningful to its creators and supporters were surveyed by combining the analysis of consumer narratives (Kozinets, 2008), through different online platforms such as information sites (KIVAFriends.org[6]), networking sites (Change.org[7], Paris KIVA Cup 2009, Facebook (n.d.), LinkedIn, n.d.) and blogging initiatives (Inside KIVA), with the analysis of KIVA's brand genealogy (Flannery, 2007; Holt, 2004)[8].

KIVA means other way of thinking in the charity field, unifying utopian ideology of *entrepreneurial charity*, which challenges the view(s) of charitable giving and poverty alleviation. This utopian ideology has been created both by joining the KIVA's creators with the media and KIVA's supporters by building up their KIVA "portfolios" and actively assuming the entrepreneurial role of "angel investors".

What is really new from Entrepreneurial and Virtuous Approach is that KIVA replaces the traditional "hands-off" charitable giving by inspiring visions of "hands-on" entrepreneurial charity. This revolutionary concept is to see the poor as an entrepreneur waiting to be found out by microloans instead of seeing the poor as a beggar. This "cultural innovation" (Holt & Cameron, 2010) means complementing the patronizing handout of traditional charity ... with a dignified "hand-up" partnership between entrepreneurial investors and borrowers.

Social Change from Organizational and Digital Dynamics Approach

What is really new in the developing countries from social and financial perspective is the social transformation from innovative "traditional" social models of finance, such as microfinance, into a digital transformation based on new organizational and digital dynamics which have come up with examples such as KIVA, a novel Internet-based, peer-to-peer micro-lending model of crowd-sourced microfinance. Indeed, until now, none as KIVA has gone so deep as on-site cooperation on group lending and none has impacted so much on the online solidarity.

Looking at our background where we have compared KIVA to other organizations (Hartley, 2010), we observe big differences with Kiva.org such as:

1. TrickleUp and Global Giving both offer robust online platforms on which to observe Internet-based cooperation, but both diverge from KIVA in the scope of their model. TrickleUp offers business development services and Global Giving caters to individual and institutional groups looking to choose between charities;
2. Acción International and Grameen Foundation are less focused on providing capital to developing world borrowers to target and mitigate global poverty. Therefore, their scope is microfinance, they are less involved in peer-to-peer micro-lending than they are in the mentorship of MFIs and credit bureaus to improve access to capital on the ground in developing nations;
3. UniThrive has moved into the space of providing Internet-based peer-to-peer, zero-interest loans, similar to KIVA, but they differs in its scope from KIVA, which is tightly restricted to Harvard students.

Despite the diverse ecosystem of Internet-based peer-to-peer micro-lending, substantive variations among these loan providers' positions KIVA as unique in its Internet-based peer-to-peer platform, membership, breadth and scope of international lender and borrowers, its zero-interest lending scope.

Founded in 2005 by Matt and Jessica Flannery, Kiva.org seeks "to connect people through lending for the sake of alleviating poverty." On Kiva.org, internationally based lenders create and personalize individual profiles. Lenders can act individually, or can join "Lending Teams" to attribute their loans to a collective campaign or to compare their joint impact with other like-interest, regional or demographic groups. Lenders capitalize their Kiva.org accounts with personal money, browse the profiles of entrepreneurs located around the world, and choose those to whom they would like to lend money. The platform is additionally built with systemic checks and balances, such as the feature that lenders cannot view or post to message boards until their account is capitalized with at least $25, the minimum lending amount. Lenders are advised on potential country, Microfinance Institute, and personal loan default potential, described respectively as "Country Risk," "Field Partner Risk," and "Entrepreneur Risk." Loans are provided globally to KIVA "Field Partners," or MFIs, located in 83 countries that then distribute money to individual entrepreneurs. Capital is provided in minimum $25 loans at zero percent-interest, though repayment to the lender comes incrementally, with small cash installments that recapitalize individual accounts. Field Partner risk is assessed by KIVA employees based on financial audits, organizational experience, and existing loan portfolio size and risk. Risk ratings do not indicate probability of default, and exposure at default, but are qualitative one-to-five star ratings that have implications on initial entrepreneur loan requests. In this way, Kiva.org additionally helps regional MFIs establish credit histories by allowing even historically poor performers to request loans to build a positive portfolio.

KIVA Field Partners are regionally located MFIs, and are often guided by the on-site expertise of centrally trained, pro-bono KIVA Fellows. KIVA has over 300 Field Partners around the world, and it is the responsibility of the MFI to screen entrepreneurs, determine whether entrepreneurs belong to borrowing groups wherein individual default is linked to group outcome, and post loan requests on KIVA. Once an entrepreneur has been approved by the MFI, they can create a KIVA profile, and specify the loan amount they require. The MFI is additionally responsible for the disbursement and collection of the loan provided to the entrepreneur, and for documenting loan impact.

Although Kiva.org is neither the first nor the only web-based peer-to-peer micro-lending institution with global focus, they are the first to pioneer zero-interest entrepreneurial lending. Indeed, Kiva.org offers an online platform in which lenders collaborate, in various sized open and closed groups, and around various topical categories, to provide affordable debt capital to entrepreneurs in need. For instance, KIVA provides dynamic Lending Team analytics, feedback, or explicit training. These analytics would visually –like Google Analytics, Google Insights for Search or Google Finance– illustrate to group members participation rates over time.

By transparently explaining the value and effectiveness of their on-site design levers, Kiva.org could help facilitate digital growth and entrepreneurial engagement, offering to the developing market a unique online platform where to provoke the social and entrepreneurial change. Information and Communication Technologies (ICTs) offer new opportunities to help the poor as potential lenders are virtually connected. The first microfinance revolution took a supply-side perspective, and showed that the poor were bankable. The second revolution highlighted demand-side concerns to meet the complex livelihood needs of the poor. As a result, there is a need to improve the design and outreach of MFIs on a continuous basis and to put these institutions in a perspective which matches demand and supply concerns (Arun, Hulme, Matin, & Rutherford, 2005).

As stated in the objectives set in this book chapter, the impact of the digital space through this case study has been important, as more than nearly USD 850 million have been given to more than 2 million borrowers at risk of social exclusion. These results are possible thanks to the use of ICTs where lenders are connected through the cyberspace and are informed in real time on the granting and repayment of their loans.

Moreover, regarding the possible replication of this ICTs-based model, the following three-fold practical solutions and lessons to be applied in other NGOs, as follows (Hartley, 2010):

1. KIVA's Top Large Closed Lending Teams have, on average, more members;
2. KIVA's Top Large Open Lending Teams provide over three times as many loans and three times as much money per loan; and
3. KIVA's Top Small Closed Lending Teams have greater per capita lending.

Indeed, these lending teams tend to transform borrowers into entrepreneurs defined by their use of economic resources to create both social and economic wealth. Given this social impact linked to solidarity, social entrepreneurship is generated and social businesses are created to reduce the risk of social exclusion.

FUTURE RESEARCH DIRECTIONS

Transnational microfinance socio-financial networks like KIVA are responding to global problems, such as poverty. Under normal circumstances, nation states are responsible for their own citizens, for implementing social policies, and providing banking and saving services. However, NGOs have taken over some of these responsibilities through services rendered by digital entrepreneurs to citizens globally. Given the global-scale financial crisis, socio-financial networks like Kiva.org play a critical role in helping bring about social change by providing economic relief to millions of people worldwide. There is a transnational level of participation in socio-financial networks like KIVA, despite not having any tax advantage for lenders whose residence is outside of the United States.

Kiva.org is one of the many transnational public spheres where donors and borrowers (digital entrepreneurs) connect in a disruptive space to participate in social and financial transactions in order to alleviate poverty as we have named *social-on-linezation* of poverty. Given its success, it is foreseeable that other Peer-to-Peer microcredit initiatives will be put into action rooted on ICT-based microfinance.

According to Yartey (2014), the concept of *transnational public sphere* can be theorized as a tangible virtual space where people congregate to develop online socio-financial entrepreneurial relationships, share goodwill through microfinance, and establish connections that have potential of lasting a lifetime in order to alleviate poverty. Therefore, the transnational public sphere is also a stage for performance; it is a public space where potential donors from the Global North provide financial relief to the poor in the Global South. It is a space where individuals and groups in Third World countries exhibit their eligibility for loans through various communication strategies.

We see that future research might be focused on the advent of a new profile of entrepreneurs, the digital entrepreneur, and its environments (Appadurai, 1990):

1. Appadurai's *technoscapes* can be conceptualized as a technological public sphere that exists in cyberspace. This space consists of computers linked together by the fabric of the Internet with various inter-nodal connections that link up local microfinance institutions to Kiva.org in a sophisticated communication network. The *technoscapes* are web 2.0 technological tools for lenders who help bring social change by improving the lives of the poor through
2. *Finanscapes* theoretically are paths that financial transactions flow to and from globally. The transfer of money through digital mediums like Kiva.org to developing countries follows paths created by finanscapes and technoscapes.

These assumptions are based on:

1. In light of present transnational globalization trends, cultural economies flounder;
2. Cultural economies also move seamlessly between ethnoscapes, technoscapes, financescapes, mediascapes, and ideoscapes; and
3. *Social-on-Linezation* to alleviate poverty (Yartey, 2014).

Globalization is bringing many First-world households in developed countries who have never been in direct contact with it the reality of poverty, so solidarity can flourish. So, and in addition to these new cultural trends, they coexist with traditional values such as solidarity and charity, following the latter a religious sense.

CONCLUSION

From Digital Entrepreneurs for Social Change by analyzing the KIVA's Case of Study, there is no doubt that an organization is well meaning while striving to bring economic relief to the poor, ending up the superiority of the Global North over the Global South. KIVA complements the informal lending practices based on family and friends. These arrangements can usually be distinguished from those of moneylenders in that there is no usury, as lenders do not receive any interest rate. KIVA lenders do not expect to gain money from their lending, as all money received is given to benefit the poor.

These disruptive practices of empowering the poor as Kiva.org has opened new ways of understanding the relationships between developing and developed countries, relationships based on equal social exchange, losing traditional forms of developed countries rooted on charity. The new wave has come beyond the microcredit frontiers by transforming the organizational dynamics based on *borrowing side* into *lending groups* and what is more, losing the financial return by moving forward to other forms of "return" as feeling the Gate's philanthropy.

Organizations, like the Grameen Bank of Bangladesh, have paved the way for social change, development of countries' entrepreneurship and empowerment of the poor through microfinance. Technology has complicated and introduced new challenges to global lending. Using online platforms by means of implementing what we have named as *social-on-linezation* of poverty or action of alleviation of poverty through entrepreneurial online platforms which impact directly on society and their change, this chapter has also showed the need for global communication processes and empowerment through online microfinance. Web 2.0 digital technologies and its complex tools make it possible for borrowers living in developing countries to receive loans, adding a new dimension to global lending, providing people who otherwise would not have had access to loan products with the support needed to improve their lives and in turn, the social change.

Finally, scholars and researchers need to continue exploring new ways in theorizing global processes of communication, and in exploring new financial and social concepts based on the flow of money, return, communication, and interrelationships in cyberspace in order to deal with this disruptive social digital entrepreneurial movement known as Kiva.org. We are sure that this first Peer-to-Peer microfinance institution in the world will be soon followed by others with the unique objective of offering a better life for the poor, following lenders a double way, either born from philanthropy (to help others through donations) or from religious beliefs (charity).

REFERENCES

Appadurai, A. (1990). Disjuncture and difference in the global cultural economy. *Theory, Culture & Society*, *7*(2), 295–310. doi:10.1177/026327690007002017

Armendáriz, B., & Morduch, J. (2007). *The Economics of Microfinance.* MIT Press.

Arun, T., Hulme, D., Matin, I., & Rutherford, S. (2005). Finance for the poor: the way forward? In Finance and Development. Surveys of Theory, Evidence, and Policy. doi:10.4337/9781845424602.00016

Barrick, M. R., & Mount, M. K. (1991). The Big Five Personality Dimensions and Job Performance: A Meta-Analysis. *Personnel Psychology*, *78*, 111–118.

Black, S. (2009). Microloans and Micronarratives: Sentiment for a Small World. *Public Culture*, *21*(2), 269–292. doi:10.1215/08992363-2008-029

Butler, J. K. Jr, & Cantrell, R. S. (1984). A behavioral decision theory approach to modeling dyadic trust in superiors and subordinates. *Psychological Reports*, *55*(1), 19–28. doi:10.2466/pr0.1984.55.1.19

Cameron, K. S., Bright, D., & Caza, A. (2004). Exploring the relationships between virtuousness and performance. *The American Behavioral Scientist*, *47*(6), 766–790. doi:10.1177/0002764203260209

Carr, J., Dickinson, E., McKinnon, S. L., & Chávez, K. R. (2016). Kiva's Flat, Flat World: Ten Years of Microcredit in Cyberspace. *Globalizations*, *13*(2), 143–157. doi:10.1080/14747731.2015.1062603

Cha, H.-J., Yang, H.-K., & Kim, Y.-H. (2016). A Node Management Scheme for Stable P2P Service in Mobile Ad-hoc Networks. *Peer-to-Peer Networking and Applications*, *9*(3), 558–565. doi:10.1007/s12083-015-0392-5

Chae, C.-J., Shin, Y. J., Choi, K., Kim, K.-B., & Choi, K.-N. (2016). A Privacy Data Leakage Prevention Method in P2P Networks. *Peer-to-Peer Networking and Applications*, *9*(3), 508–519. doi:10.1007/s12083-015-0371-x

Chun, R. (2005). Ethical character and virtue of organizations: An empirical assessment and strategic implications. *Journal of Business Ethics*, *57*(3), 269–284. doi:10.1007/s10551-004-6591-2

Collins, D., Morduch, J., Rutherford, S., & Ruthven, O. (2009). *Portfolios of the Poor: How the world's poor live on $2 a day*. Princeton, NJ: Princeton University Press.

D'Aveni, R. A. (1994). *Hyper Competition. Managing the Dynamics of Strategic Maneuvering*. New York: The Free Press.

Domen, B. (2011). Reimagining Charity: KIVA's Ideology of Entrepreneurial Charity. *Advances in Consumer Research. Association for Consumer Research (U. S.)*, *39*, 489–490.

Edward, P. (2006). The Ethical Poverty Line: A Moral Quantification of Absolute Poverty. *Third World Quarterly*, *27*(2), 377–393. doi:10.1080/01436590500432739

Facebook. (n.d.). *Kiva*. Retrieved from: https://www.facebook.com/KIVA

Flannery, M. (2007). KIVA and the Birth of Person-to-Person Microfinance. *Innovations: Technology, Governance, Globalization*, *2*(1-2), 31–56. doi:10.1162/itgg.2007.2.1-2.31

Gajjala, R., & Birzescu, A. (2011). Digital Imperialism through Online Social/Financial Networks. *Economic and Political Weekly*, *66*(13), 95–102.

Godelier, M. (1986). *The Mental and the Material*. New York: Verso.

Harris, H. (2001). Content analysis of secondary data: A study of courage in managerial decision making. *Journal of Business Ethics*, *34*(3/4), 191–208. doi:10.1023/A:1012534014727

Hartley, S. E. (2010). *Kiva.org. Crowd-Sourced Microfinance & Cooperation in Group Lending*. Harvard University's DASH Repository.

Hermes, N., Lensink, R., & Habteab, T. M. (2006). Does the Group Leader Matter? The Impact of Monitoring Activities and Social Ties of Group Leaders on the Repayment Performance of Group-based Lending in Eritrea. *African Development Review*, *18*(1), 72–97. doi:10.1111/j.1467-8268.2006.00133.x

Holt, D. (2004). *How Brands Become Icons*. Boston: Harvard Business School Press.

Holt, D., & Cameron, D. (2010). *Cultural Strategy: Using Innovative Ideologies to Build Breakthrough Brands*. New York: Oxford University Press.

Ilcan, S., & Phillips, L. (2010). Developmentalities and Calculative Practices: The Millenium Development Goals. *Antipode*, *42*(2), 844–874. doi:10.1111/j.1467-8330.2010.00778.x

Jung, H. M., Hwang, I.-S., Moon, J.-K., & Park, H.-S. (2016). A Security Monitoring Method for Malicious P2P Event Detection. *Peer-to-Peer Networking and Applications*, *9*(3), 498–507. doi:10.1007/s12083-015-0369-4

Kar, A. K. (2016). Measuring Competition in Microfinance Markets: A New Approach. *International Review of Applied Economics*, *30*(4), 423–440. doi:10.1080/02692171.2015.1106445

Kiva Cup. (2009). Retrieved from: http://KIVAnews.blogspot.com/2009/07/paris-KIVA-cup-2009.html

Kiva. (n.d.). Retrieved from: http://www.Kiva.org/about

Kozinets, R. V. (2008). Technology/Ideology: How Ideological Fields Influence Consumers' Technology Narratives. *The Journal of Consumer Research*, *34*(6), 865–881. doi:10.1086/523289

Li, G., Park, E.-M., & Park, S.-T. (2016). Effects of Social Media Usage on Country Image and Purchase Intention from Social P2P Network Perspective. *Peer-to-Peer Networking and Applications*, *9*(3), 487–497. doi:10.1007/s12083-015-0379-2

LinkedIn. (n.d.). *Kiva*. Retrieved from: https://www.linkedin.com/company/KIVA-org

Lu, K., Wang, J., & Li, M. (2016). An Eigentrust Dynamic Evolutionary Model in P2P File-sharing Systems. *Peer-to-Peer Networking and Applications*, *9*(3), 599–612. doi:10.1007/s12083-015-0416-1

Lumpkin, G. T., & Dess, G. G. (1996). Clarifying the Entrepreneurial Orientation Construct and Linking it to Performance. *Academy of Management Review*, *21*(1), 135–172.

Moody, M. (2008). Building a Culture: The Construction and Evolution of Venture Philanthropy as a New Organizational Field. *Nonprofit and Voluntary Sector Quarterly*, *37*(2), 324–352. doi:10.1177/0899764007310419

Moss, T.W., & Meyskens, M. (2015). The Effect of Virtuous and Entrepreneurial Orientations on Microfinance Lending and Repayment: A Signaling Theory Perspective. *Entrepreneurship Theory and Practice*, 27-52.

Paine, L. S. (1991). Ethics as Character Development: Reflections on the Objective of Ethics Education. In R. E. Freeman (Ed.), *Ethics and Governance* (pp. 67–88). Oxford University Press.

Payne, G. T., Brigham, K. H., Broberg, J. C., Moss, T. W., & Short, J. C. (2011). Organizational virtue orientation and family firms. *Business Ethics Quarterly*, *21*(2), 257–285. doi:10.5840/beq201121216

Prahalad, C. K., & Hart, S. (2002). The Fortune at the Bottom of the Pyramid. *Strategy & Business*, *26*, 54–67.

Ramírez, J., & Anderson, S. (2007). *Kiva Case Study*. UC Berkeley.

Roy, A. (2010). *Poverty Capital: Microfinance and the Making of Development*. New York: Routledge.

Roy, A. (2012). Subjects of Risk: Technologies of Gender in the Making of Millennial Modernity. *Public Culture*, *24*(1), 131–155. doi:10.1215/08992363-1498001

Rutherford, S. (2008). Micropensions: Old Age Security for the Poor? In *New Partnerships for Innovation in Microfinance* (pp. 241–264). Springer.

Sawyer, F. H. (1975). A conceptual analysis of empathy. *The Annual of Psychoanalysis*, *3*, 37–47.

Schwittay, A. F. (2013). Making Poverty into a Financial Problem: From Global Poverty lines to Kiva.org. *Journal of International Development*, *26*(4), 508–519. doi:10.1002/jid.2966

Wagner, L. (2002). The 'new' donor: Creation or evolution? *International Journal of Nonprofit and Voluntary Sector Marketing*, *7*(4), 343–352. doi:10.1002/nvsm.191

Wiklund, J., & Shepherd, D. (2005). Entrepreneurial orientation and small business performance: A configurational approach. *Journal of Business Venturing*, *20*(1), 71–91. doi:10.1016/j.jbusvent.2004.01.001

World Bank. (n.d.). Retrieved from: http://www.worldbank.org

Wright, T. A., & Cropanzano, R. (2004). The Role of Psychological Well-being in Job Performance: A Fresh Look at an Age-Old Quest. *Organizational Dynamics*, *33*(4), 338–351. doi:10.1016/j.orgdyn.2004.09.002

Yartey, F. N. A. (2014). Microfinance, Digital Media and Social Change: A visual analysis of Kiva.org. *Communication & Social Change*, *1*(1), 63–89.

Yunus, M. (1999). *Banker to the Poor: Micro-Lending and the Battle against World Poverty*. New York: Public Affairs.

Zahra, S., Newey, L., & Li, Y. (2014). On the frontiers: The implications of social entrepreneurship for international entrepreneurship. *Entrepreneurship Theory and Practice*, *38*(1), 137–158. doi:10.1111/etap.12061

KEY TERMS AND DEFINITIONS

Banking the Poor: It consists in providing microcredit-based financial opportunities to the poor to insert them in today's financial world.

Digital Entrepreneurial Charity: Entrepreneurs full of human and/or religious values who use Internet for alleviating the poor.

KIVA: Based on the Information and Communication Technologies (ICTs), this organization is the first Peer-to-Peer microfinance organization in the world.

Microfinance: Small loans that benefit the poorest who live in Third World countries.

Peer-to-Peer Micro-Lending: Small quantities of financial aid donated at no interest rate directly given through Internet from First and Second World lenders to Third World beneficiaries.

Social Change: The creation of fairer and values-based societies defined by a higher standard of living through microfinance and digital entrepreneurial charity.

Social-on-Linezation **of Poverty:** It is defined by the sense of internalization and awareness of poverty by the most favored population of the world by using Internet.

ENDNOTES

1 The Consultative Group to Assist the Poor (CGAP) is a consortium of 31 public and private development agencies working together to expand access to financial services for the poor, referred to as microfinance. These principles were developed and endorsed by CGAP and its 33 member donors, and further endorsed by the Group of Eight leaders at the G8 Summit on 10 June 2004.

2 For instance, in November 2010, the World Bank launched a Global Program on Consumer Protection and Financial Literacy, which focuses on shielding poor financial clients from abusive and predatory practices and on providing "consumer education."

3 Grameen Bank and Muhammad Yunus pioneered in Bangladesh what it titled the "solidarity circle."

4 There are "self-help" and "community-based organizations" that have evolved from traditional credit groups. These intermediaries are especially popular in India. While "solidarity circles" and "self-help" groups of borrowers are similar, the former involves remote oversight whereas the latter strives to create independent, locally run micro-banks that encourage borrower repayment but there is no paucity of academic and popular media coverage of *group lending* (Hartley, 2010).

5 KIVA lenders cannot receive interest for legal reasons, among other, because KIVA is a 501(c)(3) what means being a nonprofit for tax purposes.

6 KIVA Friends is perhaps the most comprehensive off-KIVA website for coordination. Touted as "A community for lenders, by lenders," its look and feel is similar to Kiva.org. On the homepage daily statistics are displayed which show the total value of loans made, number of KIVA lenders, number of countries represented, number of entrepreneurs who have received loans, percentage of loans made to women, number of KIVA Field Partners, and number of countries hosting Field Partners. Additionally, KIVAFriends.org provides links to other off-KIVA coordination sites. KIVAFriends is the most prominent platform for understanding extant online options on lending cooperation.

7 The KIVA profile on Change.org shows the support of over 1,000 individuals, fundraising, supporter biographies, blogs, embedded videos, recent news, and job postings.

8 *KIVA* means "unity" in Swahili.

Chapter 10
Universities Fostering Business Development:
The Role of Education in Entrepreneurship

Omar Alonso Patiño Castro
Universidad EAN, Colombia

Jose Emilio Jimenez Ibañez
Universidad EAN, Colombia

Catalina Lucia Ruiz Arias
Universidad EAN, Colombia

Francisco Javier Matiz Bulla
Universidad EAN, Colombia

ABSTRACT

Universities play a fundamental role in promoting entrepreneurship and directly contributing to the economic development of the country. This chapter presents the case of EAN University, an institution that from its foundation has focused on entrepreneurship as its mission. Over time the university has taken concrete actions to develop its three mission areas: training, research and social outreach. At EAN University, the entrepreneur training model is mandatory for all programs, making it part of the student's core education. In addition, the university promotes and creates a culture of entrepreneurship in its community through permanent business incubation and acceleration services, which are available to all students and graduates. EAN's accompaniment model begins with a diagnostic which determines the stage the company is in: preincubation, incubation or acceleration. The university's effort is focus on directing and accompanying companies in the preincubation and incubation stages.

DOI: 10.4018/978-1-5225-0953-0.ch010

INTRODUCTION

This chapter presents EAN University's 47-year history in the context of entrepreneurship. From its conception, EAN University, which stands for *Escuela de Administración de Negocios* and means Business Administration School, has been a benchmark institution at the national and international levels for its focus on entrepreneurship and its developments in supporting entrepreneurs and businesses.

As the business environment has changed, EAN University has adapted and created a curriculum based on entrepreneurship and enterprise development in order to empower its students.

With entrepreneurship at its core, EAN University has developed its three areas of focus: teaching, research and social outreach. All of the areas are complementary. Teaching is informed by research-based practices, and focuses on relevancy in the entrepreneurial environment. Research aims to improve the learning process for students as well as deepen their impact on the surrounding community.

Sometimes the academic world repeats itself and does not take advantage of best practices. Often institutions that follow in the footsteps of others are doomed to walk the same path and commit the same mistakes as others before them. In this sense, the purpose of this chapter is to share the experience of EAN University and generate a shared learning process that leads us to grow together as a system rather than individually.

The first part of this chapter gives a short review of the importance of entrepreneurship in the development of society and the impacts it can have on economic growth. This section brings up previous studies that analyze different countries capacity to create business and entrepreneurs. The section concludes with points about the role of universities in strengthening entrepreneurship skills in students.

On the second part, the chapter focuses on the university's historical developments. This includes philosophical changes as well as changes to the curriculum offered to each student based on their field of study. This part also discusses organizational changes and their impact on creating a greater entrepreneurial spirit within the university's academic, research and outreach communities.

ENTREPRENEURSHIP AND ECONOMIC DEVELOPMENT

In recent decades, entrepreneurship, the social and economic phenomenons associated with the creation of new companies, has been gaining importance as a development strategy. Developed and developing countries are especially looking to capitalize on a formula that creates entrepreneurship through public policy, economic blocks and multilateral entities. An example of this is the "Green Paper on Entrepreneurship in Europe," published by the European Commission in 2003.

This association between business creation and socio-economic development comes from new approaches companies bring to generating value, incorporating new technologies in their respective economic sectors and absorbing qualified human resources (Matiz, 2006). Businesses do this in addition to paying taxes and staking their chains of command.

Various studies, applied in different locations, have shown the benefits of new businesses, especially those with greater dynamism. They are a significant source for employment, innovations, creative energy, and renewal of the business sector and the region's economy (Audretsch & Thurik, 2001; Kantis et al, 2002; OCDE, 1999, 2001; Reynolds et al, 1999).

In this sense, research in entrepreneurship is still a young research area in academia (Matiz 2009; Veciana 1999; Vesper, 1996). Entrepreneurship began its development in the 1970s. Since then it has become a strategy not just for business development, but also for research and instruction in the academic arena.

The Global Entrepreneurship Monitor (GEM) arose in the 1990s as an essential part of this development. The GEM helped create a global consensus on entrepreneurship. One of the most important conclusions the committee came up with was about the clear and direct relationship between economic and social development of a nation and its ability to create and grow businesses.

Since its inception in 1999 under the leadership of Babson College (Boston - USA) and London Business School (London - UK), GEM has measured entrepreneurial activity.

In the beginning it was in only in ten countries, it passed twenty in 2001 and by 2012 was in fifty-four countries on five continents. Today GEM is the largest network of collaboration and research in the world focused on entrepreneurship.

GEM focuses on three main objectives. First, measure the level of entrepreneurship in different countries. Second, identify factors that lead to the creation of new businesses. Third, and finally, stimulate policies that positively benefit the creation and cultivation of entrepreneurship.

The main objective of this study is to, "Determine the difference in national levels of entrepreneurial activity in a large sample of countries, and evaluate the effect of entrepreneurship on economic growth in each country."

GEM's measurement tools such as Total Early-Stage Entrepreneurial Activity (TEA) correlate with macroeconomic variables such as GDP per capita. In addition, GEM analyzes the differences of entrepreneurship generated in the different countries. Through this GEM has discovered a direct relationship between higher levels of venture capital and socio-economic development of countries.

The Colombian Case

Since its inclusion in the study, Colombia has ranked as a country with high rates of Total Early-Stage Entrepreneurial Activity (TEA). Its average value between 2006 and 2013 was 23%. That means 23 out of every 100 Colombians studied by GEM were linked to the creation of businesses.

Figure 1 shows that about 6.5 million Colombians are undertaking business initiatives. This is a significant statistic when compared with values presented by countries with similar economies to Colombia's.

Among many variables, the study investigated the level of entrepreneur training and examined it over time. Figure 2 shows this development.

This evidence displays an important trend that reaffirms findings of recent research conducted in Latin America (Kantis and Others 2002). This research shows that 50% of the most dynamic entrepreneurs in the region are university graduates.

University Education and Entrepreneurship

Several studies and researchers at the international level have examined the relationship between education and entrepreneurship. The researchers examined different topics. Some aimed to demonstrate the relationship between educational levels of entrepreneurs and their propensity to start new businesses (Borjas & Bronars, 1989; Cowling & Taylor, 2001; Delmar &Davidsson, 2000; Karcher, 1998; Lafuente & Salas, 1989; Robinson & Sexton, 1994). Others associated entrepreneurs' education levels with the type of their company and its use of technology (Colombo & Delmastro, 2001; Fayolle & Ulijn, 2001;

Figure 1. Total early-stage entrepreneurial activity (TEA) 2006 - 2013
Source: GEM Colombia 2006 – 2013.

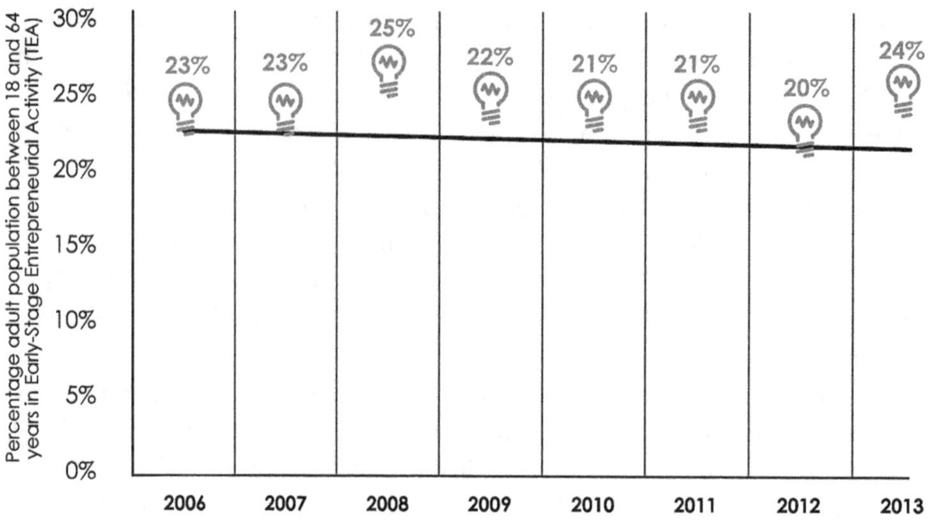

Figure 2. Level of entrepreneur training in Colombia 2006-2013
Source: GEM Colombia 2006 – 2013.

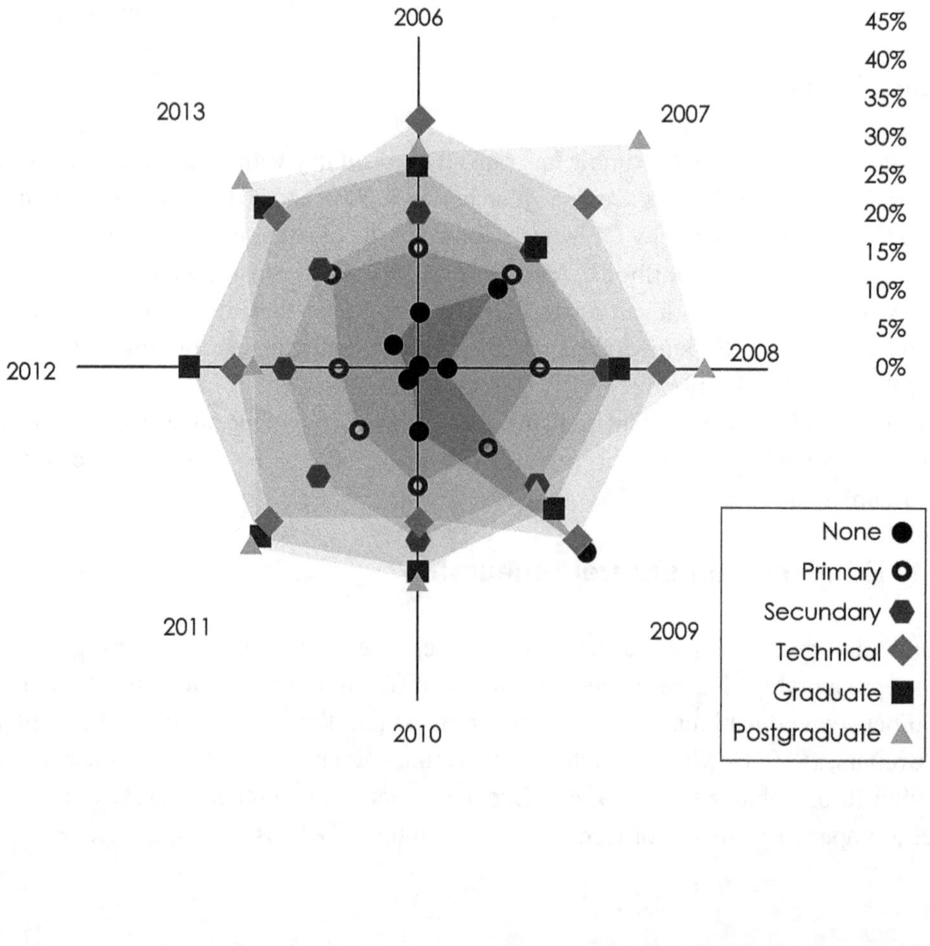

Litvak & Maule, 1976). Finally, some evaluated the impact of the universities courses and / or programs on the characteristics of its graduates businesses (Charney & Libecap, 2000; Levie et al, 2001; Lüthje & Franke, 2002; Tackey et al, 1999; Upton et al, 1995).

Every day the role of academia and especially universities becomes more important. Academic institutions can serve as an engine to develop a new generation of entrepreneurs able to use their knowledge and training to create and serve new companies. These entrepreneurs can generate value, innovation and competitiveness. For this reason, some now consider universities as incubators of entrepreneurs (Veciana, 2002).

Therefore, a significant number of universities worldwide, understanding their role in the development of entrepreneurial skills, have implemented frameworks within their academic programs that address issues related to content, business creation and cultural exchange. This system also allows students to assess the possibility of a corporate career as part of their career path.

Several authors have evidenced a significant growth of entrepreneurship programs within universities in the last 30 years. This includes the emergence of research demonstrating increasingly strong demand for content and courses created by educational entities (Kolvereid & Moen, 1997; Fayolle, 1998; Flinke & Deeds, 2001; Lüthje & Franke, 2002).

Entrepreneurial Training: EAN University's Constantly Developing Mission

Essentially, universities must do three basic activities: teaching, research and social impact. In the exercise of each of these activities, many institutions chose to provide its stakeholders a unique education that distinguishes it from other universities in the country or region.

With this thinking, EAN University, *Escuela de Administración de Negocios*, which means Business Administration School, was founded in 1967. From inception, entrepreneurship has been the university's focus. Despite Colombia's social and economic conditions at the time, entrepreneurship was not considered a relevant part of training at the university level. On the contrary, programs that had this component could suffer from negative judgement when others assessed the university's curricula.

EAN University's most important institutional goal is described in its charter:

EAN University aims to stimulate the development of the country's economy and society, through the application of principles and management techniques to create and develop businesses.

Since its inception, this mission has permeated all of the University's activities, particularly its curriculum and programs.

Since 1993, this training has appeared explicitly. At that time, the Business Administration program began offering study units in Development of Entrepreneurial Spirit, where students developed business creation skills through four courses.

The purpose of these initial courses was for students to become aware of holistic entrepreneurship. Students learned that projects are not only built with individual effort, but also with other areas. These include the person's family and the community to which they belong. From this understanding, students could generate initiatives to address specific social needs.

In addition to such training, the program offered students classes in Business Creation and Business Opportunity Analysis. Business Creation empowered students to boost entrepreneurial skills within organizations. Meanwhile Business Opportunity Analysis focused on researching environments for new options and opportunities for doing business.

Other programs accepted these units as part of their curriculum. Some were implemented as a mandatory component without a grade, in the traditional classroom environment and the virtual one.

Although the focus of the university has always been entrepreneurial training, it was not until 2003 that it became a part of all the undergraduate programs of the university. It was then that the university designated certain classes mandatory, creating a core curriculum. These subjects included, and still include, basic language training, humanities, and training in entrepreneurship.

In addition, administrators made the decision that entrepreneurship should not only be taught in undergraduate programs but should be incorporated in all graduate programs.

As part of this reform, the university created the Entrepreneur Program EANTEC®. The focus of this program is an entrepreneurial training that is flexible, relevant and research-based.

Each of these elements represents a fundamental component for the development of a competency-based curriculum.

These aspects have been utilized in the design of courses at the undergraduate and graduate level, as well as in the two learning environments -- traditional (face to face), and e learning (virtual).

This change in the structure of the curriculum allows students to take four units of study called Entrepreneurship I to IV, which provide 9 academic credits [4] in curricula with a total of 160 credits that they must accomplish.

This changed the system. Unlike what was done in the previously, this new proposal allows students to not only develop aspects of their personality, but also to strengthen their propensity towards projects focused on business creation. It also incorporates creativity and innovation into their business ideas, and students learn how to structure businesses that are durable and grow over time.

The other significant change in the new system was the inclusion of the creation of business plans. Initially, only Business Administration students had to create business plans. These plans are developed during the three first courses. By the fourth course, students would have projects ready for launch and testing in business simulators.

Finally, the program included training in business financing. With this, students could be certain that their project could access the resources needed for its implementation.

As part of the academic development of entrepreneurship, in 2008 a comprehensive review of the contents of all programs was completed. This resulted in the formulation of a new philosophy regarding how the students will be trained. The Entrepreneurship Program changed the names of its classes to Culture of Entrepreneurship I, II and III, and the last level became Business Creation. In addition, Economics and Business courses were expanded. The purpose was for students to gain macroeconomic knowledge since their businesses would definitely be affected by market forces. The change was more than the new given name. The development of the curriculum sought to inspire students to see entrepreneurship as a way of life.

As stated, the university included entrepreneurship in the undergraduate and graduate curriculum.

Since the curriculum reforms of 2002, all specialization programs have two units, one called Entrepreneurship and the other Strategic Thinking and Global Management. Each one has an intensity of 6 credits (3 each) of a total of 24 that make up each program.

The scope of the units of study varies according to the undergraduate program.

Although the focus continues to be entrepreneurship, the focus is more in the quality of the business and its importance in a society that needs job providers that create economic wealth through business creation and existing businesses growth. Training in strategic thinking and understanding globalization are factors that allow the creation of businesses that have a great impact on the country's economy.

This decision had greater weight in the design of master's degrees and doctorates. In these curriculums, EAN University included the same units designed for specializations. The purpose was to give greater emphasis to research and consultancy skills associated with the creation and management of companies.

ENTREPRENEURSHIP AS A WAY OF INTERACTING WITH THE ENVIRONMENT: CONTINUING EDUCATION

Over the years, the relationship between the University and its environment, especially in entrepreneurship, has been mediated by institutes or centers specifically created to handle this relationship. Despite being one of the fundamental objectives from the founding of the University, issues of entrepreneurship were handled by the same entities that manage formal education.

Pioneering entrepreneurial education at the national level has brought the university to explore different ways of articulating its three basic functions as well as its services in the field of administration.

Only in 1983, the university created a specialized project management and research body. This body was named *the Centro de Investigación y Desarrollo Empresarial*, CIDEAN, which means the Center for Research and Business Development.

With the passing of time, advancement of the university as well as its reputation in teaching, research and outreach grew. Therefore, the university decided to create divisions to manage each of these facets. In 1987 the university created the *Centro de Desarrollo Empresarial* (CEDEAN), wich means the Center for Business Development. This body was dependent on the Vice President's Academic office, but had unique functions such as consultancy. This left research as a function delegated to the Vice Presidency. In 1995 this center became part of the President's office, and acquired greater strategic importance in the development of the university.

In 1998, in order to maximize the relationship with graduates and to continue strengthening interactions with the business sector, the university created the Office of Managerial Development. This office remained dependent on the President, and fused the two aforementioned priorities. At this time, units related to entrepreneurship already existed in the Business Administration program, so this faculty focused on teaching.

In 2005, the university welcomed a process management model. The three basic functions of the university (teaching, research and outreach) were each given their own space in the institution's organizational structure.

With this reform, the Social Outreach Process was created. This gave a new opportunity for the CEDEAN, whose functions now included supporting the external sector in addition consultancy services. Furthermore, Social Outreach Process created various programs including Promotion of the Professional Entrepreneur, Promotion and Support of the Executive Entrepreneur, Promotion and Support of Women Entrepreneurs, Promotion and Support of University Entrepreneur, and Business Outreach, amongst others.

Only two years after its creation, the CEDEAN became part of the Vice President's Social Outreach program. The institution considered it necessary to raise the status of the Social Outreach process to balance the hierarchy. This put the program's scope and capacity of decision-making in balance with the other two processes, Teaching and Research.

Just a few months later in the same year, the Entrepreneur Program, the body responsible for all academic aspects of the university, became part of the Vice Presidency with the aim of fostering closer ties with the business sector. This allowed entrepreneurship skills to have a direct relation to the outreach function and affected teaching and academic policies regarding entrepreneurship at the University.

A new reform took place in 2012 with the creation of the *Instituto de Emprendimiento, Liderazgo e Innovación* (IDELI), Institute of Entrepreneurship, Leadership and Innovation. This institute seeks to explicitly articulate the necessary elements in the training of entrepreneurs. In addition, it is part of the Vice Presidency Social Outreach project and currently is part of the Academic Vice Presidency. IDELI is created with the function to manage the activities related to entrepreneurship both within the university and for its stakeholders.

In the exercise of its functions, IDELI manages and develops the EANTEC® model, through which all of the university's units of study are managed. EANTEC and IDELI also focus on developing research. From here, Entrepreneurship Group researches the entrepreneurial process. It also promotes the business ideas of undergraduates and graduates. Following these reforms and the acquianted experience by EAN University with the training and promotion in entrepreneurship, IDELI developed the EAN IMPACT Model, a business development strategy aimed at the pre-incubation (discovery), incubation and acceleration of businesses.

THE EAN IMPACT MODEL

Background

The model's origins are from the IN3 program (incubate, innovate, integrate), that operated starting 2001. Its objective was to make viable the ideas, business models and plans generated by students and graduates of EAN University.

With the creation of IDELI in 2013, IN3 was transformed into the *Punto de Atención al Emprendedor* - PAE, Entrepreneurs Attention Point, giving continuity to the incubation and acceleration model. The creation of IDELI and PAE was accompanied by the creation of the EAN University Capital Seed Funding, an initiative to fund EAN University entrepreneurs' projects. Along the three versions of the contest (2013-2015), the fund distributed USD $142,500 annually to 15 projects at an average of $9,500 per project.

This contest, which publically called EAN students and graduates with entrepreneurial initiatives, gave a special value to projects with sustainability components and innovation, financing investment in fixed assets, working capital, business strategy, pre-operating expenses and prototypes. In addition, it specially valued companies close to launch.

The university improved several aspects of the contest, over time. The first was the change in the application process, from filling out a business plan form to a form with a summary of the business plan. In both cases, the jury of the contest evaluated the document.

Once evaluated, participants who passed the first stage, in the first version (2013) would submit their business ideas. Meanwhile in the second version (2014), prior to evaluation, participants would participate in workshops on design thinking and business models design. Here they were given tools to improve and complete their business profiles, thus developing more structured business plans. Furthermore, in the second version, an additional filter was installed to train participants on Elevator Pitches, so they could strengthen their project for the Selection Committee. The flow of the contest in the second version is shown in Figure 3.

The strategy around accompanying the entrepreneurs, not only was to maximize the resources allocated by the University. In addition IDELI's entrepreneurs participated in a national contest with Incubatec Ufro (Chile), StartUp Essentials (LATAM) and Lever Laboratory S.A.S. as accompanying institutions, to compete for funding from the national government.

The aim of participating in this competition was to obtain resources that would improve IDELI's incubation and acceleration model, and reinforce IDELI's teachers' skills. From this, IDELI could identify opportunities for improvement and collect lessons from IN3, PAE and the Seed Capital Fund.

In the process, one of the added values was to have achieved methodologies and best practices transferred by the accompanying institutions. In addition, IDELI sought to implement the lean startup methodology, early validation strategies and strategic design. In the third version of the contest, IDELI incorporated a Bootcamp in which entrepreneurs worked on discovering the demands of specific clients. Participants also engaged in interviews in the field, developing alternative solutions that were prototyped and validated directly with clients.

The process of the Seed Capital Fund Contest does not end with the allocation of resources. From the moment in which the participants are awarded funds, a supervisor oversees the allocation of resources and actions are taken to ensure the project starts and grows.

Figure 3. Seed capital fund contest, second version, (2014)
Source: Elaborated by authors.

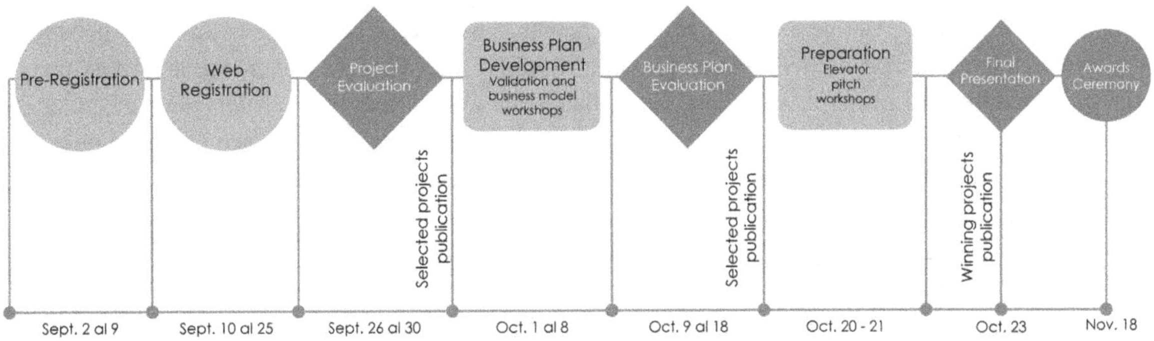

Figure 4. Seed capital fund contest, third version, 2015
Source: Elaborated by authors.

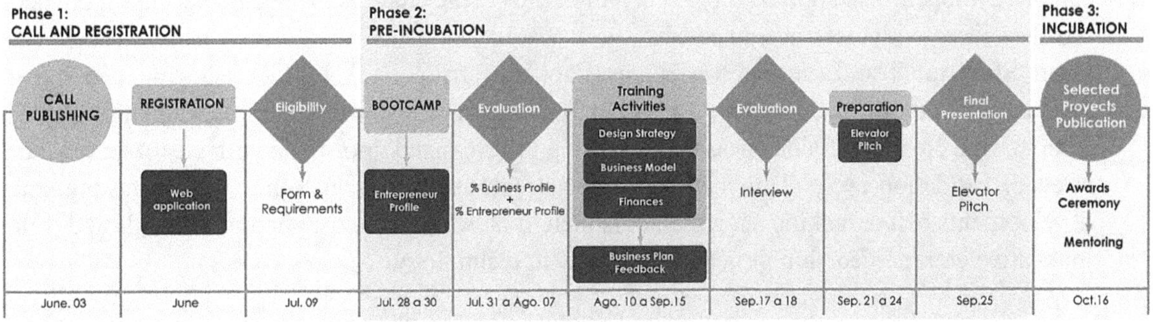

The project resulted with the improvement and strengthening of IDELI's ability to incubate and accelerate, with the conceptualization of a new model of services for entrepreneurs called EAN IMPACT program. It focuses on incubating and supporting entrepreneurial projects to get started and grow, not only through the university's equity funds, but also through other existing funds and other services provided by different organizations and the entrepreneurial ecosystem of Colombia and Latin America.

The built model uses as a methodological basis, implements tools to solve complex problems, and uses creative techniques to generate ideas. These tools are explained below:

- **Design Thinking:** Stanford University (USA) in partnership with the University of Potsdam (Germany), created the Hasso Plattner Institute of Design, also known as d.school, a meeting place for students of different academic programs at Stanford. Here students can learn Design Thinking to solve complex and dynamic problems using creative tools. Professors David Kelley and George Kembel, whom saw the creative processes of artists and designers as an opportunity to innovate in the business world, launched the initiative. In addition, Design Thinking decreases risk during the prototype phase by testing the products and services before going to market and marketing.

Another major contribution of Kelly and Kembel with Design Thinking is the multi-disciplinary professional profiles that can participate in the development of the methodology. This experience enriches the results and improves possible solutions to problems due to the different points of view of the participants.

- **Design Process:** Ambrose and Harris (2010) propose a sequential seven steps design process based on Design Thinking. The process starts with the definition of the problem or situation where the project starts. The next step is based on research to determine the context or background on which the defined problem develops. In the third step of the process is where the creativity emerges in the form of ideas for solutions. From there the process moves to prototyping and selecting the best prototypes. The penultimate step is the implementation and delivery of the solution that responds to the initial problem. Finally, there are lessons learned and feedback to help improve future processes.
- **Human-Centered Design:** The HCD (Human Centered Design) is one of the methodologies designed by Stanford University. It has been developed and implemented by consulting design firms; for example, IDEO and Frog Design employed it in their product development projects and their consultancy services for other companies. This methodology puts users (people) at the beginning and end of the design process. In this sense, the observation of human actions, behaviors and desires, are the starting points for detecting opportunity areas where entrepreneurial projects can be developed. The application of this tool improves the chances of success because it connects business ideas with real market needs.
- **Lean Startup:** The Lean Startup is an adaptation from LEAN Manufacturing and LEAN Production methods made by Eric Ries (2013). It is a process based on rapid validation that accelerates the creation of a business by measuring progress and improving with customer feedback at every validation cycle. This method helps entrepreneurs to reduce the risks by testing small developments before making a big market launch. It is one of the most popular methods to develop innovative entrepreneurship projects, especially in technology.

Figure 5. Design process
Source: Ambrose and Harris (2010).

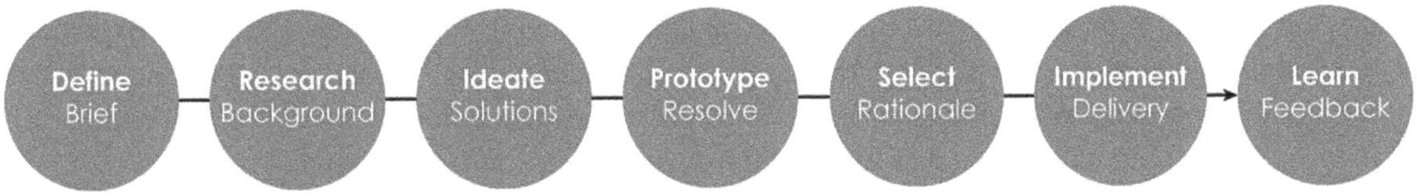

Origins of The EAN IMPACT Program

EAN IMPACT seeks to integrate the learning process that takes place in the university, from the core subjects in entrepreneurship, with a program of pre-incubation, incubation and acceleration of businesses. This program aims to strengthen business ideas from the Eanista community with innovative components and competitive advantages.

The implementation of such a program requires the active participation of the institution in giving feedback to new business ideas to help form and consolidate the new businesses.

At this point, EAN IMPACT seeks to bring to market valuable business ideas by structuring and validating business plans and models likely to be incubated and transformed into sustainable enterprises with high growth potential.

The main objective of the program is to generate a permanent process of pre-incubation, incubation and acceleration of business initiatives through services in technical and soft skills development. This ensures the creation of companies with high growth potential and entrepreneurial leaders who are committed to the social, economic and environmental sustainability of the country.

To meet this objective, entrepreneurs should be offered services to help them design and develop business models, with emphasis on aspects of sustainability in three areas: economic, social and environmental. Furthermore, entreprenuers need help to define strategies to enhance their growth sales and job creation. Along this mentoring process, enterprises with sustainable business models and high growth potential must be identified, to support them in acquiring and managing financial resources for its consolidation and growth. Additionally, identifying and articulating the different actors of the entrepreneurial ecosystem in the country will help EAN entrepreneurs take advantage of the services and benefits of such ecosystem.

Figure 6. EAN IMPACT general program scheme
Source: Elaborated by authors.

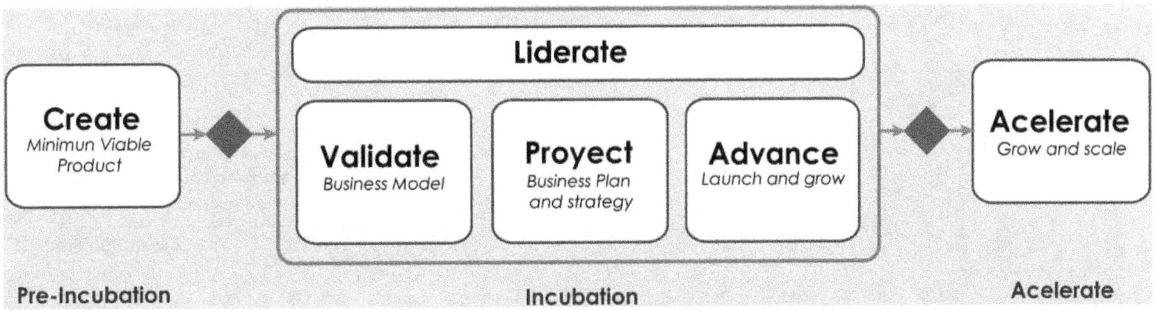

Description of the Model

The EAN IMPACT program focuses on providing support services to entrepreneurs through Pre - Incubation, Incubation and Acceleration paths. Starting with an identification and diagnosis process, entrepreneurs who apply to the program are directed to different routes depending on the degree of development of their initiatives and the evidence of sustainability and innovation factors in their proposals.

The pre-incubation path is constituted mainly by the Creation phase. Here, through methodologies of design thinking, lean startup and early validation, entrepreneurs redefine their business idea, by identifying problems and opportunities in the market based on a thorough analysis of the context and potential users. In addition, the generation of alternative solutions and conducting experiments allow entrepreneurs to validate their ideas and define a value proposition that serves as a starting point to develop a business model.

The incubation path is based on accompaniment by specialized tutors. These instructors help entrepreneurs validate their business models and develop skills in the areas of marketing, organization, operations and finance. From this, EANists create and consolidate sustainable enterprises with high growth potential.

The acceleration path seeks to articulate companies that are in a mature stage with the entrepreneurial ecosystem, after analyzing the needs of each company through an enterprise diagnosis method.

Figure 7. EAN IMPACT program scheme – routes
Source: Elaborated by authors.

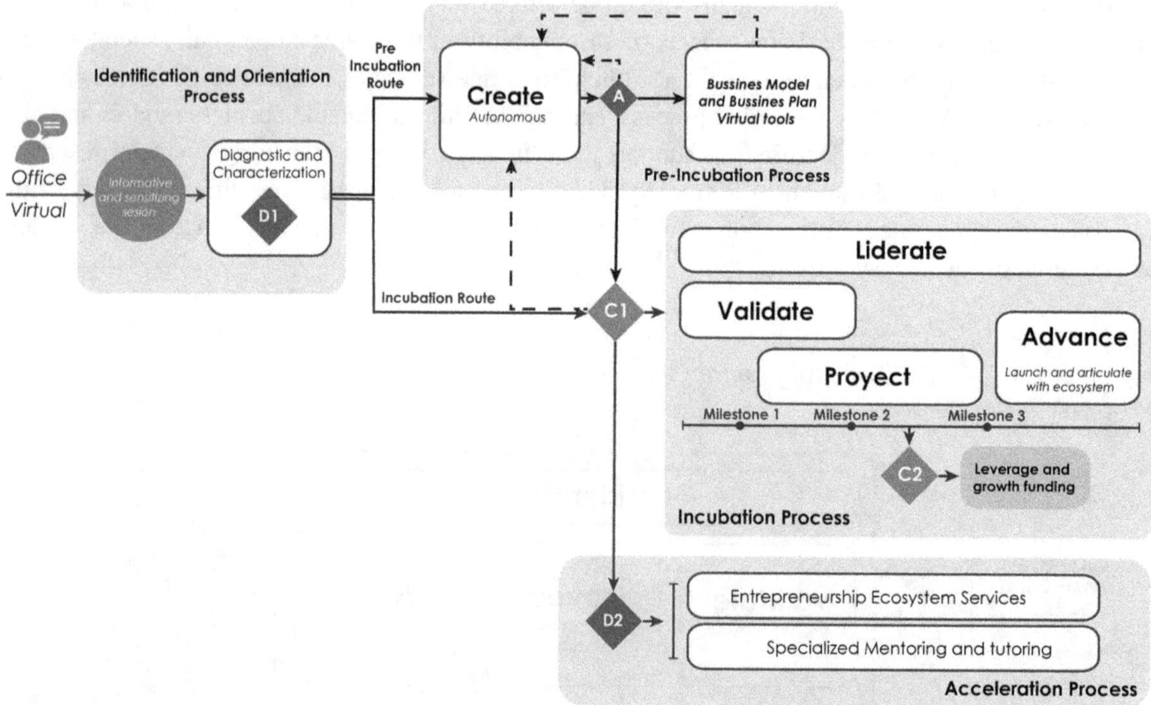

Model Operation

Part 1: Identification and Orientation Process

This process diagnoses and routes the business initiatives that will join the program. It is important to note that this process includes projects that are in all stages--from business ideas, to business models, to business plans and operating companies.The identification and orientation process includes entrepreneur sign up, attending briefings, and filling out registration and characterization forms. With these forms the program coordinator and mentors can diagnose whether the project is an idea, a model, a business plan or a growing business and are the basis to identify the current status of the project submitted and resolve any technical or conceptual issues. The process ends by directing each project to one of the following processes: pre-incubation route or incubation Committee.

Specific Activities in the Identification and Orientation Process:

- **Informative and Sensitizing Session:** Conducted once a month to socialize the program components and to sensitize participants with the concept of Sustainable Entrepreneurship.
- **Entrepreneur's Registration:** Registration form which contains personal data, type of relationship with the university (Active student – Graduate), and project basic data, among others.
- **Project Diagnostic:** Diagnostic form, were the major components of the business idea are described (description of the business idea, business model and validation, business plan and business aspects in general).

In the process of characterization it is important to understand the clarity, relevance and real opportunity of the proposed business idea in relation with the market and the environment. Furthermore, to get to know in detail the business model, the maturity of the idea and the market validation.

Figure 8. Identification and orientation process
Source: Elaborated by authors.

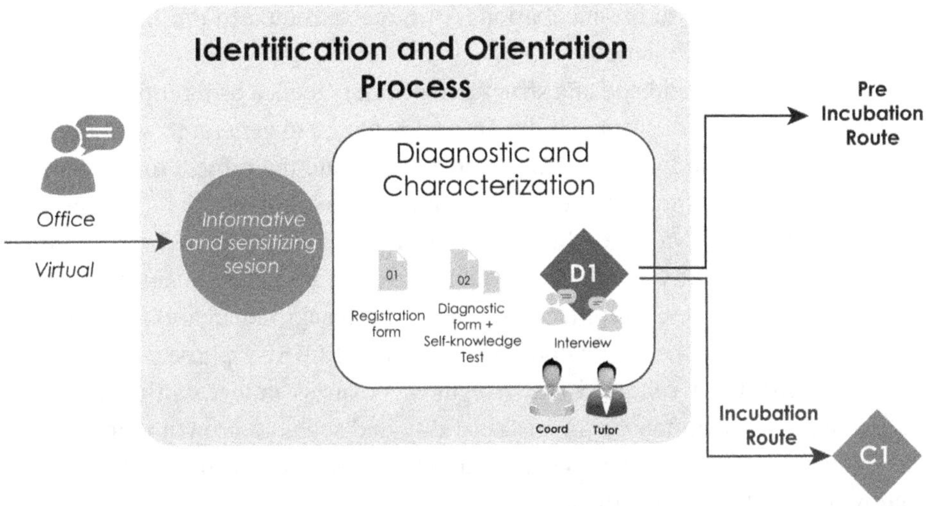

- **Self-Knowledge Test to Examine Entrepreneur's Soft Skills:** Registered entrepreneurs should develop the *color test*. This test is designed to enable people to understand more about their personality and is based on the *Theory of Colours* by Ned Hermann; Hermann theorized that humans' ability to develop critical preferences leads us to have a tendency to develop certain skills more than others.

Upon completion of the test a graph in which behavioral trends are observed is presented.

- **Interview:** Once the forms have been delivered and evaluated, if it is not clear enough with the information presented to define the route, the entrepreneur must attend an interview. The purpose of the interview is to validate the diagnosis of the project and identify weaknesses or strengths in the soft skills of the entrepreneur, which have already been evidenced in the self-knowledge test.

The final concept of the interview determines whether the entrepreneur must be submitted to the incubation committee or is directed to the pre-incubation process.

- **Incubation Committee:** Once the evaluation process of diagnosis and final interview is completed, the projects routed to the committee are presented. The committee reviews the diagnosis and opinion issued in the interview to validate projects addressing the appropriate next step in the process.

According to the concept and evaluation issued by the incubation committee projects are addressed to the next process.

Here are the concept's criteria:

1. If the entrepreneur either fails to present a validated and structured business, lacking sustainability components in its 3-axis component, or if the entrepreneur lacks the potential to generate a scalable business model with detailed growth potential the project is routed to pre-incubation. Companies that are already in place, but do not have a differentiated value proposition, obvious sustainability components, or the potential to generate a scalable business model and with detailed growth potential may also be routed to pre-incubation. All projects routed to pre-incubation are invited to participate in the CREATE phase.
2. If the entrepreneur has a validated and structured business idea, a differentiated value proposition, with obvious components of sustainability and the potential to generate a scalable business model with growth potential, the Incubation Committee will route the project to incubation, and assign the project a coach with who will oversee the process through the VALIDATE phase.
3. If the entrepreneur has a validated business model and a differentiated value proposition, with components of sustainability and the potential to generate a scalable business model and with detailed growth potential, the Committee will route the project to incubation, and assign the project a coach who will oversee the incubation process in the PROJECT phase.
4. If the entrepreneur has a validated and structured business model, with a differentiated value proposition, and with components of sustainability and scalable growth potential the committee will assign the project a coach who will structure the work plan during the ADVANCE phase and also who will begin the process of coordination with other players in the ecosystem.

Table 1. Identification and orientation process resources, products, and indicators

Resources Needed	Physical infrastructure to support entrepreneurs
	Forms
	Program coordinator
	Administrative assistant
	Tutors
	Incubation Committee
Products	Registrations to the EAN IMPACT program
	Completed diagnostic forms
	Route concepts: Pre-incubation path or presentation to the Incubation committee.
	Incubation Committee Concepts
	Tutor assignment
Indicators	Number of entrepreneurs who attend information and sensitizing sesions
	Number of entrepreneurs registered
	Number of entrepreneurs diagnosed
	Number of businesses in the pre-incubation route
	Number of incubated companies presented to the Incubation committee and routed to the Incubation process.
	Total sales earned by incubated companies early in the process.

5. If the entrepreneur has a company with an ongoing structure, a differentiated value proposition, obvious components of sustainability, and the potential to continue its escalation and growth, the committee will route the project to the acceleration process.

Part 2: Pre-Incubation Route

Preincubation looks to structure and to define a business idea, through identifying problems and market opportunities. It also develops ideas that respond to these problems and opportunities, and conducts experiments that allow entrepreneurs to validate their ideas and define a value proposition that helps to develop their business model.

The preincubation process starts in the CREATE phase, which begins with the exploration of opportunities and the redefinition of problems. It continues with the process of idea generation for alternative solutions. Then it turns to the selection of ideas to be prototyped and validated. To conclude, a final idea is structured and a value proposition is defined. It is important to note that in order to complete the CREATE phase the company must have the endorsement of a mentor to perform again the diagnostic process at the identification and selection stage.

In the event that a project completes the CREATE phase and is not endorsed to be routed to the incubation process, the entrepreneur can choose to repeat the CREATE phase or request tools for autonomous work for the development of his or her business model and business plan, if necessary (in this case there is no support from tutors).

Stage 0: Create Phase

The Create phase is a mixed program that both trains and accompanies entrepreneurs through 9 virtual modules. These modules are autonomous and have spaces to interact with specialized tutors who help them in the process. In addition, these modules help students test their business ideas in a group setting. Here, interaction between entrepreneurs in the class is established to foster collaboration.

Once entrepreneurs have completed all modules, they engage in a final presentation session, where specialized mentors conduct the assessment and issue a decision. If the mentor OKs the development process, the entrepreneur completes the diagnostic process again in the identification and selection stage and presents to the incubation committee. If denied approval, the entrepreneur can repeat pre-incubation or utilize virtual tools for model development and business plan writing (without tutors).

The CREATE program is based on methodologies of design thinking, brand driven innovation, lean startup methodologies and early validation.

The modules are shown in Table 2.

Part 3: Incubation Route

Incubation according to UBI Global Organization, seeks to facilitate the process of entrepreneurship and help companies that are in their early stages. This is achieved through a systematic process that includes direct support services (training, consulting) and creating leverage to access venture capital (UBI Global).

Business incubation is a dynamic process in the development of a company. Incubators strengthen young firms, helping them survive and grow during the startup period in which they are most vulnerable. In addition, incubators provide management assistance, access to financing and provide critical feedback or technical assistance. They also offer shared office services, access to equipment, flexible leases and expandable space (NBIA, 2001).

Moreover, incubation is a term that covers a wide variety of processes. All the processes help reduce the percentage of problems in the early years of a company and accelerate growth of companies. This empowers companies to potentially generate substantial employment and sales (UKBI).

Figure 9. Pre-incubation process
Source: Elaborated by authors.

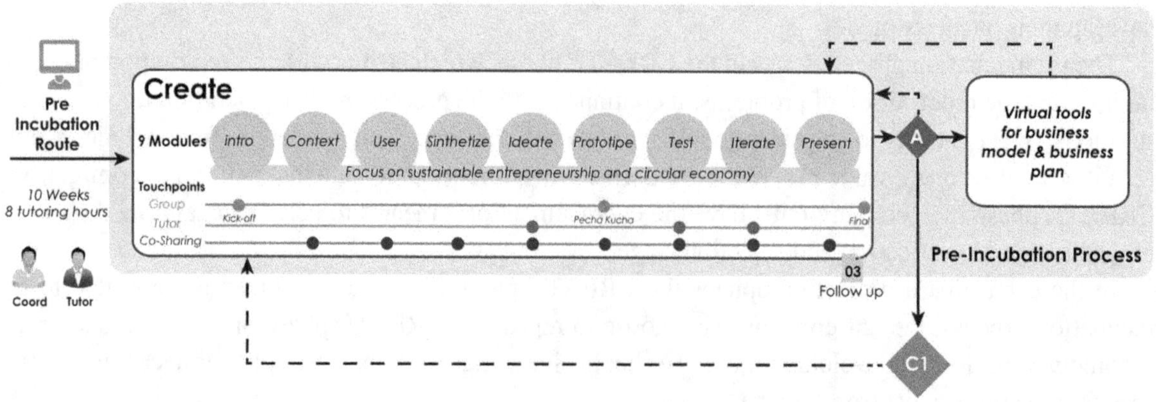

Table 2. Modules in CREATE phase

Introductory Module	Provides propaedeutics or preliminary information in which the overall objective of the program is presented. Students comprehend market needs to identify how their product can succeed.
Context Module	Here participants look to understand their initial business idea. Initially this comes from the inner vision of the entrepreneur. Later it comes from the external view of the market, and concludes in constructing a hypothesis.
User Module	The user becomes the focus of the products or services offered by the entrepreneur. Entrepreneurs comprehend users' needs through fieldwork.
Synthesis Module	Entrepreneurs convert data obtained from fieldwork done in the previous module into valuable information. They use different methodologies to gain insight, synthesize needs and build a user experience.
Idea	Creativity is at the heart of this module. It seeks to develop and select ideas through the use of techniques to develop a value proposition for users.
Prototyping Module	Entrepreneurs validate ideas through the prototyping of products and services, reducing the risks of entrepreneurship.
Testing Module	With the developed prototype, the next step is to test to ensure the effectiveness of the product or service
Iteration Module	This sections focuses on analyzing the results of the previous modules, and making necessary adjustments, even going back to revise previous modules.
Presentation Module	Finally participants present the final results of the process
Leadership and Sustainability Learning Module	During the preincubation phase learning soft skills and sustainability is emphasized. In each module tools for reflection and developing soft skills for entrepreneurship and sustainability concepts and guidance will be included, and their completion will be verified by the tutor. In addition, entrepreneurs complete an assessment to continue their personal reflection. (16 personalities development test - MBTI). https://www.16personalities.com

Table 3. Pre-incubation route resources, products, and indicators

Resources Needed	Monitoring platform
	Virtual modules material
	Interaction sessions
	Module Templates to be filled: Intro, context, user, synthesize, idea development, prototyping, testing, iterate and presentation
	CREATE phase tutors
Products	Structured and validated ideas, with a value proposition defined
Indicators	Number of entrepreneurs who start the CREATE process.
	Number of entrepreneurs who complete the CREATE phase and present to the incubation committee.

EAN University's business incubator, materialized in the EAN IMPACT program, accompanies students and graduate entrepreneurs to help them formulate and validate models and business plans in order to create and develop sustainable enterprises with high-impact potential.

The incubation process begins with a value proposition, which allows the identification and validation of the business model in the VALIDATE Phase. During the PROJECT Phase the business model is structured in the following areas: commercial, technical and operational, organizational, financial and investment. Incubation ends with the implementation of the operational plan in the ADVANCE Phase. Additionally, the entrepreneur will receive tools for reflection and development of soft skills in the LIDERATE Phase.

The incubation process begins with the endorsement of the incubation committee which assigns the project a tutor who designs the work plan and determines the project's tutoring needs and the appropriate specialized topics. During the VALIDATE, PROJECT and LIDERATE Phases, the tutor will track progress to achieve the implementation of defined work plan and meet the milestones established with the entrepreneur. It also is expected that during the validation process the first sales projection is generated.

Phase 1: Validate

In this phase of the incubation process the entrepreneur formulates and validates the business model using lean startup and early validation methodologies; this includes tools such as lean canvas, rapid prototyping, development of viable minimum products and business model canvas.

The final product must be a validated business model, showing articulation and coherence between the proposal and the market value. This validation process should be documented as it is the input for the projection of the business model and the company.

Phase 2: Project

This phase advises the entrepreneur in screening of his or her business model. This is to detail the operation of the company, prepare it for launching, and manage resources within the enterprise ecosystem.

For the projection of the business model, the business model validated in the previous phase and the documentation used in the validation process are used as input.

Figure 10. Incubation process
Source: Elaborated by authors.

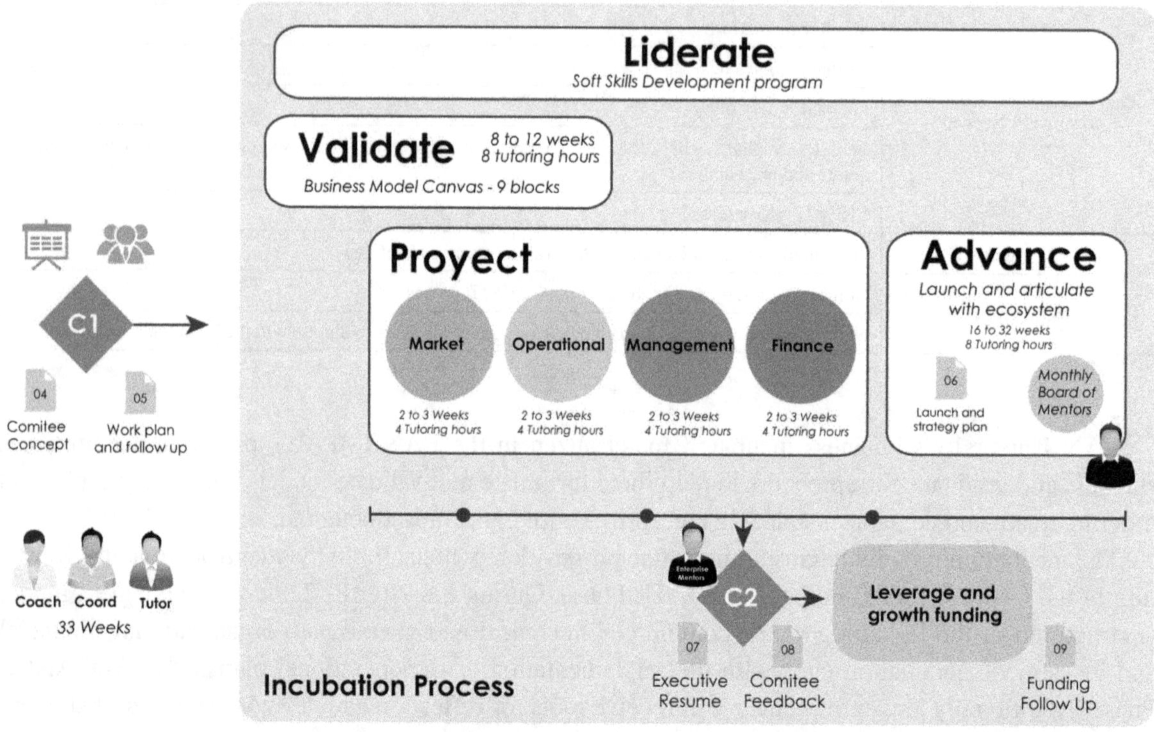

Aspects to consider for the projection of the business model are the market components, production, administrative components, and finally a financial component to assess the profitability of the company. With the development of these aspects, the entrepreneur will have the information to write his or her business plan for fundraising and investment, as the case may require.

1. **Commercial Component:** The commercial component is essential in the PROJECT phase, it is conducted first as it serves as input to develop the following components. The commercial component is defined as a test to clarify market characteristics, size and share of the company, and to describe the strategies that the company will continue to use to position itself in the market. To distill and define all essential business information is a guarantee of success, as the information available to the sales force and the company generally presents an overview of both customers and marketing channels, allowing the company to take concrete actions to reach results. In the commercial component the entrepreneur should develop the following aspects:
    - 1.1. Sector Analysis:
        - 1.1.1. Sector characterization,
        - 1.1.2. Entry and exit barriers.
    - 1.2. Market Research and Analysis:
        - 1.2.1. Market segmentation,
        - 1.2.2. Consumer description,
        - 1.2.3. Market risks and opportunities.
    - 1.3. Competitive analysis,
    - 1.4. Price analysis,
    - 1.5. Marketing strategy:
        - 1.5.1. Product or service concept,
        - 1.5.2. Distribution strategy,
        - 1.5.3. Price strategy,
        - 1.5.4. Promotion strategy,
        - 1.5.5. Communication strategy.
    - 1.6. Service strategy,
    - 1.7. Marketing Mix Budget plan,
    - 1.8. Sales projections.
2. **Operational Component:** Once a definition of a market exists, specially the possible demands for the product or service, an operational study is conducted. The operational study is based on the size and characteristics of the infrastructure and processes needed, and describes the methods, techniques and technology used to achieve the production of goods or services to be marketed. For most businesses, this section is critical because production and suppliers' management that come before the manufacturing of a product or service are key to achieving operational profitability.

The most important areas of the operational component are: labor, materials and infrastructure, equipment and processes. These areas provide the company part of the competitive advantage, which are critical when submitting a plan to an investor.

The simplest way to document the operations of a company is to think in a linear process that can be described in a series of sequential tasks to produce a product or deliver a service. Once the task list is complete, each must determine what is needed to perform each task and who should perform.

The importance of knowing in detail each step of the production process is to develop strategies that are based on intelligent decisions about what processes to do internally and what should and can be outsourced. The development of the operating component applies to both manufacturing companies and trade and service companies.

The development of the operational component should respond to questions such as:

- How long does it take to produce a unit or a predetermined set of units?
- What measures should be defined to receive customer feedback to your product or service? This is if it has been assigned time to create and test prototypes, pricing mechanisms, or delivery mechanisms.
- What are the installed capacity and responsiveness to higher demand flows, or processes or steps that will be required to do when a sale of the product or service is conducted?

In the operational component the following aspects should be developed:

- 2.1. Product or service data sheet
- 2.2. Process description
- 2.3. Requirements and necessities
    - 2.3.1. Technological characteristics
    - 2.3.2. Raw materials and supplies
- 2.4. Operational localization
- 2.5. Production plan
    - 2.5.1. Order processing and inventory control
    - 2.5.2. Operation scalability
- 2.6. Production capacity
- 2.7. Quality control plan
- 2.8. Research and development
- 2.9. Purchase plan
- 2.10. Production cost
- 2.11. Infrastructure
- 2.12. Labor required

3. **Administrative Component:** This component addresses different themes ranging from the legalization and formalization of the business, to the formation of work teams. To start up a business, team experience and skills are incredibly valuable; these variables are taken into account by investors, partners and funders.

This should ensure the creation of a complementary and multidisciplinary team aligned to the business. It should give a description of their profiles, required academic and professional experiences, and functions.

In addition, you should review and document the procedures and time related to the formalization of the company; this will ensure the launch operates under a legal framework. It is necessary to follow the procedures, permits and legal issues that would otherwise delay the formalization of the company. Entrepreneurs must also clearly identify the regulations governing the business, as well as the entities that should generate these processes for formalization.

On the other hand, it proposes what will be the type of business organization; this should be the result of a process of discussion with the Chamber of Commerce, which is the entity where a company must sign a formal registration. Here entrepreneurs can analyze differences in legal forms that exist to formalize the business, and understand the associated costs, fiscal and tax implications and responsibilities arising from the legal form chosen by entrepreneurs to formalize the company.

The preparation of the above two components, both commercial and operational, facilitate identifying the need for human resources for both managerial and operational processes; therefore they are essential to the development of this component.

This component should consider the number of jobs to be generated, considering that in either case, the entrepreneur must be part of the team. According to the enterprise type, they should contemplate the type of contracts and other benefits. Every job position should have a description of functions, as well as the experience needed, profiles and responsibilities.

Having a qualified work team permits:

- Deploy the business strategy,
- Improve competitiveness and productivity.

The administrative component should develop the following aspects:

- 3.1. Strategic Analysis
    - 3.1.1. Corporate values
    - 3.1.2. Mission
    - 3.1.3. Vision
    - 3.1.4. SWOT Analysis
- 3.2. Organizational Structure
    - 3.2.1. Job Profiles and functions
    - 3.2.2. Organizational Chart
    - 3.2.3. Recruitment and remuneration scheme
    - 3.2.4. Corporate government scheme
- 3.3. Legal Aspects
    - 3.3.1. Legal structure and type of society
    - 3.3.2. Special regimes
- 3.4. Administrative Costs
    - 3.4.1. Personal Costs
    - 3.4.2. Start Up Costs
    - 3.4.3. Annual Administration Costs
- 3.5. Support Organizations and partners

4. **Financial Component:** The financial study and analysis of a business model can be defined as a quantitative summary and analysis of the results of the other components of the screened business model. Financial simulation can be deployed with various programs or software that are easy to use, but the really important simulations are the sources of the information and the interpretation of results. The administrative component should develop the following aspects:
    - 4.1. Financial Projections
        - 4.1.1. Sales forecast

4.1.2. Expenses budget
4.1.3. Income projections
4.1.4. Cash flow statement
4.1.5. Projected balance sheet
4.1.6. Breakeven analysis
4.2. Financial indicators
4.3. Financial references
4.4. Financial evaluation

Phase 3: Liderate

This is a cross-phase in the incubation process and can be develop in parallel with the Validate, Project, and Advance Phases. At this stage, entrepreneurs develop autonomously nine thematic modules where they will find theoretical material, inspiration and instructions for personal reflection.

During the development of the modules, entrepreneurs have group tutoring sessions with a Leadership Specialist Tutor: Introductory Session (Group), Intermediate Session (Group) and Final Session (Group).

Modules and topics to be addressed are shown in Table 4.

Phase 4: Advance

This is the last phase of the incubation process in which the launch of the start-up is monitored. Such monitoring will be held under a system of governance, assigning a board of directors to each of the companies. In this board, Mentors are invited to support entrepreneurs with crucial decision-making.

In this phase, the tutor and the entrepreneur devise a plan for the implementation of the project, based on components developed in the PROJECT phase.

During the PROJECT and ADVANCE phases the tutor can nominate a Start-up to the Development Committee that will evaluate the project by analyzing two documents submitted by the entrepreneur: the executive summary and the investment plan.

The projects presented to the development committee shall meet the following requirements:

- Be nominated by the tutor.
- Have validated the nine blocks of the business model.

Table 4. Modules and topics of LIBERATE phase

Self-Awareness	Optimism, exploration, wonder, contemplation, existentialism, self-image, visualization, empathy
Purpose	Call, meaning, sense
Communication	Listening, immaterial communication
Humble Leadership	Compassion, sharing, humility
Transparency	Trust, honesty, integrity, balance
Resiliency and Intuition	Managing uncertainty, smell, creativity, flexibility
Perseverance	Commitment
Control	Sharing power, delegate
Negotiation	Problem solving, decision making

- Have developed at least one component of the projection of the business model to justify a specific financial need.
- Demonstrate commitment to the program and evidenced soft skills.

The Committee must submit a concept about the project, identifying areas for improvement for the entrepreneur and the project. Additionally, it must check whether the enterprise meets the criteria for capital leverage and make appropriate recommendations about available funds that the entrepreneur can apply to.

Table 5. Incubation route resources, products, and indicators

Resources Needed	Monitoring Platform
	Tutors (Coaches and specialized ones)
	Mentors
	Workplan (VALIDATE, PROJECT and ADVANCE)
	Articulation with ecosystem plan
	Committee executive summary
	Committee evaluation
	Funds record
	PROJECT phase components documentation
	LIDERATE phase Reflections
Products	Validated business model
	Projection of business model (four components)
	Milestones and completed work plan
	Development committee concepts / Capital funds orientation
	Business start up
Indicators	Acceptance rate: number of entrepreneurs who apply versus the number of entrepreneurs accepted into the incubation process
	Number of jobs created by entrepreneurs
	Sales achieved by entrepreneurs at different levels of the process
	Number of entrepreneurs who have received funds in the previous year
	Amount of investments secured by entrepreneurs
	Number of commercial and investment events attended by entrepreneurs
	Number of investors with the intention to invest in the projects
	Number of accompanying mentors
	Number of national allies in the entrepreneurial ecosystem
	Number of international allies in the entrepreneurial ecosystem
	Survival rate: Number of incubated entrepreneurs in operation in the third year after the launch

Part 4: Acceleration

Given the cycle of business development within the framework of the Colombian Public Policy, the role of universities, and the participation of different entrepreneurial actors, the EAN University offers its students and alumni a service to articulate entrepreneurs with the ecosystem, according to the needs of each company.

At this stage, the company maximizes its value added through internationalization and innovation, and focuses its efforts towards consolidating sustained growth. It does so by seeking strategic partners and new customers via networks of national and international contacts and attracting capital.

Accelerating business involves increasing execution speed scalability and achieving direct results. It also means optimizing the commercial, administrative, financial, technological and human resources involved in the company management, and aligning them to a series of triaged objectives and business strategies (Quiroga et. al, 2014, p. 18).

The acceleration process begins with the development of a detailed business assessment, based on which an articulation plan with the entrepreneurial ecosystem is developed (services and offers from other entities in the ecosystem). Additionally, projects will have the option for mentoring and specialized tutoring, which will be provided in some cases by students from different academic backgrounds from various EAN University programs.

Process

The process of accelerating business is focused on entrepreneurs who come to the program for specific advice in different areas or those who have emerged as part of the incubation process, and come back for support for accelerating the company.

The process has different stages, depending on whether it is a company in operation or one that comes from the incubation process with EAN IMPACTA prorgam.

Figure 11. Acceleration process
Source: Elaborated by authors.

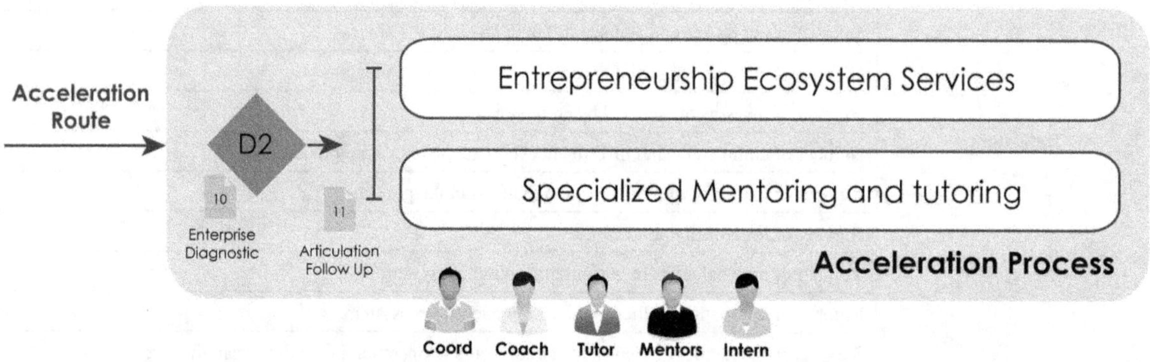

Companies in Operation

Entrepreneurs with companies incorporated into the acceleration process and companies that require acceleration services must comply with the following requirements*:

- Being registered in the chamber of commerce,
- Average annual sales of USD48000,
- Generates at minimum three jobs per year,
- Minimum 2 years of operation.

To start this process, the employer fills out the registration and characterization forms, which must contain the concept of a specialized tutor indicating that it is a company that meets the selection requirements and can be routed to the acceleration process.

In the case of companies that, in respect to the tutor, do not have a clear idea of their needs in the acceleration process to apply for a business assessment, this process identifies the strengths and weaknesses of the company and proposes strategic solutions that can bolster its growth potential.

The business assessment will be made using the model "Modernization for Management Organizations" (MMGO®), which by its general nature is equally applicable in business services, commercial or manufacturing. Its main feature is the level of detail with which it handles in the situational analysis and identification of activities that make up the route change, improvement and conversion to the best practices in the organization. It works systematically, facilitating the establishment of priorities for management modernization projects, and proposing descriptors for activities and elements that the company must have to define the routes that optimize these key elements. Highlighted aspects in a restructuring process in the management of SMEs are (Pérex-Uribe, 2009):

- Monitoring and understanding of the environment,
- Strategic direction,
- Knowledge Management,
- Communication and information,
- Market management,
- Imports and exports,
- Financial management,
- Operations management,
- Human resource management,
- Cultural and organizational structure,
- Associativity,
- Social and environmental responsibility.

According to the result obtained in the diagnostic process, the assessment must define a clear strategic route with the different actors of the entrepreneurship ecosystem. It is necessary to prioritize the weakest areas of the company and the direction the employer should take to strengthen these areas. This process also it provides information about the entrepreneurial ecosystem services.

To define the articulation route, the tutor must identify the five most critical areas resulted from the MMGO® diagnosis, for the organization's development and growth according to the business model.

The entrepreneur and tutor generate a monitoring plan in which the impact that the actions have had on the development of the company are evaluated.

Start-Up Coming from the Incubation Process

Companies that finished the incubation process with EAN IMPACT program that are prepared to continue to the acceleration process, must have an approval from the Development committee. These companies should also meet the requirements set for the *Companies in Operation*. In line with the strategic plan of the company, the articulation plan with the different actors of the ecosystem is defined and is monitored by a tutor.

MMGO® diagnostic process and articulation strategic route formulation example:

The table shown below is an example of a MMGO® comprehensive report, which constitutes the basis for the definition of the strategic route.

In this case, the critical areas in which organizations should implement an organizational intervention plan are:

- Human Resource Management,
- Financial Management,
- Social Responsibility,
- Exports,
- Organizational Structure.

In this sense, the strategic route must contain the five areas mentioned above. It must do this without losing sight of the employer's obligation to strengthen the processes that showed good results.

To define specifically weak issues, tutor should refer to the report of each area and identify critical issues again. For example:

In this case the areas to strengthen are:

- Product and service differentiation,
- Business export,
- Outward oriented talent.

In this sense, the workplan should be oriented to:

- Strengthen the value proposition of the product,
- Design an export strategy,
- Design a program for employees training and a plan to aquire skilled personnel in foreign trade.

The same exercise is made with each of the identified and prioritized activities.

Each strategic activity redefined must have tools and strategies for their development as well as management indicators.

An example of the Strategic Route should be as shown in the Table 6.

Figure 12. MMGO® report
Source: Pérez et al (2009).

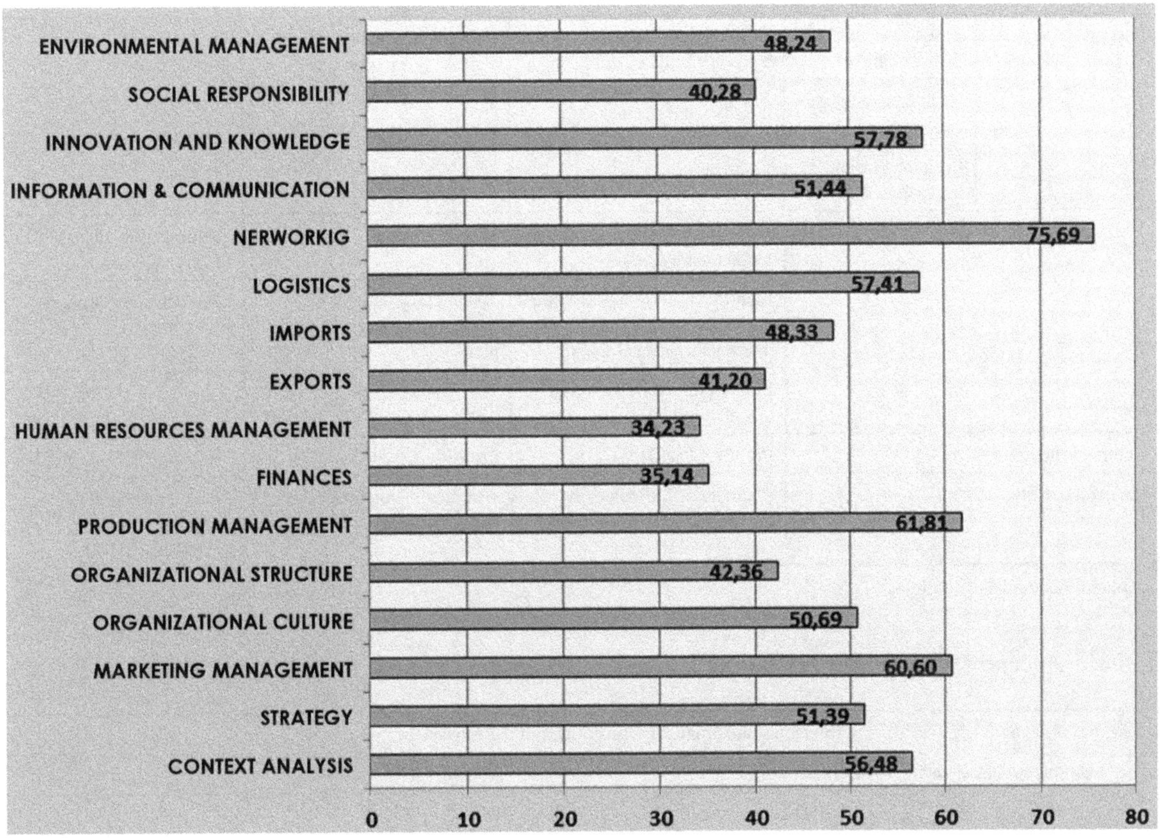

Figure 13. MMGO® exports report
Source: Pérez et al (2009).

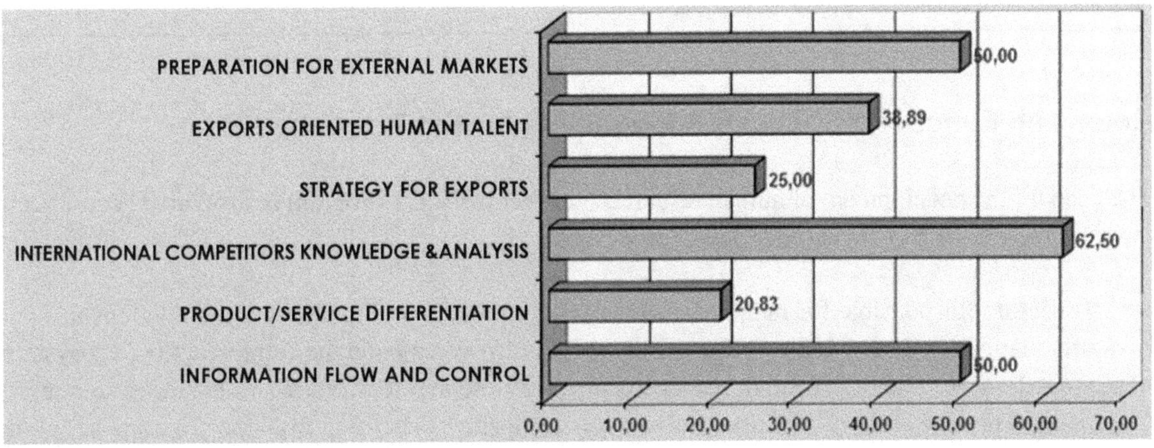

Table 6. Strategic route

Critical Area	Strategic Activities	Strategic Objectives		
Exports	Product or service differentiation	Develop and validate a product with a clear value proposition for the international market	Design and implement a strategy for company internationalization	Train 50% of employees in foreign trade
	Business strategy to export			
	Personnel oriented for foreign trade			
Tools		Analysis of international market tendencies	Market research	Training agreement with SENA
		Identification of opportunities in the company's sector and in the international market	Technical assistance in designing strategic export plan	Conduct training course for exports
		Prototyping, validation and business model		
Actors of the Ecosystem		PROCOLOMBIA	PROCOLOMBIA	SENA / PROCOLOMBIA
Indicators of Transaction		(Number of Validated and prototyped products/1)x 100	1 Strategic plan for design and implementation	(Number trained employees/ No. Total employees) x 100

Source: Elaborated by authors.

Table 7. Acceleration process resources, products, indicators

Resources Needed	Agreements and alliances with entities in the ecosystem
	Formats
	MMGO Diagnostic tools
Products	Companies clearly diagnosed and prioritized for business acceleration
	Clearly defined Strategic Route tailored to each company
	Articulation EAN-Eecosystem to support entrepreneurs.
Indicators	Number of enterprises articulation strategic route versus number of companies diagnosed with MMGO.
	Established connections of employers/businesses with actors in the ecosystem
	Strategic indicators for each company

Roles and Functions of Those Responsible for the Program

The need for the development of human resources in EAN IMPACT program is structured according to the roles the team will play in the different processes.

- **Director:** Responsible for monitoring the program process and facilitating the relationships and articulation with other actors and areas of the EAN University and the entrepreneurial ecosystem.
- **Coordinator:** Offers executive management for all the services offered to businesses/entrepreneurs. This role reviews, monitors and controls companies benefiting from the program, assigning responsibilities to tutors and management within the University and with the ecosystem to facili-

tate the service. The coordinator should monitor the implementation process and carry out the actions necessary to ensure the sustainability and the program improvement.

- **Administrative Assistant:** Responsible for monitoring and deploying administrative processes to facilitate the services delivery to entrepreneurs, supports the program coordinator.
- **Tutors:**
    - **Coach:** Tutor delegated to the accompaniment of a project from inception to finish in the incubation process. They will also be allocated to companies in the acceleration process to accompany the diagnosis and monitor the strategic plan.
    - **Specialized Tutors:** Expert on specific topics of the pre-incubation and incubation processes, to accompany entrepreneurs in the design, development and validation of its business models, the construction of the business plan and the implementation of the company.
    - **Mentors:** Businessperson who is linked to the program and voluntarily accompany entrepreneurs to established monitoring boards and other activities. They oversee the entrepreneur's strengthening and consolidation as a business man.
    - **External Consultants:** these are those specialized professionals who are not part of the University but are intended to support the entrepreneur in certain topics that are crucial for their venture, in very specialized fields or those in which the University lacks specialized human resources.
    - **Incubation Committee Participants:** integrated by the program coordinator, IDELI Director, a coach and a specialized tutor if necessary. The committee have to review the diagnosis forms and the interview results, which should contain a technical concept of a specialized tutor if necessary, and define the route that each entrepreneur should take in the EAN IMPACT program.
    - **Development Committee Participants:** Integrated by IDELI director, a tutor different from the project coach (from any faculty), two executives of the University (Vice-Rectors, general ward members or members of the Board of Governors), and at least one mentor. This committee will meet once a month if there are at least four (4) entrepreneurs nominated.

CONCLUSION

The modern dynamics of change and development demand entrepreneurs to develop skills that go beyond the technical aspects of a business; the entrepreneur must combine the knowledge of how to do something, with the necessary bases, and also be able to manage the company and plan for the future, from the inception of the company.

Accordingly, entrepreneurship programs or units are obliged to stay constantly updated and generate spaces and tools for entrepreneurs to facilitate the success of their initiatives. In this sense, the university has the responsibility to provide the necessary theoretical basis, and either by itself or in partnership, provide practical business tools required to strengthen the enterprises.

That is why from the university in particular, the entrepreneurship training and support model should be constantly evolving, aligning with a changing local and global context, and developing new methodologies to learn from the individual experiences of every entrepreneur.

Furthermore, training in entrepreneurship should be practical (hands on) and as a fundamental element, followed by concrete actions to promote entrepreneurship, enabling relevant connections with the ecosystem to leverage and collaborate with entrepreneurial initiatives.

In the same way that there is a commitment from educational institutions to the processes of entrepreneurship, there must be tangible support and actions from the ecosystem, offered in a clearly structured and purposeful way; otherwise the offering is disassociated from the educational engine and fail to leverage resources needed to fulfill the value promised and to foster the role of universities in the entrepreneurial ecosystem.

The educational processes in general, but particularly institutions of higher education, do not have the option to stay out of these initiatives. In cases in which there is funding support, it is welcome, but in cases in which there is not, the responsibility to support these initiatives is no less. It is the same society that faces the same challenges that demand solutions; while entrepreneurship is part of the solution, it is not the only one.

Finally, as stated by Patino (2016), the responsibility of eradicating social problems such as poverty and business informality, is shared between different actors in the social fabric: the government, society in general, and education institutions. These institutions, like EAN University, are expected to provide both theoretical knowledge and an approach to the challenging reality in which entrepreneurs live, to remove the obstacles so that entrepreneurial initiatives and business developments can truly impact our world.

ACKNOWLEDGMENT

We would like to express our special thanks to Nicholas Moran Brown and Lincoln Howard Mitchell. They have contributed to the present chapter by translating it from Spanish to English.

REFERENCES

Ambrose, G., & Harris, P. (2010). *Basics Design 08: Design Thinking*. London: Bloomsbury Publishing.

Audretsch, D., & Thurik, R. (2001). *Linking Entrepreneurship to Growth*. OECD.

Borjas, G., & Bronars, S. (1989).Consumer Discrimination and Self-Employment. *Journal of Political Economy*, *97*(3), 581–605. doi:10.1086/261617

Charney, A., & Libecap, G. (2000). *Impact of entrepreneurship education*. Kauffman Center for Entrepreneurial Leadership.

Colombo, M., & Delmastro, M. (2001). Does InternetMake a Difference? *Small Business Economics*, *16*(3), 177–190. doi:10.1023/A:1011127205758

Cowling, M., & Taylor, M. (2001). Two DifferentSpecies? *Small Business Economics*, *16*(3), 167–175. doi:10.1023/A:1011195516912

Delmar, F., & Davidsson, P. (2000). Where do they come from? Prevalence and characteristics of nascent entrepreneurs. *Entrepreneurship and Regional Development*, *12*(1), 1–23.

Fayolle, A. (1998). *Teaching of entrepreneurship: Outcomes from an innovative experience*. Internationalizing Entrepreneurship Education and Training Conference, Oestrich-Winkel, Alemania.

Fayolle, A., & Ulijn, J. (2001). *Comparing entrepreneurial and innovation cultures: The European perspective of French, German and Dutch engineers, some empirical evidence about their technology versus market orientation*. The Future of Innovation Studies Conference, Eindhoven, The Netherlands.

Finkle, T., & Deeds, D. (2001). Trends in the Market for Entrepreneurship Faculty. *Journal of Business Venturing*, (16): 613–663.

Free Personality Test. (2016). Retrieved on 20th may 2016 from https://www.16personalities.com/free-personality-test

Kantis, H., Ishida, M., & Komori, M. (2002). *Empresarialidad en economías emergentes: Creación y desarrollo de nuevas empresas en América Latina y el Este de Asia*. Banco Interamericano de Desarrollo, Departamento de Desarrollo Sostenible, División de Micro, Pequeñas y Medianas Empresas.

Karcher, B. (1998). *Does Gender Really Matter? The influences of gender and qualifications on self-employment and their implications on entrepreneurship education*. Internationalizing Entrepreneurship Education and Training Conference, Oestrich-Winkel, Alemania.

Kolvereid, L., & Moen, O. (1997). Entrepreneurship among business graduates: Does a major in entrepreneurship make a difference? *Journal of European Industrial Training*, *21*(4-5), 154–157.

Lafuente, A., & Salas, V. (1989). Types of entrepreneurs and firms: The case of newspanish firms. *Strategic Management Journal*, *10*(1), 17–30. doi:10.1002/smj.4250100103

Levie, J., Brown, W., & Steele, L. (2001). How entrepreneurial are Strathclyde Alumni? In International Entrepreneurship: Researching New Frontiers Conference. Universidad de Strathclyde.

Litvak, I., & Maule, C. (1976). Some perspectives. *Journal of International Business Studies*, *7*(1), 31–38. doi:10.1057/palgrave.jibs.8490692

Lüthje, C., & Franke, N. (2002). *Fostering entrepreneurship through university education and training: Lessons from Massachusetts Institute of Technology.Second Annual Conference of European Academy of Management, Suecia.*

Matiz, F. (2006). Emprendimiento Como Un Pilar para La Competitividad de las Naciones. *Revista EAN*, (59), 119-129.

Matiz, F. (2009). Investigación en Emprendimiento, un reto para la Construcción de Conocimiento. *Revista EAN*, (66), 169-182.

NBIA. (2001). *Best Practice in Action: Guidelines for Implementing First Class Business Incubation Programs*. NBIA.

OECD. (1999). *Fostering Entrepreneurship*. Paris: OECD.

OECD. (2000). *A new economy? The changing role of innovation and information technology in growth*. Paris: OECD.

OECD. (2001). *Entrepreneurship, Growth and Policy*. OECD DSTI/IND.

Patiño, O. (2016). Key Aspects to Develop Long-Term Microfinance and Financial Inclusion for Social Development. In Handbook of Research on Social Entrepreneurship and Solidarity Economics (pp. 260–281). Hershey, PA: IGI Global.

Pérez-Uribe, R. (2009). *Modelo de Modernización para la Gestión de Organizaciones MMGO*. Bogotá: Universidad EAN.

Quiroga, R. et al. (2014). *Panorama del ecosistema del emprendimiento de Bogotá*. IMPACT-A.

Reynolds, P., Camp, S., Bygrave, W., Autio, E., & Hay, M. (1999). *Global Entrepreneurship Monitor: Executive Report*. Kauffman Center for Entrepreneurial Leadership/Babson College/London Business School.

Ries, E., & San Julián, J. (2013). *El método lean startup: cómo crear empresas de éxito utilizando la innovación continua*. Barcelona: Deusto.

Robinson, P., & Sexton, E. (1994). The effect of education and experience on self-employment success. *Journal of Business Venturing*, *9*(2), 141–157. doi:10.1016/0883-9026(94)90006-X

Tackey, N., Perryman, S., & Connor, H. (1999). *Graduated business star-ups*. The Institute for Employment Studies, Reino Unido. UK Business Incubation (UKBI). Retrieved from: www.ukbi.co.uk

Upton, N., Sexton, D., & Moore, C. (1995). *Have we made a difference? An examination of career activity of entrepreneurship majors since 1981. In Trabajo presentado en la Entrepreneurship Research Conference*. Babson College.

Varela, R., Moreno, J., & Bedoya, M. (2015). *Global Entrepreneurship Monitor 2006 – 2013*. Cali: Universidad ICESI.

Veciana, J. M. (1999). Creación de empresas como programa de investigación científica. *Revista Europea de Dirección y Economía de la Empresa*, *8*(3), 11–36.

Veciana, J. M. (2002). *Función Directiva*. Alfaomega, Universidad Autónoma de Barcelona.

Vesper, K. H. (1996). *New Venture Experience*. Seattle, WA: Vector Books.

Chapter 11
E-Learning Solution for Enhancing Entrepreneurship Competencies in the Service Sector

Constanţa-Nicoleta Bodea
Bucharest University of Economic Studies, Romania

Stancu Stelian
Bucharest University of Economic Studies, Romania

Radu-Ioan Mogos
Bucharest University of Economic Studies, Romania

ABSTRACT

The chapter proposes an e-Learning solution for the entrepreneurship education, based on several simulation modules integrated into a classical learning management system. The originality of the approach is that the solution is domain independent and applied advanced IT technologies, such as agent-based simulations and extended graphical support. Using this e-learning solution, the students can learn how to choose characteristics/aspects for particular type of business and how important is each of them according specific criteria; how to set realistic values for different characteristics/aspects of the business, how a business scenario can be changed in order to fit better to the business context (business reality), modeled through by the scenario pattern and how to assess/evaluate business scenarios.

DOI: 10.4018/978-1-5225-0953-0.ch011

INTRODUCTION

The increasing role of the service sector for the economic growth, competitiveness and employment was revealed by studies which were carried on in the last 10-15 years (de Jong, Bruins, Dolfsma & Meijaard, 2003; Howells & Tether, 2004; Tether, 2005; Jose Silva et.al, 2011). Considering the importance of this sector, it becomes important to identify the main factors of the innovative capability of the service organizations. Entrepreneurship, as a major source of innovation allows these organizations to undertake their role (Harper, 2003; Barth, Yago & Zeidman, 2006; Vicens & Grullón, 2011; Ragalie et al., 2007). The term of entrepreneurial innovative capability was introduced by Jose Silva et.al (2011) to integrate the various components arising from the innovation process, namely product innovation, process innovation, organizational innovation and marketing innovation. Entrepreneurial innovative capability varies from one organization to another, and is determined by a number of factors, the most important one being the individual competencies, as the application of knowledge, skills and abilities in order to achieve the desired results. Besides the individual competencies, there are team competencies, that address the collective performance of individuals joined toward a purpose and the organizational competencies, which address the strategic capabilities of a self-sustaining unit of people (IPMA, 2015).

By enhancing the individual competencies, the entrepreneurial education and training has a positive impact on the entrepreneurial mindset of young people, on their intentions towards entrepreneurship, their employability and finally on their role in society and economy (Audretsch & Thurik, 2002), (Valerio, Parton & Robb, 2014). E.U. (2012) concludes that the objectives of entrepreneurship education should be the improvement of the young people's entrepreneurial competences, and increasing the role of entrepreneurs in the economy and society at large. European Commission launched several official documents, such as *Entrepreneurship 2020 Action Plan* and *Rethinking Education Communication*, to emphasize the need of embedding the entrepreneurial learning in all levels and sectors of education, including the non-formal one. Before leaving compulsory education, all young people should benefit from a practical entrepreneurial experience (E.U., 2013).

The conceptual framework developed by Global Entrepreneurship Monitor (Amorós & Bosma, 2014) includes the following key entrepreneurial conditions: entrepreneurship education and training (the extent to which competences for creating or managing business are developed by the education and training system at all levels), the availability of financial resources, the presence and quality of governmental programs, R&D transfer (the extent to which national research and development will lead to new commercial opportunities and is available to SMEs), commercial and legal infrastructure (the presence of property rights, commercial, accounting and other legal and assessment services and institutions that support or promote SMEs), market entry regulation (for assuring market openness and dynamics), physical infrastructure and cultural and social norms.

Although the need for entrepreneurial education and training is growing, there are still many obstacles to its development. One of these obstacles is represented by the controversy on the competences repository which should be taught in the entrepreneurship programs and courses and the extent to which they are really teachable. Entrepreneurship competences are often considered to be only a subset of managerial competences and, for this reason, extensive managerial subjects are included in the entrepreneurial educational programs. But, entrepreneurship can occur outside the corporate sphere, making the managerial competences as being in some context less important. Even if there are common elements, specific entrepreneurial competencies can be still identified and firstly considered by every educational program or course on entrepreneurship.

Caird (1992) identified as the most important entrepreneurship competences the communication and career skills, knowledge, attitudes and personality variables. Gibb (1998) considers that entrepreneurial skills include intuitive decision making, creative problem solving, managing interdependency on a know-who basis, ability to conclude deals, strategic thinking, project management, time management, persuasion, selling, negotiation and motivating people by setting an example. The DUBS model (Cotton & Gibb, 1992) structures the entrepreneurial competences into four dimensions: ideas (the corresponding competences are: opportunity seeking, investigation and creativity), planning (it includes planning and problem solving), doing (it refers to the risk taking, autonomy, commitment, persistence and initiative) and self-awareness (it encompasses the self-awareness, self-confidence, initiative and motivation). Van der Kuip & Verheul (2003) define the following types of entrepreneurial competences: achievement motivation, need for autonomy, creativity, initiative, risk, opportunity seeking or recognition, goal setting, self-awareness, internal locus of control and persistence.

The study on the effects and impact of entrepreneurship programmes in higher education, which was undertaken by the Entrepreneurship Unit of the European Commission, the Directorate-General for Enterprise and Industry, in 2012 (E.U., 2012), has structured the entrepreneurial competences into the following categories: attitude (sense of initiative, risk propensity, self-efficacy, need for achievement, structural behavior), skills (creativity, analysis, motivation, networking, adaptability), and knowledge (understanding role entrepreneurs, knowledge of entrepreneurship). Among these competences, initiative and understanding role entrepreneurs are the most important one.

The competence model promoted by IPMA (2015) includes 29 competence elements, organized in three competence areas:

- **People Competencies:** They define the personal and interpersonal competencies. There were identified the following people competencies: self-reflection and self-management; personal integrity and reliability; personal communication; relations and engagement; leadership; teamwork; conflict and crisis; resourcefulness; negotiation; result orientation.
- **Practice Competencies:** They define the technical aspects of managing projects, programmes, and portfolios. There were identified the following practice competencies: design; requirements, objectives and benefits; scope; time; quality; finance; procurement and partnership; plan and control; risk and opportunities; change and transformation; select and balance.
- **Perspective Competencies:** They define the contextual competences that must be navigated within and across the broader environment. There were identified the following perspective competencies: strategy; governance, structures and processes; compliance, standards and regulations; power and interest; culture and values.

Entrepreneurs should continue develop their ability to define and implement new areas of work, with more integration across projects and more focus on the long term benefits. "The professionals of tomorrow will work in distributed environments with overlapping and often conflicting stakeholder interests; shaped by real time data and performance management tools, challenged with too much information and not enough communication, and judged by their ability to deliver products or services that align with short and long term strategies, to deliver benefits" (IPMA, 2015). For these reasons, the competence profile of the entrepreneurs is changing over time, by redefinition of the competence elements required by the modern entrepreneur.

There are many approaches for developing competencies, the individuals or the organization adopting some of them based on the preferences, the situation and the availability of resources. The interactions between individual, collective and organizational competence development offer different approaches to the development of individual competencies. Competence development are achieved mainly through education and training. But it also happens when individuals perform activities according to their specific roles and thus acquire new knowledge, skills and abilities. They interact with others and could ask for respective knowledge, could exchange experiences and / or support each other in performing the activities in a project, programme or portfolio. A community of practice is one example for individuals interacting in a formal or an informal way and collectively developing their competences. The individual could use a community of practice to facilitate learning through discussing, experimenting and reflecting all kinds of practical issues. It is also a means to feed information back to the embedding organization that could make use of the lessons learned in other projects. According to (IPMA, 2015), the main approaches for developing competencies are the following:

- **Education and Training:** Attending a seminar, lectures, and training sessions, where the trainer delivers specific know-how. This could be done through a presentation, interactions between the participants and the trainer as well as using case studies, group exercises, and simulation games.
- **Self-Development:** Reading books, standards, case studies and articles helps to gain knowledge, reflecting on the application in practical situations and deriving learnings from that. Other ways of self-development are studying, experimenting, trying things out or learning by doing. The latter helps to gain experience in a certain context or to develop certain skills.
- **Peer-Development:** Reflecting with colleagues on how things are going, asking for a feed-back on the own performance and ways to improve it. Learning partners from different disciplines could help to see a situation from a different angle and lever the development to the benefit of both peers, e.g. one through the questions asked and the other through the insights provided.
- **Coaching and Mentoring:** Getting feedback, advice and support by a coach, leader or mentor whilst performing certain activities or striving to develop specific competences. Typically, a coach, leader or mentor is an experienced person that does not deliver direct answers, but challenges the individual through questions that draw the attention to certain aspects and re-quires finding an adequate answer.

In order to achieve the competence development, several prerequisites should be considered and if possible fulfilled. Firstly, the actual situation and the target state of the individual competencies should be known. Secondly, there should be access to expertise (e.g. know-how, experts) and sufficient resources (e.g. budget, time). It is important to create a culture, in which the development of competences is perceived to have added value.

Developing personal and interpersonal competencies and strengthening personal attitudes required by the entrepreneurship through education and training may be a very difficult task (Bodea, Mogoş & Dascălu, 2012). It is acknowledged that students should experience the concept of entrepreneurship, rather than just learn it in conventional manner (Gibb, 2002). According to (Van der Kuip & Verheul, 2003), the learning approach should be process and student-oriented, with teachers acting as facilitators rather than experts, organized in flexible and interactive sessions with negotiated learning objectives and multidisciplinary problem focus. The student-centered and competence-based programs increase the students' employability (Starta-Etan, 2003; Kohn, 1997). The students should be engaged with what they are doing, in order to achieve a deeper understanding of the domain. Passman (2001) studied the benefits of the student-centered model of teaching, and found out that the learning process is improved if the students have to make decisions and to manage their consequences.

The personal and interpersonal competencies for entrepreneurship cannot be taught in a classical educational and training setting, using the common methods for knowledge transfer. Instead, the entrepreneurship should be taught in an active and experiential way, stimulating students to think creatively and to act entrepreneurial. Van der Kuip and Verheul (2003) argue that for facilitating entrepreneurial learning, the principles of theory of generative learning and experience-based learning should be systematically applied. The theory of experience-based learning (learning by doing) promotes the active involvement of the students in real-life problem-solving situations with a high personal relevance (Wittrock, 1990; Kourilsky,1996). For facilitating the entrepreneurial learning, the learning environment should stimulate students to get involved in different activities, by experience in real-life context, risk taking and making mistakes, by creative problem solving, by feedback through social interaction, by role playing. The learning environment should help students to discover themselves how to learn, only support them in designing their own learning process. Entrepreneurship does not necessarily involve a specific school subject; but rather it requires a way of teaching, with experiential learning and project work playing the main role (E.U., 2013).

The chapter discusses the role of education and training in strengthening entrepreneurship and the challenges for entrepreneurship teaching. After the identification of the most important methods and tools for entrepreneurship teaching, the authors propose an e-learning solution for the entrepreneurship education and training. The proposed solution is built upon the learning management system (LMS) implemented in the Bucharest University of Economic Studies to deliver the online programmes. Some additional modules, mainly for simulation and risks analysis were integrated into the existing LMS.

METHODS AND TOOLS FOR TEACHING ENTREPRENEURSHIP

Several methods and tools are applied for teaching entrepreneurship (Arasti, Falavarjani & Imanipour, 2012; E.U., 2013; Fayolle, 2007). Simulation is considered one of the most important learning methods in entrepreneurship, due to its effectiveness in providing valuable learning experiences. Developing competences through simulation and gaming (board or computer games) is achieved by the interactions and behaviors of individuals shown in such a setting. Often simulations and games represent a mix of approaches, e.g. enabling self-development combined with peer-development and coaching in a training environment. It could also be helpful to combine these approaches based on the previous experiences, the stage of development an individual is in or the possibilities an organization is having in its circumstances.

Simulations performed during the training sessions are usually referred as educational simulations, in order to differentiate them from other simulation activities, such as: experiments for decision support, entertainment and imitation (Orehn, 2009). Simulations began to be used in business education in late 1950s (Gredler, 1996) and, progressively, the usage of games, simulations and case-based models became a well-established paradigm, especially for marketing and strategic policy issues. Educational simulations are considered as one of the most effective methods for teaching higher order skills, such as: analysis, synthesis, creation of knowledge (Leger, et al., 2011). Well-designed educational simulation allows learners to test different problem-solving strategies, to experience the consequences of their decisions and to adjust their actions accordingly, basis for an accelerated competency development process, with a deeper understanding of domain (O'Neil, Wainess & Baker, 2005). Galloway (2009) analyzes the usage of different simulation experiences, such as the role playing, standardized cases, partial task simulation, complex task simulation, integrated simulation, and full mission simulation. He considers the post-simulation debriefing process as being the most important phase of the educational simulation because helps learners to clarify and integrate the simulation experience with previous knowledge.

Adopting simulation as an educational strategy could require a considerable financial effort. The cost of the simulation tools for the simulation labs/learning systems can range from a small to a considerable expense. Glavin (2008) proposes to use the performance pyramid, the base of the pyramid being the factual or didactic knowledge and at the higher levels the skill competencies in order to evaluate the appropriate use of simulation. If the learner is required to show or demonstrate a skill, simulation will likely be an appropriate learning tool.

The simulation labs are preparing students in medical education for practice environment despite limited experience on real cases (Fettig, Robertson & Priem, 2013). For running simulations individually in the simulation lab, the students have to have basic informatics knowledge, skills and attitudes (commitment). The usage of the Elsevier simulation solutions allows getting benefits, such as the consolidated use and familiarity of the students with the publisher sites, access to the e-books and outside assignments online. Also, it reinforces idea of research and learning through technology. In the simulation lab, the students utilize embedded links to websites, able to discuss reasons why sites differ, and explore immediate availability of information. The education materials are also immediately available so students can review information and gain confidence prior to review with the patients. The simulation labs are equipped with simulators/simulation packages and different artefacts emulating the real environments.

Schindehutte (2006) considers the following simulations packages as relevant for entrepreneurship education and training:

- **Threshold Entrepreneur:** (www.prenhall.com/threshold) It develops abilities for start-up enterprises (planning, organizing and controlling the organization). It is available in two versions: Team Entrepreneur (student team) and solo Entrepreneur.
- **The Business Disc:** (http://online.sfsu.edu/~cmgaglio/BizDiscSimulation.PDF) Students are supposed to develop a business plan for a new start-up company and the software make a simulation for one year of operation.
- **GoVenture:** (www.goventure.net) Students are entrepreneurs setting up a new company. Students choose the Quick Start feature, if they want to be assigned to a particular simulated company or choose to start a business of their own choice.
- **VCommerce:** (http://misweb.bus.msu.edu/acc890/resources/vc-EntreGuide.pdf) A small group of entrepreneurs will design and implement some companies that use the web to market goods and services.
- **Venture.SIM:** (http://simventure.co.uk/index.php) The simulation is used in entrepreneurial finance to study alternative strategies, to assess financial needs, to assess risk and uncertainty of cash flows, and to compare different contract terms.
- **Entrepreneur:** (http://home.att.net/~simulations/entre2.htm) The simulation involves the takeover and continuing operation of a retail clothing store. The students act as a management team.

The simulation packages are frequently integrated into existing learning management systems (LMSs) in order to allow the users to define courses with simulations. The Elsevier's Learning Simulation System is a relevant example of LMS integrating simulations. Elsevier offers a powerful e-learning platform that makes the continuing education process simple to use and administer. In 2015, Elsevier announced the personalized learning solution, named Sherpath, during the National League for Nursing Education Summit (Elsevier, 2015). Sherpath represents a nursing educational platform which integrates content, assessment and simulations personalized to each program, course and student. With a seamless, responsive and optimized platform, it provides a tailored learning path at the individual level, guiding the student to success on formative and summative assessments and ensuring a simple and engaging process of genuine learning. Sherpath uses data, analytics and adaptive techniques to track students' interactions with content, assessments and simulations while providing a highly focused learning path for each student. It empowers faculty to easily customize the learning modules to their specific course outcomes. Programs have now real-time, learning analytics to monitor their program against accrediting body standards and student success towards licensure.

Many education and training providers already use an LMS in their organization. Simulation facilities can be integrated, if required. The integration of the simulation facilities can be done at course level or even at higher level (for example, at the platform level). The difference resides in the flexibility offered to the platforms users to add simulations in the educational experiences. When the simulation is considered only at the course level, the simulators are transparent for the user, they only are asking to the course management component to add a specific element (simulation) in the course. When the simulation is integrated at a higher level, the platform users have directly access to the simulators and can run simulation sessions inside, but also outside the regular courses activities.

E-LEARNING SOLUTION FOR ENTREPRENEURSHIP EDUCATION

An e-learning solution was developed in the Bucharest University of Economic Studies (BUES), initially as an independent e-learning platform (Bodea et al, 2015). The e-learning platform is intended to support the development of students' entrepreneurial competences in identification of business opportunities and taking decisions about the product/service to be produced, the start-up strategy and the marketing and selling strategy. The e-learning platform is domain-independent and has three main components:

1. **Module for the Identification of the Business Opportunities:** The component assists students to learn how to identify business opportunities, by using several techniques, such as: screening the web and reading business materials, contacting to people and asking questions The business opportunities identification module is a web-based component, which recommends web documents related to students' business profiles: it was designed using JAVA technologies in NetBeans IDE and can be accessed via any browser.
2. **Module for Developing Business Scenarios:** The component assists students to learn to define a start-up strategy and a marketing strategy. An entrepreneur can decide to produce an existing product/service for an existing market, a new product/service for a new market, a new product/service for an existing market or an existing product/service for a new market. In order to decide what product or service to provide, the entrepreneur should consider many factors, including the idea's market potential, the competition, financial resources, risks and its own skills and interests.

Then, the entrepreneur should decide the start-up strategy, meaning innovation, differentiation, or niche specification. Another important decision is about the marketing and selling strategy to be adopted. The start-up and marketing strategies are considered as parts of business scenario, together with other characteristics of the business initiative. The simulation module is a multi-agent system, with three types of actors and agents (entrepreneur/student, teacher and platform).

3. **Module for Risk Analysis:** The risk analysis module assists the students to quantify and to understand better the effect of risk factors upon their decisions, using the Monte Carlo simulation method. The aim of the simulation is to settle the business strategy and the marketing conditions. The module is linked with the business parameters list identified in the previous components and allows the students to perform successive simulations in order to improve their decisions The simulation consist of running several scenarios with the same target parameter value and different optimistic, most probable and pessimistic values. For each scenario, the student will obtain a current probability to achieve the target value, the probability distribution and the cumulative probability curve. Monitoring the probability trend will help the student to improve entrepreneurial abilities for the selected business scenario in risk conditions and to take better decisions.

After the platform was fully development, it was decided to integrate the components of this platform as modules of LMS already in use in BUES. This integration was decided in order to assure an easier utilization and a large scale adoption of the solution by the university programmes. The BUES utilizes a Moodle-based LMS for promoting the blended-learning approach in all its programmes. The LMS is accessible from the home page of the university (Figure 1) or direct at the address http://online.ase.ro (Figure 2).

The main users are the students and the didactical staff. When a teacher logon on, she/he has access to the list of the courses (Figure 3), being able to update her/his profile (Figure 4) and to update the course format (Figure 5) and resources (Figure 6).

Figure 7 presents the LMS architecture after the integration of the simulation modules for entrepreneurial education.

The simulation modules are integrated at a higher level, in order to allow to the LMS users to have directly access to these modules and to run simulation sessions as part of the courses, but also as separate activities.

The main simulation techniques applied are the agent-based simulation (for the module for developing business scenarios) and Monte Carlo (for the risk analysis module). The agent-based simulation is done by a multi-agent system, with three types of actors and agents (entrepreneur/student, teacher and platform). The evaluation of business scenario is performed as a negotiation process, using the tree representation of the objective which should be achieved, the aspects and criteria defining the business scenario. The Monte Carlo simulation method is linked with the business parameters list and allows the students to improve their decisions. For a chosen parameter, the student will select a risk scenario which will be saved into the database. Each risk scenario is characterized by a target value for the chosen parameter and another three values: optimistic, most probable and pessimistic. The parameter target value is defined by the student taking into account the business strategy. The optimistic, most probable and pessimistic values are defined based on the identified risks. The Monte Carlo method (Bodea & Purnus, 2012) first generates artificial variable values, using a random number generator uniformly distributed in the interval [0, 1] and the associated cumulative distribution function. Then, the Monte Carlo method uses the obtained results to extract values from the probability distribution that describes the behavior of

Figure 1. Login page for BUES's LMS
From the home page of the university.

the stochastic variable. For discrete stochastic variables, the list of possible values and the corresponding probabilities form a discrete probability distribution. In the terminology of probability theory, one can note the stochastic variable as X, xi being a particular value of the variable X. The probability that the value of a variable X equals to xi is denoted as P(X = xi) = P(xi). The probability that the value of variable X exceeds a certain value xi is called the cumulative distribution function and it is denoted as F(xi). The most common theoretical discrete probability distributions are the discrete uniform distribution, the binomial distribution and the Poisson distribution. The risk analysis module can use one of the following distributions: triangular, beta, or log-normal. The simulation consist of running several scenarios with the same target parameter value and different optimistic, most probable and pessimistic

Figure 2. Login page for BUES's LMS (direct)

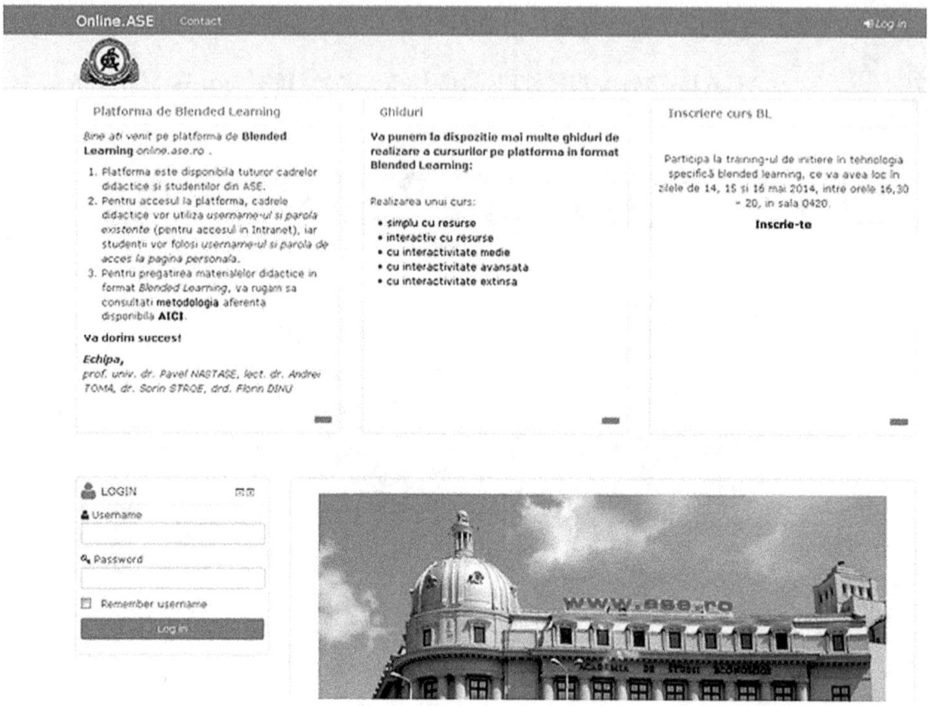

Figure 3. The course list

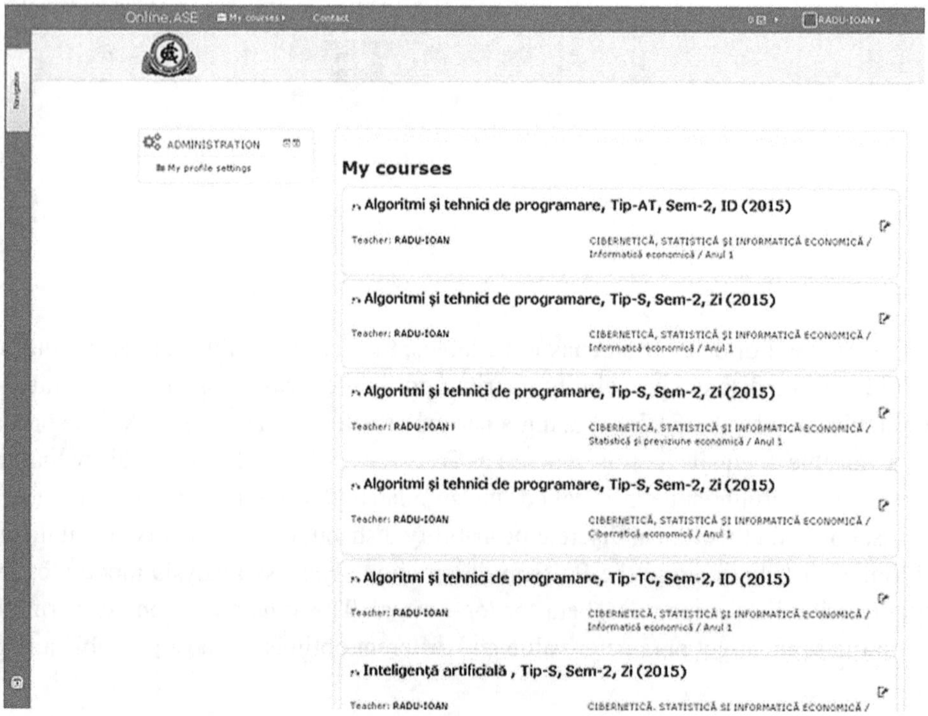

Figure 4. The options for updating the user profile

Figure 5. Updating the course format and resources (course format)

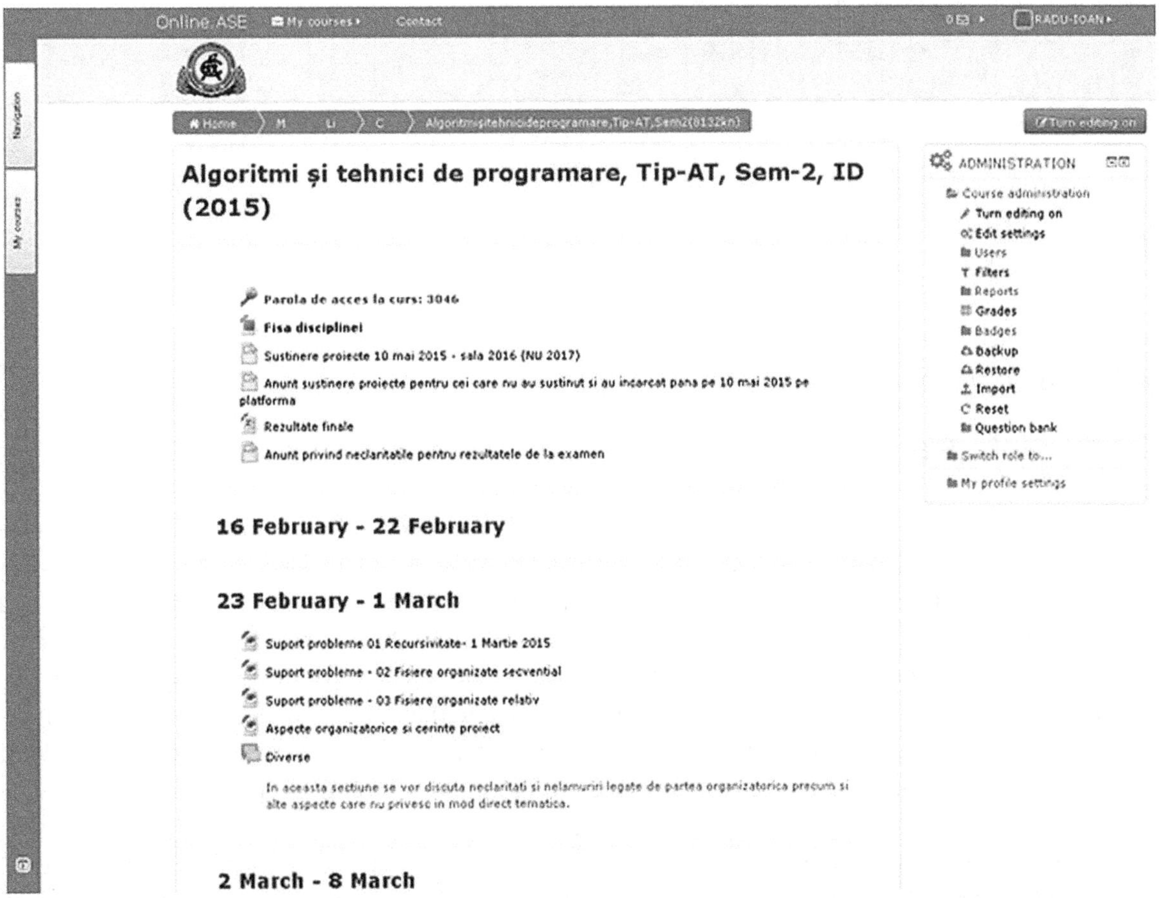

Figure 6. Updating the course format and resources (resources)

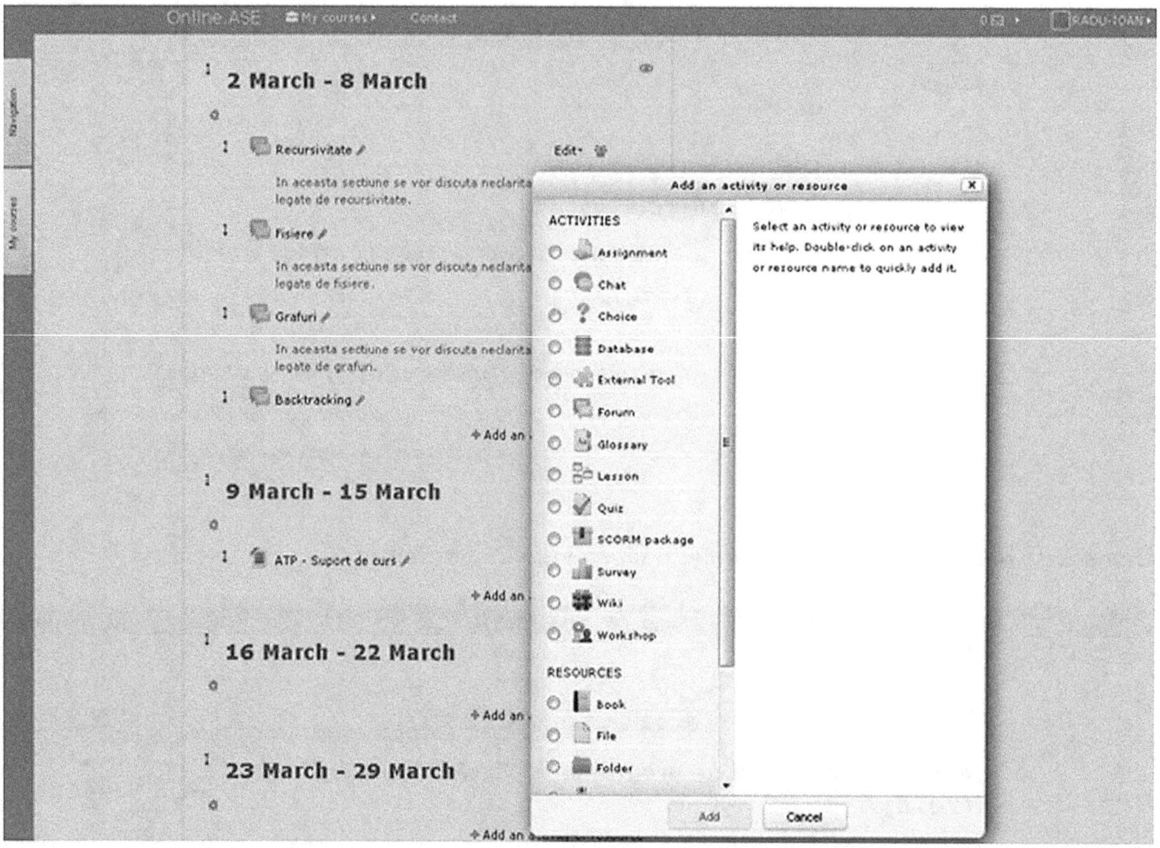

values. For each scenario, the student will obtain a current probability to achieve the target value, the probability distribution and the cumulative probability curve.

Experimental Results

The first experiments started in 2014, for the standard LMS features. The Blended Learning platform (online.ase.ro) become soon an indispensable means facilitating the collaborative interactions between teachers and students. Usefulness of the platform is confirmed by the increasing number of students and teachers who accessed it so far. Since October 2015 until January 2016, the platform was accessed by 8,400 unique users, of which 8184 were students and 216 teachers (25% of the entire didactical staff). This activity has shown the ability of the teachers to stimulate the students engajement in different didactic activities according to the needs and academic expectations that they have. Even if only 25% of the teachers used the platform on the period of the experiment, they were in contact with approximately 40% of the students enrolled in the university's programmes.The average number of unique users who daily accessed the platform is 630, the maximum being reached in January 2016 during the exam period.

Figure 6 presents the number of unique users in the period October 2015 – January 2016. The peak was attended in January, when 6215 unique users were active on the platform.

Figure 7. The architecture of the LMS with simulation modules for entrepreneurial education

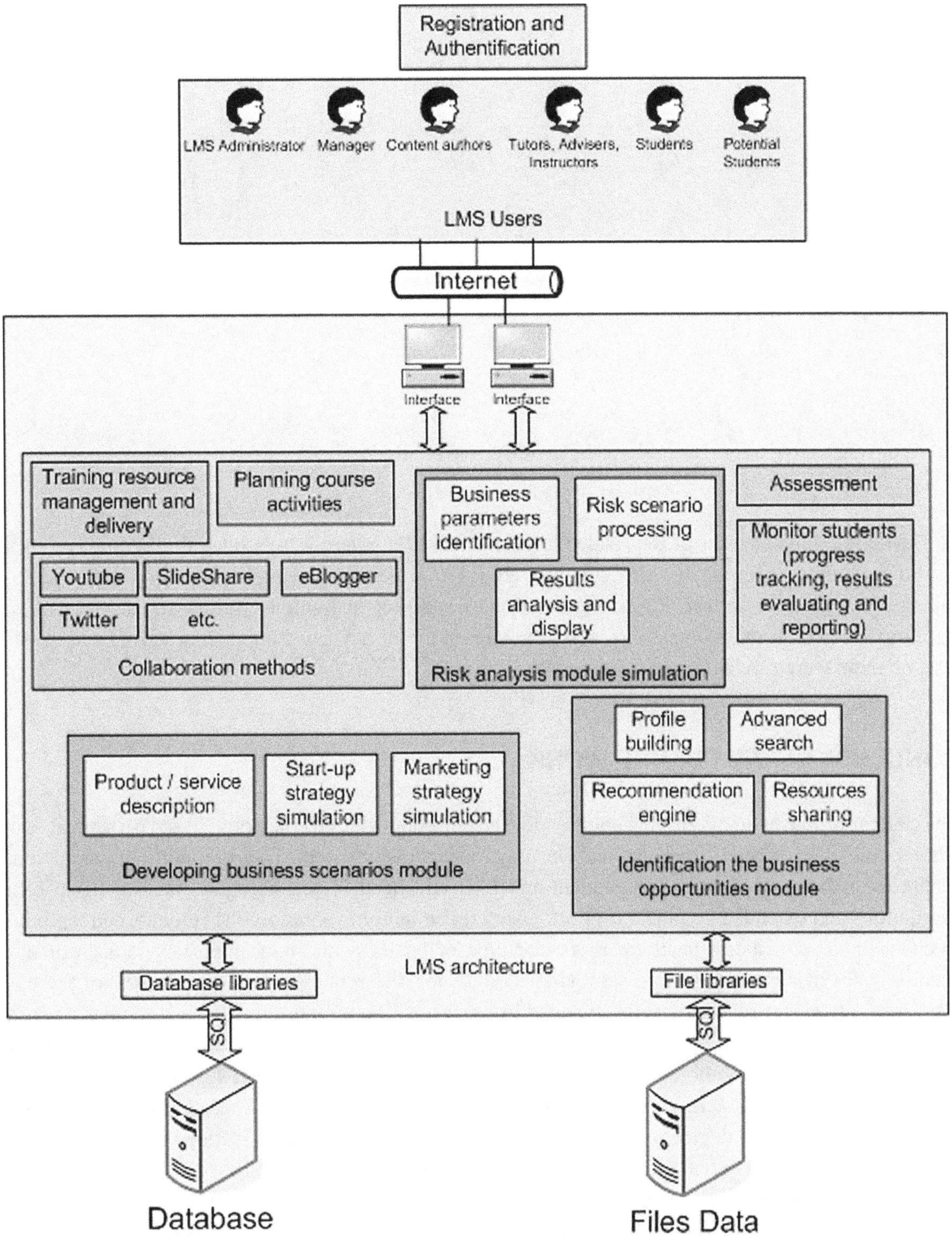

Figure 8. The number of unique visitors on the blended learning platform
Report of the Rector, February 2016, page 29, www.ase.ro.

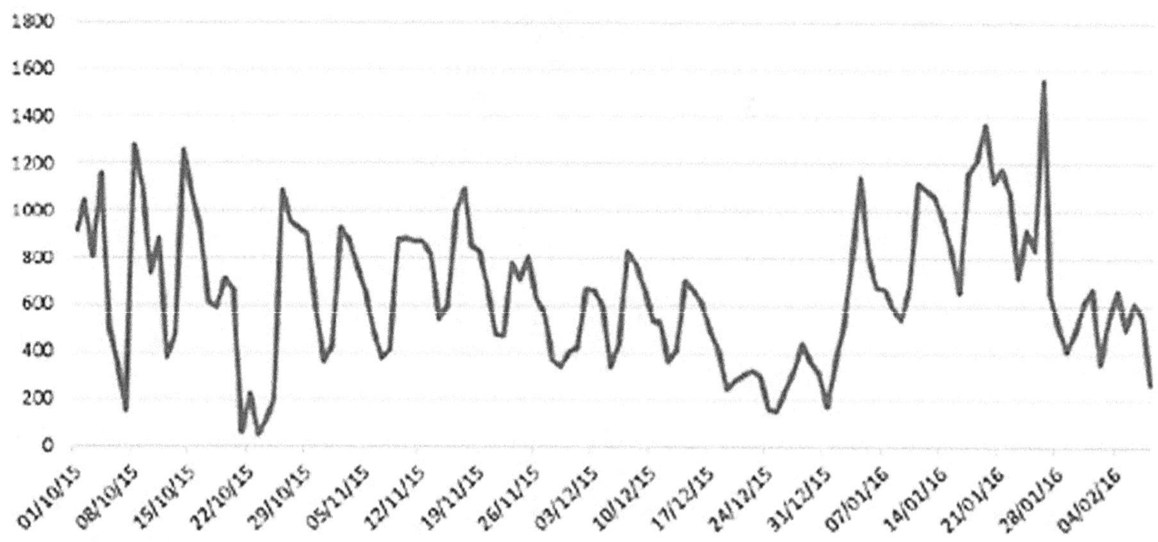

During the fall semester of the academic year 2015-2016, there were loaded on the platform 8745 teaching resources (projects, lectures, seminars etc.), 197 professors had published online tests, and 3678 students had taken these tests. Some methodologies and practical guides on how to use the platform were made available on online.ase.ro. Also, the integration of the new components started and the first tests on them were run, with promising results.

CONCLUSION AND FUTURE WORK

The chapter proposes an e-Learning solution for the entrepreneurship education, based on several simulation modules integrated into a classical learning management system. The originality of the approach is that the solution is domain independent and uses advanced IT technologies, such as agent-based simulations and extended graphical support. Using this e-learning solution, the students can learn how to choose characteristics/aspects for particular type of business and how important is each of them according specific criteria; how to set realistic values for different characteristics/aspects of the business, how a business scenario can be changed in order to fit better into the business context (business reality), modeled through by the scenario pattern and how to assess/evaluate business scenarios. The main strengths of the solution are the integration in a learning management system environment and the extended simulation activities which can be designed and undertook. The main limitations of the solution are the reduced degree of detail and the limitation in regard of information volume and sources types that are accessed in real time (to assure the efficiency of the operation).

As future work, we consider enriching the functionalities of the simulation modules, mainly by assuring a collaborative environment for simulations and adding a data mining component for extracting knowledge about the evolution of the business environment in time, analysing the scenario templates defines in different moments of time. Also, the motivational aspect will be included in the simulation environment.

REFERENCES

Amorós, J. S., & Bosma, N. S. (2014). Global Entrepreneurship Monitor 2013 Global Report. Global Entrepreneurship Research Association, Universidad del Desarrollo, Santiago, Chile.

Amway Global Entrepreneurship Report (AGER). (2013). *Encouraging Entrepreneurs – Eliminating the Fear of Failure*. GfK Nuremberg.

Arasti, Z., Falavarjani, M. K., & Imanipour, N. (2012). A Study of Teaching Methods in Entrepreneurship Education for Graduate Students, *Higher. Educational Studies*, *2*(1).

Audretsch, D. B., & Thurik, A. R. (2002). *Entrepreneurship: Determinants and Policies in a European-US Comparison*. Kluwer Academic Publishers. doi:10.1007/b109395

Axtell, R. (2000). *Why agents? On the varied motivations for agent computing in social sciences*. Working paper 17. Center on Social and economic dynamics, Brookings Institution.

Barth, J. R., Yago, G., & Zeidman, B. (2006). *Barriers to Entrepreneurship in Emerging Domestic Markets: Analysis and Recommendations*. Milken Institute. Retrieved from www.milkeninstitute.org

Bodea, C. N., Mogoş, R. I., & Dascălu, M.-I. (2012). How e-Learning Experiences Enhance the Social Presence in Community of Practices: an Empirical Analysis. In C. El Morr & P. Maret (Eds.), *Virtual Community Building and the information Society: Current and Future Directions* (pp. 75–120). Hershey, PA: Information Science Reference. doi:10.4018/978-1-60960-869-9.ch005

Bodea, C.N., Mogoş, R.I., Dascălu M.-I, Purnuş, A., & Ciobotar, N.G. (2015). Simulation-Based E-Learning Framework for Entrepreneurship Education and Training. *Amfiteatru Economic*, *17*(38).

Bodea, C. N., & Purnus, A. (2012). Project Risk Simulation Methods – A Comparative Analysis. *Management & Marketing Challenges for the Knowledge Society*, *7*(4), 565–580.

Caird, S. (1992). Problems with the identification of enterprise competencies and the implications for assessment and development. *Management Education and Development*, *23*(1), 6–17.

Cotton, J., & Gibb, A. A. (1992). *An Evaluation Study of Enterprise Education in the North of England, Enterprise and Industry Education Unit*. Durham City, UK: Durham University Business School.

De Jong, J., Bruins, A., Dolfsma, W., & Meijaard, J. (2003). *Innovation in services firms explored: what, how and why?* Zoetermeer: EIM, Business and Policy Research.

Elsevier. (2015). *Next-Generation Personalized Learning Solution at NLN Education Summit*. Press Release. Retrieved from http://www.elsevier.com/about/press-releases/clinical-solutions/elsevier-to-introduce-next-generation-personalized-learning-solution-at-nln-education-summit-2015,-sept.-30-oct.-2

EU. (2012). *Effects and impact of entrepreneurship programmes in higher education*. Entrepreneurship Unit, Directorate-General for Enterprise and Industry European Commission.

EU. (2013). *Entrepreneurship Education – A Guide for Educators*. Directorate-General for Enterprise and Industry European Commission.

Fayolle, A. (2007). *Handbook of Research in Entrepreneurship Education, A General Perspective* (Vol. 1). Edward Elgar Publishing Limited. doi:10.4337/9781847205377

Fettig, K., Robertson, L., & Priem, S. (2013). *Informatics in the Simulation Lab; Learn how faculty at Minnesota State College – Southeast Technical in Red Wing and Winona, MN have developed full integration of informatics into Sim Lab*. Minnesota Simulation Conference 2014. Available at http://www.riverland.edu/mnsim/Informatics%20in%20the%20Classroom%20with%20notes.pdf

Galloway, S. J. (2009). Simulation Techniques to Bridge the Gap Between Novice and Competent Healthcare Professionals. *Online Journal of Issues in Nursing*, *14*(2), 31.

Gibb, A. A. (1998). Entrepreneurial core capacities, competitiveness and management development in the 21st century. *Internationalizing Entrepreneurship Education and Training Conference*. European Business School.

Gibb, A. A. (2002). Creating conducive environments for learning and entrepreneurship – living with, dealing with, creating and enjoying uncertainty and complexity. *Industry and Higher Education*, *16*(3), 135–148. doi:10.5367/000000002101296234

Glavin, R. (2008). When simulation should and should not be in the curriculum. In R. Kyle & W. Murray (Eds.), *Clinical simulation: Operations, engineering and management*. New York: Elsevier, Inc. doi:10.1016/B978-012372531-8.50017-0

Gredler, M. (1996). Educational games and simulations: a technology in search of a (research) paradigm. In Technology (Vol. 1, pp. 521–540). MacMillan. Retrieved from http://www.aect.org/edtech/ed1/17/index.html

Harper, D. A. (2003). *Foundation of Entrepreneurship and Economic Development*. New York: Routledge. doi:10.4324/9780203380215

Hindle, K. (2002, June). A grounded theory for teaching entrepreneurship using simulation games. *Simulation & Gaming*, *33*(2), 236–241. doi:10.1177/1046878102332012

Howells, J., & Tether, B. (2004). *Innovation in Services: Issues at Stake and Trends, Final Report*. Brussels: Commission of the European Communities.

IPMA. (2015). *IPMA Individual Competence Baseline ICB4, International Project Management Association*. IPMA.

Jose Silva, M., Sousa, G., Moreira, J., & Simões, J. (2011). Innovation Activities in the Service Sector: Empirical Evidence from Portuguese Firms. *Journal of Knowledge Management, Economics and Information Technology*, (6).

Kohn, A. (1997). Students don't "work", they learn. *Education Week*, (Sept): 3.

Kourilsky, M. L. (1996). *Mini-Society*. Kauffman Center for Entrepreneurial Leadership, E.M. Kauffman Foundation.

Macal, C., & North, M. J. (2009). Agent-based modeling and simulation. In *Proceedings of the 2009 Winter Simulation Conference*. IEEE. doi:10.1109/WSC.2009.5429318

Macal, C., & North, M. J. (2010). Tutorial on agent-based modeling and simulation. *Journal of Simulation*, *4*, 151–162.

O'Neil, H. F., Wainess, R., & Baker, E. L. (2005). Classification of learning outcomes: Evidence from the computer game literature. *Curriculum Journal*, *16*(4), 455–474. doi:10.1080/09585170500384529

Orehn, T. (2009). Modeling and simulation: a comprehensive and integrative view. In Agent-based Simulation and Systems Engineering. Wiley-VCH, GmbH&Co, KgaA.

Owen, S., Anil, R., Dunning, T., & Friedman, E. (2011). *Mahout in Action*. Manning Publications.

Passman, R. (2001). Experiences with student-centered teaching and learning in high-stakes assessment environments. *Education*, *122*(1), 189–199.

Ragalie, S., Neagu, C., Muscalu, M. S., Russu, C., Badileanu, M., Bulearca, M., & Neacsu Dalu, I. et al. (2007). Modalitati de imbunatatire a productivitatii la nivel de organizatie, sectorial si national, *Journal of Industrial Eonomics. Centre for Industrial Economics and Services*, *5*(4), 7–14.

Saaty, T. L. (1994). How to make a decision: The analytic hierarchy process. *Interfaces*, *24*(6), 19–43. doi:10.1287/inte.24.6.19

Schindehutte, M. (2006). *Entrepreneurship Simulations*. Accessed September 2, 2014, http://old.ied.econ.msu.ru/cmt2/lib/c/198/File/simulations.pdf

Starta-Etan Expet Group. (2003). *Higher education and research for the ERA: Current trends and challenges for the near future*. Bruxelles, Belgium: European Commission, DG Research.

Tether, B. (2005). Do Services Innovate (Differently)? Insights from the European Innobarometer Survey. *Industry and Innovation*, *12*(2), 153–184. doi:10.1080/13662710500087891

Valerio, A., Parton, B., & Robb, A. (2014). *Entrepreneurship Education and Training Programs around the World: Dimensions for Success*. Washington, DC: World Bank. doi:10.1596/978-1-4648-0202-7

van der Kuip, I., & Verheul, I. (2003). *Early Development of Entrepreneurial Qualities: the Role of Initial Education. SCALES-paper N200311*. EIM Business and Policy Research.

Vicens, L., & Grullón, S. (2011). *Innovation and Entrepreneurship: A Model Based on Entrepreneur Development*. Inter-American Development Bank. Retrieved from http://www.iadb.org

Wittrock, M. C. (1990). Generating processes of comprehension. *Educational Psychologist*, *24*(4), 345–376. doi:10.1207/s15326985ep2404_2

ADDITIONAL READING

Brew, L. S. (2008). The role of student feedback in evaluating and revising a blended learning course. *The Internet and Higher Education*, *11*(11), 98–105. doi:10.1016/j.iheduc.2008.06.002

Brown, M., Anderson, B., & Murray, F. (2007). *E-learning policy issues: Global trends, themes and tensions*, ICT: Providing choices for learners and learning. Proceedings ASCILITE Singapore, http://www.ascilite.org.au/conferences/singapore07/procs/brown-m.pdf

Casanova, D., Holmes, B., & Huet, I. (2009). *Aiding Academics to move from knowledge management to knowledge creation: Conceptualisation of a Personal Academic Environment (PAE* Proc. MICTE, Research, Reflections and Innovations in Integrating ICT in Education, Badajoz, Spain, 2009 Chen, L.-J, Chen C.C., & Lee W.-R (2008). *Strategic Capabilities and Innovation Intensity. Journal of Service Science and Management, 1*, 111–122.

Commission of the European Communities. (2005). *Proposal for a Recommendation of the European Parliament and of the Council on Key Competences for Lifelong Learning*. Brussels, Belgium: European Commission.

Cruz Natali, A. C., & de Almeida Falbo, R. (2002). *Knowledge Management in Software Engineering Environments*, available at: www.inf.ufes.br/~falbo/download/pub/Sbes2002.pdf 9.

Dittmann, L. (2004). *Ontology-based Skills Management*, Work Report no. 22, Institute for Production and Industrial Information Management, University of Duisburg-Essen, april 2004.

Garcia, A. C. B., Kunz, J., Ekstrom, M., & Kiviniemi, A. (2003). *Building a Project Ontology with Extreme Collaboration and VD&C, CIFE Technical Report #152*. Stanford University.

Garofano, C., & Salas E. (2005). *What influences continuous employee development decisions?*, Human Resource Management Review, Nr. 15, 2005, pg. 281–304

Grau, N., & Bodea, C. N. (2013). *ISO 21500 Project management Standard*. Shaker Verlag GmbH.

Gremmen, H., & Potters, J. (1997). Assessing the efficacy of gaming in economic education. *The Journal of Economic Education, 28*(4), 291–303. doi:10.1080/00220489709597934

Hitt, M. A., Ireland, R. D., & Lee, H. U. (2000). Technological learning, knowledge management,firm growth and performance: An introductory essay. *Journal of Engineering and Technology Management, 2000*(3-4), 231–246. doi:10.1016/S0923-4748(00)00024-2

Kanellopoulos, D., Kotsiantis, S., & Pintelas, P. (2006). Ontology-based learning applications: a development methodology, *Proceedings of the 24th IASTED International Multi-Conference Software Engineering*, February 14-16, Innsbruck, Austria.

Klassen, K., & Wiloughby, K. A. (2003). In class simulation games: Assessing student learning. *Journal of Information Technology Education, 2*, 1-13.

Kröll, M. (2007). Limits and possibilities for competence development in firms. European Conference of Educational Research (ECER). Ghent. Retrieved December 12, 2009, from http://www.vet-research.net/ECER_2007/sep21/session_10c/985/985_pdf/attach/Kroll_ecer2007_p985.pdf

Lundin, J. (2005). *Talking about Work: Desiging Information Technology for Learning in Interaction*. Gothenburg: Gothenburg Studies in Informatics.

Marimuthu, M., Arokiasamy, L., & Ismail, M. (2009). Human Capital Development and Its Impact on Firm Performance: Evidence from Developmental Economics. *The Journal of International Social Research*, *2*(8), 265–272.

Markulla, M. (2006). *Creating Favorable Conditions for Knowledge Society through Knowledge Management, eGovernance and eLearning,* available: http://www.fig.net/pub/monthly_articles/june_2006/june_2006_markkula.pdf, 2006.

Raymond, L., Croteau, A. M., & Bergeron, F. (2009). *The Integrative Role of IT in Product and Process Innovation: Growth and Productivity Outcomes for Manufacturing*, Enterprise Information Systems, 11th International Conference, Berlin Heidelberg: Springer-Verlag, pg. 27-39 doi:10.1007/978-3-642-01347-8_3

Santos, J. (2002). Developing and implementing an internet-based financial system simulation game. *The Journal of Economic Education*, *33*(1), 31–40. doi:10.1080/00220480209596122

Stergiou, N., Georgoulakis, G., Margari, N., Aninos, D., Stamataki, M., Stergiou, E., & Karakitsos, P. et al. (2009). Using a web-based system for the continuous distance education in cytopathology. *International Journal of Medical Informatics*, *78*(12), 827–838. doi:10.1016/j.ijmedinf.2009.08.007 PMID:19775933

Suikki, R., Tromstedt, R., & Haapasal, H. (2006). Project management competence development framework in turbulent business environment. *Technovation*, *26*(5-6), 723–738. doi:10.1016/j.technovation.2004.11.003

Turner, J. R. (1996). International Project Management Association global qualification, certification and accreditation. *International Journal of Project Management*, *14*(1), 1–6. doi:10.1016/0263-7863(96)88794-1

Vannakrairojn, S. (2003). *E-learning Standard,* www.su.ac.th/html_broadcast/2.ppt

KEY TERMS AND DEFINITIONS

Agent-Based Simulation: A modeling technique used to model complex systems composed of interacting, autonomous agents. It is an alternative simulation approach to system dynamics, discrete-event for modeling systems. Agent-based simulation refers to a model in which the dynamic processes of agent interaction are simulated repeatedly over time.

Educational Simulation: The simulation applied in the education processes, to motivate student to engage in problem solving, hypothesis testing, experiential learning, schema construction and development of mental models. Simulation technologies are powerful educational tools which are becoming more widely used because of their effectiveness in providing relevant learning experience and to the increasing availability of the simulation resources, such as: computers, network bandwidth, and software simulator.

E-Learning: A type of distance education in teaching-learning interaction is mediated by an environment set up by new information and communication technologies, in particular the Internet. Internet is both the material environment, as well as the communication channel between the actors involved.

Learning Management System (LMS): A system having the following functions: enrolment and administration of learners; courses and curriculum planning in order to meet the learner's requirements (teaching material, trainer), lessons delivery, development of learner progress reports, communication support, testing and assessment of knowledge.

Simulation: An act of imitating the behavior of a physical or abstract system, such as an event, situation or process that does or could exists. Or: The execution over time of models representing the attributes of one or more entities or processes. Or: A simulation is the implementation of a model over time.

Simulation Based Teaching: A teaching process which include simulations. To get the simulation going, the following components should be developed: the simulation objectives, the simulation scenario, and the game mechanism (the simulation rules).

Chapter 12
Online Facial Symmetry and Achieving Success in a Digital Entrepreneurship World

Spoaller F. Dorin
Babeș-Bolyai University, Romania

ABSTRACT

The second most used page on a newly found website from which a client might buy an item is the ABOUT US page. Using real images of real employees has an overall positive effect on the outcome of that visit. However, it is not always as simple as that. Subtle differences in the content of the pictures may have more of an impact than previously thought. Online advertising has long been considered as a media where creativity is not needed, with the advertising agencies focusing on a few basic rules for displaying ads: call to action, numbers, colors, human figures or symbols and so on. The results are analyzed only on their conversion rate value, this being one of the main reasons why online advertising is being regarded as the future of advertising worldwide. This study is focusing on the Click Through Rate (CTR) impact of different ads using different Facial Symmetry Indices (FSI) on the human figures used. The hypothesis proposed is that a person with a higher FSI, noticeable only at a subconscious level, will generate more clicks on an online ad.

DOI: 10.4018/978-1-5225-0953-0.ch012

INTRODUCTION

Online advertising has taken off in the recent years, having been adopted by the industry as the new ideal method of reaching out to potential clients and having them interact with certain brands. It has seen a constant rise in the budgets allocated to it year after year as the online media consumption is increasing at a steady pace, overtaking offline consumption back in 2013 (Global Web Index 2013).

The problem with online advertising is that it lacks the creativity factor that traditional advertising was using in order to differentiate between brands. Online advertising is characterized by a handful of repeating elements that rarely allow any form if creativity. But where offline advertising fails and online advertising excels is in the brand interaction area. And when you have brand interaction online, you have raw data that you can use to improve your advertising, thus obtaining more brand interaction between your product and the potential client. Which in turn generates even more data and this cycle goes on until you are certain that the ads that you placed online are maximizing your overall profit.

There are many aspects of online advertising that surpass the results that offline advertising obtains. Despite explaining some of them in the chapters to come, one that the author of this paper would mention now and the one that is the pillar of this paper is the ability to implement AB testing at a fraction of the costs if it was a technic used in offline advertising. You could literally have multiple online versions of the same advertisement, and observe which is the version that generates more user interaction, thus eliminating the low performing versions in the process.

The lack of creativity and the ability to use cost effective AB testing has led the advertising industry to seek out new and improved methods of achieving user interaction with a product in the online medium. What the author of this paper is proposing is that the FSI, or Facial Symmetry Index, of a person, even if it's at a subconscious level, will generate different engagement levels from end-users when presented with this aspect in online ads.

To be more precise, the author is proposing that a higher FSI will generate more interaction than a lower one, with a more concise explanation being that people with symmetrical faces will generate more interaction than the ones with asymmetrical faces, at a purely subconscious level of acknowledgement.

This chapter is focusing on the click through rate (CTR) impact of different ads using different facial symmetry indexes (FSI) on the human figures used. The hypothesis proposed is that a person with a higher FSI, noticeable only at a subconscious level, will generate more clicks on an online ad than a person with a lower FSI.

The experiment used in this study shows a clear difference in CTR between a lower and a higher FSI, with women being more likely to click on an advertisement if the span between the two FSI is lower and men only clicking more on one type of FSI if the difference is at the maximum proposed without exiting the subconscious level of the viewer.

A higher FSI is causing a higher CTR with some demographics and it can be implied that the trust expectation of having a higher FSI pictures of employees in a new business environment can have a positive impact in the outcome of that new business. With the online advertising market becoming more influential with every passing year, the focus of advertisers with shift from the basic methods used in the present to new and subtle ways of gaining the engagement of the viewer. The facial symmetry index of the human figures used in this ads will have to be one criteria to be taken into account as it can provide a higher CTR and, as a result, a higher conversion rate.

Term Explanation

The most exciting aspect of online advertising campaigns is that you can use AB testing in order to improve and adapt an ad to perform at its maximum potential. AB testing is, essentially, the same ad run at least twice, to the same targeted audience, but with two changed variables. In online advertising these variables might be different wording, different icons, different people, different social status, hair color, skin color and so on. What makes this testing exceptional in online advertising is the fact that you can do it with a reduced budget, as opposed to offline advertising where this can't be done easily or, in the few cases that it can be done, the cost of it will rapidly rise. Imagine doing an AB testing on a television ad. The cost of production and post-production will double, the cost of media planning will also double and so on.

Facial symmetry is not new to science, but it is something that has never been taken into account in online advertising. The author of this paper believes that the facial symmetry of an individual is the total number of anatomically symmetrical markers in ones' facial aspect. More details regarding these aspects are found in the methodology of the experiment implemented to prove or disprove the hypothesis of the paper.

The facial symmetry index is the grading received by an individual as a result of the facial symmetry markers identified and analyzed in an early stage. A higher facial symmetry index results in the individual having few to none marker measurement variations, whilst a low facial symmetry index results in the individual having as many marker measurement variations as humanly possible, without leaving the subconscious observable ability of the viewer. To be more precise, a low facial symmetry index will be taken into consideration only if it does not result in the individual being tagged as one with a facial handicap. One example of this aspect of the paper would be the former Prime Minister of Romania, Victor Ciorbea. The former Prime Minister is suffering from a clear facial handicap which would result in a very low facial symmetry index, which, in turn, would make him ineligible for this experiment.

By markers, the author of this paper refers to the following: at the eye level there are two measurements taken into account, left corner of the eye and the right corner of the eye, and their distance from the central T zone shape of the face. The next marker taken into account is the cheek bone, followed by the nostril area. The last marker taken into account is the corner of the mouth.

Literature Review

Facial symmetry and the effects that it has on other individuals has been a hot topic in the science community over the last two decades. It has produced a number of papers built around the idea that a more symmetrical face has many implications for the person whose face it is, implications raging from attractiveness to more stable and better paid job opportunities. One of these papers is "Facial symmetry and the perception of beauty" (Rhodes et al. 1998) in which the authors link a perfect facial symmetry to a greater perception of beauty in both males and females. This result is further correlated with a study of perceived better health in individuals with a higher facial symmetry index (Jones, 2001).

These studies prove that an asymmetrical face is detrimental to an individual, but at the same time they prove that a human response, at an unconscious level, is more likely to be positive when confronted with a symmetrical face index of a previously unknown individual. Furthermore, a study published in 2011 showed that facial asymmetry is more common in the general public, and that there is a general understanding of the fact that a higher facial symmetry index is linked to a higher perceived beauty, an ideal that leads to many cosmetic surgeries whose sole purpose is to achieve this goal (Cheong & Lo, 2011).

In advertising facial symmetry is missing in the strategic decisions leading to the publishing of an ad. This is, in part, a consequence of casting actors that will be perceived by the public as beauty icons, and there has been a study that linked successful Hollywood actors to a higher facial symmetry index than the general public (Grammer & Thornhill, 1994).

What is present in advertising is the need to cast male actors with pronounced jaw lines, a factor linked with a higher degree in testosterone and perceived masculinity (Chia, 2008).

Asymmetry is present not only at the facial area, but all over the human body. There is no other organ as important as it is asymmetrical in its understanding of the world as the human brain. In a study published in 1966, "Dual functional asymmetry of the brain in visual perception" (Kimura, 1966), the author established that the brain perceives the environment around it differently depending on the hemisphere that is responsible for the perception. It seems that asymmetry is hard coded not only in our ability to assess and understand the world around us, but also in the small and barely understood realm of one of the most complex creations in the Universe, the human brain.

A more in depth analysis was made in 1973 to assess the nature of the possibility that asymmetry is present even in the ability of the human brain to hear differently with one ear of another. It was established, in the study "Ear asymmetry in perception of emotional non-verbal stimuli" (Carmon & Nachshon, 1973) that as a direct result of the asymmetry found at the brain level, humans have a dominant hemisphere that leads to a dominant ear. This proves that our perception of the world is influenced by internal, but also external stimuli.

Facial symmetry, however, involves many automated systems that are difficult to understand if not researched properly. A study about the systems involved is facial recognition, "Automatic recognition and analysis of human faces and facial expressions" (Samal et all, 1992), has showed that there are many issues and steps that need to be taken into consideration when assessing the ability of a person to recognize a face: detection, identification, analysis, classification, these are all parts of the same system that is involved with facial recognition. Whenever a face is recognized, the number of factors that lead to this recognition is often overwhelming to a layman. This is why the whole process is being put in place at a subconscious level, rather than at a conscious one.

One study that researched the effect of different level lesions on the brain and how they impacted an individual's ability to perceive and recognize a facial aspect is "The neuropsychology of face perception: beyond simple dissociations and functional selectivity" (Atkinson & Adolphs, 2011). It provides a clearer understanding of the way the brain works in recognizing faces and perceiving emotion, thus proving that it is possible for a human brain to identity different levels of symmetry, or asymmetry, on a facial area of an individual. Since the author of this study uses photography as a mean to gather information and data regarding how and if facial asymmetry is perceived by the general public, a study about facial recognition and photography is needed to understand this phenomenon. Thus, the study published in 1974, "Effects of uniqueness, sex of subject, and sex of photograph on facial recognition" (Carroo, 1974) is used to establish that women tend to have a higher recognition ability of other women facial attributes, whereas men tend to identify women and other man equally. This result will have great implications in regards to the final results of this study.

The author of this paper intended to achieve the identification of facial symmetry or asymmetry in others by means of pattern recognition and photo manipulation. In order to achieve this, the study written by Moss JP et all, "Methods of three dimensional analysis of patients with asymmetry of the face." (Moss

JP et all, 1991), was found to be very useful in assessing the overall ability of a computer to identify and establish a definitive result in the field of facial recognition, facial patterns and facial asymmetry. Despite being applied in the field of medical traumas, the study offers a great insight into the ability of others to perceive general asymmetry, or asymmetrical facial aspects, in other individuals. A direct result of this study is the acknowledgement that many, if not all of us, have different levels of facial asymmetry. A study published in 2001, "Functional asymmetry in the human face: perception of health in the left and right sides of the face." (Verner, 2001), studied the involuntary reaction of individuals when faced with different levels of facial asymmetry. This not only proves that attractiveness, successfulness and other aspects of our lives are influenced by our facial traits, but that we are also the ones that, at a subconscious level, determine these traits in other individuals as they are doing to us.

The last two studies analyzed by the author of this paper are directly related to the final purpose of it, "Estimating face pose by facial asymmetry and geometry" (Hu et all, 2004), and "Visual asymmetry in perception of faces" (Gilbert & Bakan, 1973). The first one analyses the best approach to determine the facial asymmetry index of an individual, with the author of this paper having taken into consideration the implications of the results of that paper, and the second one being one of the most well-known studies in the field of facial recognition and facial symmetry, having tackled the necessity of seeing the right side of the face of an individual as the dominant one, given that it resembles the whole of the individuals face more that the left side of the face, if taken out of context and digitally processed.

FACIAL SYMMETRY: ASPECTS, NOTIONS, GRADING

Determining the symmetrical or asymmetrical aspect of an individuals' face requires key placement of markers on key locations. This study uses the placement of 17 markers to determine what the author of this paper calls "Vertical Facial Symmetry / Asymmetry" or "VFS/VFA" and "Horizontal Facial Symmetry / Asymmetry" or "HFS/HFA."

Asymmetry is given by two factors: the first one is the alignment of two identical but opposing markers, as related to the center one; the second one is the length of the two opposing markers, as related to the center one. Vertical misalignment is rarely noticed, but an important factor in assessing and perceiving ones' beauty. Horizontal misalignment is usually related to a deformity or handicap of an individual.

Figure 1. Different stages of measuring the FSI of an individual: pre-adjusting vs. post-adjusting

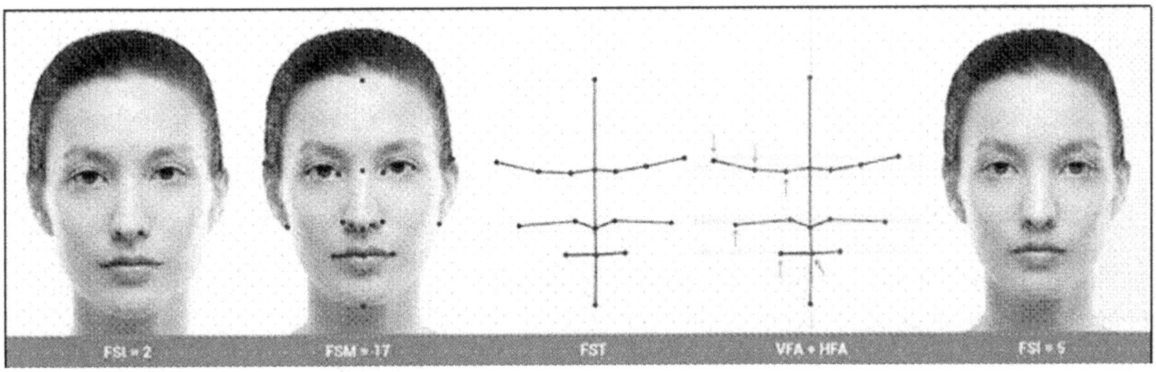

The grading of an individuals' FSI takes into consideration the micro-grading on each level, then combines them to assess and provide a final grading, from 1 to 5, with 1 being a complete asymmetrical face, while 5 being a complete symmetrical face. The study does not take into account any FSI = 1 because it tends to be associated with an asymmetry noticeable at a conscious level.

In the above figure the author of this paper started from a FSI of 2. The placement of the 17 markers, followed by the tracing of the symmetry details, showed a complete horizontal asymmetry and a partial vertical asymmetry. It also showed a multitude of different lengths between identical but opposing markers. The final image is the digitally altered version with a given facial symmetry index of 5.

FSI: Facial symmetry index,
FSM: Facial symmetry markers,
FST: Facial symmetry tracing,
VFA: Vertical facial asymmetry,
HFA: Horizontal facial asymmetry,
VFI: Vertical facial index,
HFI: Horizontal facial index.

The initial symmetry index assessment is calculated by analyzing the vertical and horizontal symmetry indexes and creating a final grading from 1 to 5, with 1 being the lowest and 5 being the highest.

$$FSI = \frac{VFI + HFI}{2}$$

The vertical facial index is given by the number of misalignments of the 3 central markers in the median line of the individuals face.

The horizontal facial index is given by the number of misalignments of the 15 markers placed along the horizontal lines and the existence of different lengths between identical but opposite markers.

Both the VFI and the HFI have a minimum grade of 1 and maximum grade of 5.

$2 \leq \text{VFI} + \text{HFI} \leq 10$

$\text{VFI} = 3 - \text{VFMM}$

VFMM = Vertical Facial Misaligned Markers.

$1 \leq \text{VFI} \leq 3$

If $\text{VFI} = 1 \rightarrow 1 \leq \text{HFI} \leq 9$

If $\text{VFI} = 3 \rightarrow 1 \leq \text{HFI} \leq 7$

The vertical facial index is more important in its relation with the horizontal facial index because of its impact on the handicap aspect of the individuals face.

$$HFI = 7 - \frac{HFMM}{HFMD}$$

HFMM = Horizontal Facial Misaligned Marker,
HFMD = Horizontal Facial Marker Differential.

Since you can have a marker differential without having misaligned markers, but you can't have misaligned markers without having marker differential:

HFMD < 0; HFMM ≤ 0

The maximum number of lines allowed for the horizontal facial misaligned markers is 6. See Figure 2.
The maximum number of lines allowed for the horizontal facial marker differential is 6. See Figure 2.
This leads up to the final form of the FSI:

$$FSI = \frac{\left[\left(3 - VFMM\right) + \left(7 - \left(\frac{HFMM}{HFMD}\right)\right)\right]}{2}$$

This will be representing the logic behind the implementation of the experiment for determining if a higher FSI has any impact over the CTR in online advertising.

Figure 2. HFMM vs. HFMD comparison

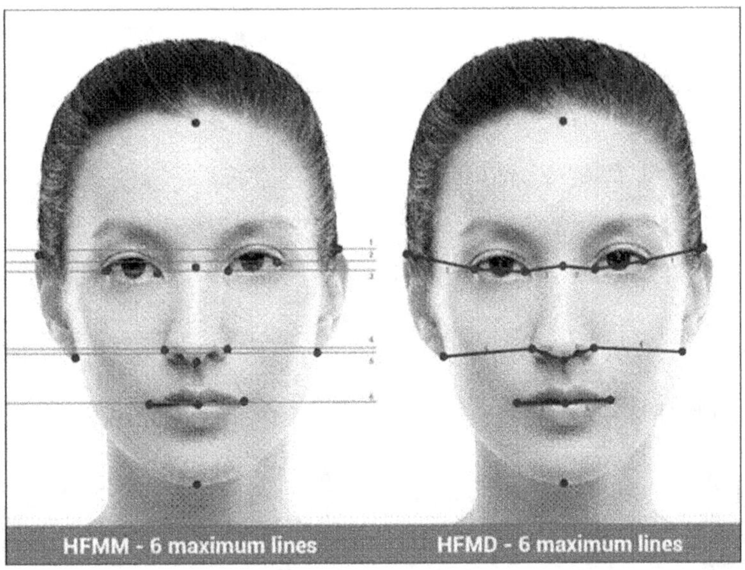

Implementation and Experiment

The author of this paper wanted to prove that a higher FSI will lead to a higher CTR. The experiment implemented was planned specifically to answer this question. The strategic thinking behind it had 2 steps: the first one was generating relevant results through AB testing methods, while the second one was trying to confirm the first results, also through AB testing methods.

The network used to implement this experiment was Facebook. The reason behind this choice involves its dominating market share in Romania regarding social media. It currently has over 7 million active accounts and it allowed the author of this paper to target a wider audience than would be possible to do on any other social media or business site in Romania. The images used (see Figures 3 and 4) in the AB testing had the following elements strategically chosen:

1. **The Background:** The choice of the background took into consideration the need to eliminate any other potential factors that could lead to the invalidation of the final results. The CTR should only be caused by the protagonists of the ads and not by the background elements. This is why the

Figure 3. Step 1, AB testing, FSI = 2; FSI = 5

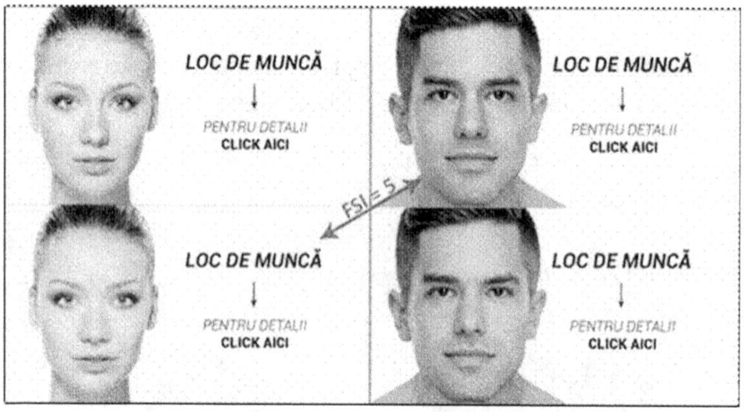

Figure 4. Step 2, AB testing, FSI = 2; FSI = 5

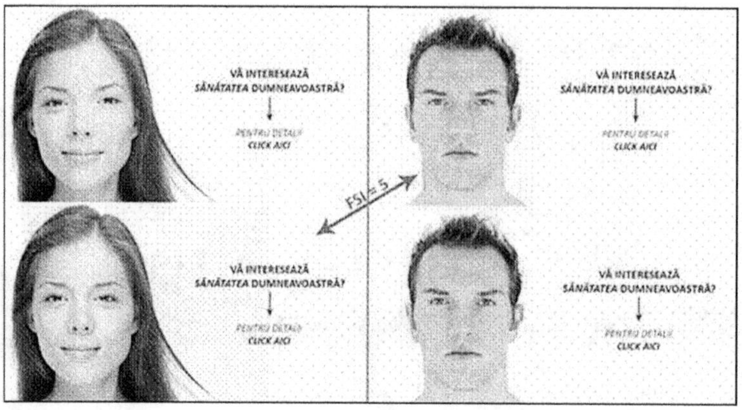

author of this paper chose a simple, minimalistic, white background. This color allows the viewers to focus on the protagonists of the ads. If the author used any other stronger color, the viewers' attention would have been distracted from the protagonists and the CTR would not have been as relevant as a white background.

2. **The Message:** The message used in the ads needed to be simple and concise, but at the same time it needed to represent an expressed interest of the wider audience targeted. The two most sought after ideals of the Romanian people are the financial status and the health status (Ioan Mărginean, 2010). This is the reason behind the message used in the first step of the experiment, "JOB OFFER. CLICK HERE FOR DETAILS", and also in the second step of this experiment, "INTERESTED IN YOUR HEALTH? CLICK HERE FOR DETAILS."
3. **FSI of the Protagonists:** The chosen FSI of the protagonists is 2 and 5. The reason behind these choices is the need to keep the protagonists at an attractive level, in order to keep the CTR symmetry related. Given the level of perceived attractiveness related to the level of symmetry that an individuals' face has, an FSI of 1, compared to an FSI of 5, would lead the viewer to choose between an attractive, perfectly symmetrical face and a less attractive, asymmetrical face of the same individual. If the results would be in favor of the FSI of 5, the validity of this result in relation to the hypothesis of this paper would be questionable.

An FSI of 2 is perceived at a conscious level as being of the same attractiveness as the FSI of 5. A higher CTR of the FSI=5 ad will be strictly related to the perceived subconscious symmetrical aspect of the individuals' face.

This is the reason behind the decision to use FSI = 2 and FSI = 5 on both steps of the experiment. Both individuals' faces are slightly neutral, but at the same time both of them are showing a smile in order. The reason behind this is the need for the viewer to focus on the faces and not only on the message.

4. **Implementation:** The general targeted audience used is the commercial targeted audience by the media: 18-60 year olds, both females and males.

The author of this paper wanted to see the reaction of both sexes. This is why the following strategy was used:

5. **Females Targeted:**
    a. Females targeted → Female ad → Symmetrical face (FSI = 5)
    b. Females targeted → Female ad → Asymmetrical face (FSI = 2)
    c. Females targeted → Male ad → Symmetrical face (FSI = 5)
    d. Females targeted → Male ad → Asymmetrical face (FSI = 2)
    e. Males targeted Males targeted → Female ad → Symmetrical face (FSI = 5)
    f. Males targeted → Female ad → Asymmetrical face (FSI = 2)
    g. Males targeted → Male ad → Symmetrical face (FSI = 5)
    h. Males targeted → Male ad → Asymmetrical face (FSI = 2)

Step 2

Step 2s role is for the author of this paper to obtain confirmation of the results of Step 1. While maintaining certain aspects of Step 1, Step 2 will change others for the results need to be validated, but not under the identical conditions.

What will be kept the same:

The FSI method will not be changed. In order to prove the results at Step 1, the FSI should continue to be FSI = 2 and FSI = 5 for the AB testing.

The targeted audience will not be changed. The author of this paper want to confirm a set of results obtained while targeting this audience and that would not be possible if it changes.

The background will not change. There is no reason to change this element as the reasoning for using it has not changed.

What will not be kept the same:

The message will be changed. This is to provide a wider understanding of the results of Step 1 and how they relate to the FSI and not the message.

The protagonists will be changed. This is to eliminate the possibility of viewer bias towards one of the protagonists used at Step 1.

Result Data: Step 1

As mentioned before, the author of this paper used Facebook Ads in order to implement the AB testing needed. There were two targeted audiences, females of 18 to 60 years and males of 18 to 60 years, what advertisers call a "commercial audience."

This experiment had a total reach of 160 000 people on each step. The market share of anti- advertising applications like AdBlock and AdBlock Plus, in 2013, is 20% (PageFair, 2013), making the final number closer to 130 000 people. The author of this paper takes this variable into account and corrects the final results. It is worth a mention the fact that Facebooks average CTR (Click Through Rate) for paid advertisement is 0.11%-0.16% according to 2015 internal Facebook data.

What these results prove is that women respond to a higher FSI both when presented with a male protagonist and with a female protagonist. The CTR obtained is slightly higher than the average due to the importance of it in regards to the average Romanian.

Table 1. Females targeted with FSI = 2 and FSI = 5, male and female protagonist

Targeted Audience	FSI and Sex	CTR	CTR w/ Adblock	Comparison
Females	F – F = 2	0.19 %	0.23%	-
	F – F = 5	0.26%	0.31%	+36.9%
	M – M = 2	0.21%	0.25%	-
	M – M = 5	0.25%	0.30%	+20%

Table 2. Males targeted with FSI = 2 and FSI = 5, male and female protagonist

Targeted Audience	FSI and Sex	CTR	CTR w/ Adblock	Comparison
Males	F – F = 2	0.25%	0.30%	+14.2%
	F – F = 5	0.22%	0.26%	-
	M – M = 2	0.34%	0.40%	+13.3%
	M – M = 5	0.30%	0.36%	-

What these results prove is that men respond to a lower FSI both when presented with a male protagonist and with a female protagonist. The CTR obtained is slightly higher than the average due to the importance of it in regards to the average Romanian.

Result Data: Step 2

In step 2 the experiment yielded the following results, under the testing condition mentioned in the previous chapter.

Despite having a changed message and despite getting a lower CTR than on Step 1, the results above confirm the general trend of women responding to a higher FSI both in male and female protagonists.

The results are consistent with the ones obtained at step 1, despite having a lower overall CTR, the trend for the men is to identify and interact with asymmetrical faces more that the opposite, at least at a subconscious level.

Table 3. Females targeted with FSI = 2 and FSI = 5, male and female protagonist

Targeted Audience	FSI and Sex	CTR	CTR w/ Adblock	Comparison
Females	F – F = 2	0.14%	0.17%	-
	F – F = 5	0.18%	0.21%	+28.5%
	M – M = 2	0.21%	0.25%	-
	M – M = 5	0.26%	0.31%	+23.8%

Table 4. Males targeted with FSI = 2 and FSI = 5, male and female protagonist

Targeted Audience	FSI and Sex	CTR	CTR w/ Adblock	Comparison
Males	F – F = 2	0.17%	0.20%	+21.4%
	F – F = 5	0.14%	0.17%	-
	M – M = 2	0.27%	0.32%	+12.5%
	M – M = 5	0.24%	0.29%	-

CONCLUSION

The hypothesis of this paper was if the symmetrical or asymmetrical index of an individual has any implication on the click through rate when displayed in an online advertisement. Despite having contradicting results when comparing women responses with those of men, the general consensus on the matter is that they both react to different facial symmetry indexes. This, in turn, proves that the initial hypothesis is correct.

A more profound analysis of the data is necessary to fully understand the reasons behind the contradictory behavior of men and women when confronted with the same elements of this experiment. As an educated guess, the author of this paper believes that women are under more pressure from the society to achieve perfection in regards to the perceived beauty aspect of their image and thus are more likely to respond to a perfectly symmetrical face rather than one slightly asymmetrical, even if the difference is noticeable at the subconscious level. At the same time, men are more technologically able individuals, more accounted with the digitally altered perfection available in the online media, thus reacting to what their subconscious identifies as being more natural than a perfectly symmetrical face.

In regards to the results of the experiment, women react and click, on average, 32% more on an advertisement involving a perfectly symmetrical females' face, and, on average, 21.9% more on an advertisement involving a perfectly symmetrical males' face. Men, on the other hand, react and click, on average, 17.7% more on an advertisement involving a slightly asymmetrical females' face and, on average, 12.9% more on an advertisement involving a slightly asymmetrical males' face.

If applied on the market, given the fact that the total online advertising revenue for 2014 was 49.5

Billion dollars (IAB, 2014), the overall growth in revenue would add up to approximately 9.5 billion dollars, on the United States market alone.

Facial symmetry has long been linked to different perceived responses in the communication sciences over the last few decades, but this study proves that it may be a factor to be taken into consideration in the advertising industry also, especially in the growing online advertisement industry.

It is clear that given the results of this paper a digital entrepreneur should use images that promote itself and its business that have a higher facial symmetry index in order to maximize the chances of success. It is also clear that the decisions that we make in regards to the understanding and perceiving the success of others is made entirely by our subconscious mind and by a set of criteria that it is easy to understand only at an academic level.

The images used on websites, in online advertising, on social media platforms, should be carefully screened before publishing, as they tend to have a higher impact onto the business side of our lives than we like to believe. We are being analyzed and profiled by the powerful processing power of others subconscious and not solely by our achievements and well-groomed overall aspect.

REFERENCES

Atkinson, A. P., & Adolphs, R. (2011). The neuropsychology of face perception: Beyond simple dissociations and functional selectivity. *Philosophical Transactions of the Royal Society of London. Series B, Biological Sciences*, *366*(1571), 1726–1738. doi:10.1098/rstb.2010.0349 PMID:21536556

Carmon, A., & Nachshon, I. (1973). Ear asymmetry in perception of emotional non-verbal stimuli. *Acta Psychologica*, *37*(6), 351–357. doi:10.1016/0001-6918(73)90002-4 PMID:4767606

Going, M., & Read, J. D. (1974). Effects of uniqueness, sex of subject, and sex of photograph on facial recognition. *Perceptual and Motor Skills*, *39*(1), 109–110. doi:10.2466/pms.1974.39.1.109

Cheong, Y.-W., & Lo, L.-J. (2011). Facial Asymmetry: Etiology, Evaluation, and Management. *Chang Gung Medical Journal*, *34*, 341–351. PMID:21880188

Chia, M. (2008). The Aetiology, Diagnosis and Management of Mandibular Asymmetry. *Ortho Update*, *1*, 44–52.

Gilbert, C., & Bakan, P. (1973). Visual asymmetry in perception of faces. *Neuropsychologia*, *11*(3), 355–362. doi:10.1016/0028-3932(73)90049-3 PMID:4792186

Global Web Index. (2013). Retrieved from https://www.globalwebindex.net/

Grammer, K., & Thornhill, R. (1994). Human facial attractiveness and sexual selection. *Journal of Comparative Psychology*, *108*, 233–242. doi:10.1037/0735-7036.108.3.233 PMID:7924253

Hu, Y. (2004). *Estimating face pose by facial asymmetry and geometry*. Retrieved from http://www.ifp.illinois.edu/~hu3/FaceTracking_PoseEstimation/Documents/Head%20pose%20estimation-CameraReady.pdf

IAB. (2014). *Internet Advertising Revenue*. Retrieved from http://www.iab.net/media/file/IAB_Internet_Advertising_Revenue_FY_2014.pdf

Jones, B. C. (2001). Facial symmetry and judgments of apparent health. Support for a "good genes" explanation of the attractiveness–symmetry relationship. *Evolution and Human Behavior*, *22*, 417–429. doi:10.1016/S1090-5138(01)00083-6

Kimura, D. (1966). *Dual functional asymmetry of the brain in visual perception*. Retrieved from http://www.ling.upenn.edu/courses/Fall_2013/ling520/Kimura1.pdf

Mărginean, I. (2010). Calitatea vieții în România: Present și perspective. *Revista Calitatea Vie*, *2*.

Moss, J. P. (1991). Methods of three dimensional analysis of patients with asymmetry of the face. *Proceedings of the Finnish Dental Society. Suomen Hammaslaakariseuran Toimituksia*, *87*(1), 139–149. PMID:2057480

PageFair. (2013). *The Rise of AdBlocking*. Retrieved from http://downloads.pagefair.com/reports/the_rise_of_adblocking.pdf

Rhodes, G., Proffitt, F., Grady, J. M., & Sumich, A. (1998). Facial symmetry and the perception of beauty. *Psychonomic Bulletin & Review*, *5*(4), 659–669. doi:10.3758/BF03208842

Samal, A., & Iyengar, P. A. (1992). Automatic recognition and analysis of human faces and facial expressions. *Pattern Recognition*, *25*(1), 65–77. doi:10.1016/0031-3203(92)90007-6

Verner, R. (2001). Functional asymmetry in the human face: Perception of health in the left and right sides of the face. *Laterality*, *6*(3), 225–231. doi:10.1080/713754415 PMID:15513171

Compilation of References

Abdullah, M. (1993). Konsep malu dan segan orang Melayu (Self-consciousness and the shame concept among Malay people). In A. H. Othman (Ed.), Psikologi Melayu (pp. 244-294). Kuala Lumpur, Malaysia: Dewan Bahasa dan Pustaka.

Abid, A. A., Rahim, M. M., & Scheepers, H. (2011). Experienced benefits and barriers of e-business technology adoption by SME suppliers. *BIMA Publishing Communications of the IBIMA*, *2011*, 1–11. doi:10.5171/2011.791778

Acs, Z. J., & Amorós, J. E. (2008). Entrepreneurship and competitiveness dynamics in Latin America. *Small Business Economics*, *31*(3), 305–322. doi:10.1007/s11187-008-9133-y

Acs, Z. J., & Audretsch, D. B. (1990). *Innovation and Small Firms*. The MIT Press.

Acs, Z. J., & Storey, D. J. (2004). Introduction: Entrepreneurship and economic development. *Regional Studies*, *38*(8), 871–877. doi:10.1080/0034340042000280901

Adamczyk, S., Bullinger, A. C., & Möslein, K. M. (2012). Innovation contests: A review, classification and outlook. *Creativity and Innovation Management*, *21*(4), 335–360. doi:10.1111/caim.12003

Adidam, P. T., Banerjee, M., & Shukla, P. (2012). Competitive intelligence and firm's performance in emerging markets: An exploratory study in India. *Journal of Business and Industrial Marketing*, *27*(3), 242–254. doi:10.1108/08858621211207252

Afuah, A., & Tucci, C. L. (2000). *Internet business models and strategies: Text and cases*. McGraw Hill Higher Education.

Agarwal, S., & Dutta, T. (2015, December). Tanusree Dutta. Neuromarketing and consumer neuroscience: Current understanding and the way forward. *Decision*, *42*(4), 457–462. doi:10.1007/s40622-015-0113-1

Ahmad, N. H. (2007). *A cross-cultural study of entrepreneurial competencies and entrepreneurial success in SMEs in Australia and Malaysia* [Ph.D Unpublished doctoral thesis]. University of Adelaide, Australia.

Ahmad, N. H., Abdul Halim, H., & Mohamed-Zainal, S. R. (2010). Is entrepreneurial competency the 'silver bullet' for SME success in a developing nation? *Interdisciplinary Journal of Contemporary Research in Business*, *2*(1), 217–236.

Ahmad, N. H., & Ramayah, T. (2010). Is entrepreneurial competency and business success relationship contingent upon business environment? A study of Malaysian SMEs. *International Journal of Entrepreneurial Behaviour & Research*, *16*(3), 182–203. doi:10.1108/13552551011042780

Ahmed, M., Chowdhury, A.S.M.R., Ahmed, M., & Rafee, M.M.H. (2012). An advanced survey on cloud computing and state-of-the-art research issues. *International Journal of Computer Science Issues, 9*(1).

Aliyu, M. S., Rogo, H. B., & Mahmood, R. (2015). Knowledge Management, Entrepreneurial Orientation and Firm Performance: The Role of Organizational Culture. Asian Social Science, 11(23). 10.5539/ass.v11n23p140

Ambrose, G., & Harris, P. (2010). *Basics Design 08: Design Thinking*. London: Bloomsbury Publishing.

Amorós, J. S., & Bosma, N. S. (2014). Global Entrepreneurship Monitor 2013 Global Report. Global Entrepreneurship Research Association, Universidad del Desarrollo, Santiago, Chile.

Amorós, J. E., Fernández, C., & Tapia, J. (2012). Quantifying the relationship between entrepreneurship and competitiveness development stages in Latin America. *International Entrepreneurship Management*, *8*(3), 249–270. doi:10.1007/s11365-010-0165-9

Amway Global Entrepreneurship Report (AGER). (2013). *Encouraging Entrepreneurs – Eliminating the Fear of Failure.* GfK Nuremberg.

Appadurai, A. (1990). Disjuncture and difference in the global cultural economy. *Theory, Culture & Society*, *7*(2), 295–310. doi:10.1177/026327690007002017

Arasti, Z., Falavarjani, M. K., & Imanipour, N. (2012). A Study of Teaching Methods in Entrepreneurship Education for Graduate Students, *Higher. Educational Studies*, *2*(1).

Argyris, C., & Schon, D. A. (1974). Theory in action. *The Journal of Transformative Studies.*

Argyris, C. (1976). Single-loop and double-loop models in research on decision making. *Administrative Science Quarterly*, *21*(3), 363–375. doi:10.2307/2391848

Argyris, C., & Schôn, D. (1987). Reasoning, action strategies, and defensive routines: The case of OD practitioners. *Research in Organizational Change and Development*, *1*, 89–128.

Ariely, D., & Berns, G. S. (2010, April). Neuromarketing: The hope and hype of neuroimaging in business. *Nature Reviews. Neuroscience*, *11*(4), 284–292. doi:10.1038/nrn2795 PMID:20197790

Aris, M. N. (2007). *SMEs: Building block for economic growth.* Retrieved from: http://www.statistics.gov.my/portal/images/stories/files/journal/smes.pdf

Arkko-Saukkonen, A., & Merivirta, M. (n.d.). *Creative Steps-Työpajan Eteneminen Askelittain.* Creative Steps. Kemi-Tornio University of Applied Sciences.

Armendáriz, B., & Morduch, J. (2007). *The Economics of Microfinance.* MIT Press.

Arthurs, K. E. (2012). *Creative cities, creative spaces, and urban policy: the impacts of regulations on artist-run centres and independent arts spaces in Vancouver, British Columbia.* Academic Press.

Arun, T., Hulme, D., Matin, I., & Rutherford, S. (2005). Finance for the poor: the way forward? In Finance and Development. Surveys of Theory, Evidence, and Policy. doi:10.4337/9781845424602.00016

Asghari, R., & Gedeon, S. (2010). Significance and Impact of Internet on the Entrepreneurial Process: E-Entrepreneurship and Completely Digital Entrepreneurship.*Proceedings of the 4th European Conference on Innovation and Entrepreneurship, Academic Conferences Limited*, 70.

Ashurst, C., Cragg, P., & Herring, P. (2011). The role of IT competences in gaining value from e-business: An SME case study. *International Small Business Journal*, 1–19. doi:10.1177/0266242610375703

Asian Development Bank. (2014). *Asia SME finance monitor 2013.* Mandaluyong City, Philippines: ADB.

Astrachan, J. H., Zahra, S. A., & Sharma, P. (2003). *Family-sponsored ventures.* Presented at the First Annual Global Entrepreneurship Symposium, New York.

Astrachan, J., & Jaskiewicz, P. (2008). Emotional returns and emotional costs in privately-held family businesses: Advancing traditional business valuation. *Family Business Review*, *21*(2), 139–150. doi:10.1111/j.1741-6248.2008.00115.x

Atkinson, A. P., & Adolphs, R. (2011). The neuropsychology of face perception: Beyond simple dissociations and functional selectivity. *Philosophical Transactions of the Royal Society of London. Series B, Biological Sciences*, *366*(1571), 1726–1738. doi:10.1098/rstb.2010.0349 PMID:21536556

Audretsch, D. B., Bönte, W., & Keilbach, M. (2008). Entrepreneurship capital and its impact on knowledge diffusion and economic performance. *Journal of Business Venturing*, *23*(6), 687–698. doi:10.1016/j.jbusvent.2008.01.006

Audretsch, D. B., & Thurik, A. R. (2002). *Entrepreneurship: Determinants and Policies in a European-US Comparison*. Kluwer Academic Publishers. doi:10.1007/b109395

Audretsch, D., & Thurik, R. (2001). *Linking Entrepreneurship to Growth*. OECD.

Autio, E., Kenney, M., Mustar, P., Siegel, D., & Wright, M. (2014). Entrepreneurial innovation: The importance of context. *Research Policy*, *43*(7), 1097–1108. doi:10.1016/j.respol.2014.01.015

Avendaño, J., Kelly, L., Treviño, R.N. & Madero, S. (2009). A Family Based Competitive Advantage: Handling Key Success Family Factors in Mexican Family Firms. *Cuadernos de Administración*, *22*(39), 191-212.

Axtell, R. (2000). *Why agents? On the varied motivations for agent computing in social sciences*. Working paper 17. Center on Social and economic dynamics, Brookings Institution.

Ayyagari, M., Demirgüc-Kunt, A., & Maksimovic, V. (2011). Firm innovation in Emerging Markets: The Role of Finance, Governance and Competition. *Journal of Financial and Quantitative Analysis*, *46*(6), 1545–1580. doi:10.1017/S0022109011000378

Baars, B. J., Ramsøy, T. Z., & Laureys, S. (2003, December). Brain, conscious experience and the observing self. *Trends in Neurosciences*, *26*(12), 671–675. doi:10.1016/j.tins.2003.09.015 PMID:14624851

Bachman, R., & Schutt, R. K. (2014). *The practice of research in criminology and criminal justice* (5th ed.). Thousand Oaks, CA: Sage Publication, Inc.

Bakar, L. J. A., & Ahmad, H. (2010). Assessing the relationship between firm resources and product innovation performance: A resource-based view. *Business Process Management Journal*, *16*(3), 420–435. doi:10.1108/14637151011049430

Balwin, J.R., & Johnson. (1996). Business Strategies in more and less innovative firms in Canada. *Research Policy*, 25.

Barbero, M. I. (2008). Business History in Latin America: A Historiographical Perspective. *Business History Review*, *82*(3), 555–575. doi:10.1017/S0007680500082647

Bargh, J. A. (2002, September). Losing Consciousness: Automatic Influences on Consumer Judgment, Behavior, and Motivation. *The Journal of Consumer Research*, *29*(2), 280–285. doi:10.1086/341577

Barkin, E. (2013, July). The prospects and Limitations of Neuromarketing. *Customer Relationship Management*.

Barnes, D., Clear, F., Romano, D., Harindranath, G., Harris, L., & Rae, A. (2012). Web 2.0 and micro-businesses: An exploratory investigation. *Journal of Small Business and Enterprise Development*, *19*(4), 687–711. doi:10.1108/14626001211277479

Barnes, D., Hinton, M., & Mieczkowska, S. (2004). Avoiding the fate of the dotbombs: Lessons from three surviving dotcom start-ups. *Journal of Small Business and Enterprise Development*, *11*(3), 329–337. doi:10.1108/14626000410551582

Baron, R. A., & Ensley, M. D. (2006). Opportunity recognition as the detection of meaningful patterns: Evidence from comparisons of novice and experienced entrepreneurs. *Management Science*, *52*(9), 1331–1344. doi:10.1287/mnsc.1060.0538

Baron, R. A., & Markman, G. D. (2000). Beyond social capital: How social skills can enhance entrepreneurs' success. *The Academy of Management Executive*, *14*(1), 106–116.

Barr, C. J. (2001). The Digital Enterprise: How to Reshape Your Business for a Connected World. *Journal of the American Society for Information Science and Technology*, *52*(12), 1084–1085. doi:10.1002/asi.1167

Barrick, M. R., & Mount, M. K. (1991). The Big Five Personality Dimensions and Job Performance: A Meta-Analysis. *Personnel Psychology*, *78*, 111–118.

Bartes, F. (2010). Competitive intelligence – Tool obtaining specific basic for strategic decision making TOP management firm. *Acta Universitatis Agriculturae et Silviculturae Mendelianae Brunensis*, *LVIII*(6), 43–50. doi:10.11118/actaun201058060043

Barth, J. R., Yago, G., & Zeidman, B. (2006). *Barriers to Entrepreneurship in Emerging Domestic Markets: Analysis and Recommendations*. Milken Institute. Retrieved from www.milkeninstitute.org

Batjargal, B. (2005). *Internet entrepreneurship*. Networks and performance of internet ventures in China.

Bercea, M. D. (2012). *Anatomy of methodologies for measuring consumer behavior in neuromarketing research*. Retrieved January 2014, from http://www.lcbronline.com/index_files/proceedingsemc12/12emc023.pdf

Bercea, M. D. (2013). *Quantitative versus qualitative in neuromarketing research*. Munich Personal RePEc Archive. Retrieved June 2015, from http://mpra.ub.uni-muenchen.de/44134/

Bergeron, P. (2000). Government approaches to foster competitive intelligence practice in SMEs: A comparative study of eight governments.*Proceedings of the ASIS Annual Meeting, 37*, 301-308.

Berman, S. J. (2012). Digital transformation: Opportunities to create new business models. *Strategy and Leadership*, *40*(2), 16–24. doi:10.1108/10878571211209314

Bernhardt, D. C. (1994). 'I want it fast, factual, actionable'—tailoring competitive intelligence to executives' needs. *Long Range Planning*, *27*(1), 12–24. doi:10.1016/0024-6301(94)90003-5

Bernstein, D. (2009). The Intercloud: Cloud Interoperability at Internet Scale. *Sixth IFIP International Conference on Network and Parallel Computing*.

Bhat, J. M., & Shroff, B. P. (2014). Entrepreneurial Value Creation in the Cloud: Exploring the Value Dimensions of the Business Model. In TDIT 2014, IFIP AICT 429, (pp. 301–310).

Bhupatiraju, S., Nomaler, Ö., Triulzi, G., & Verspagen, B. (2012). Knowledge flows – Analyzing the core literature of innovation, entrepreneurship and science and technology studies. *Research Policy*, *41*(7), 1205–1218. doi:10.1016/j.respol.2012.03.011

Bilton, C., & Cummings, S. (2010). *Creative strategy: Reconnecting business and innovation*. Chichester, UK: John Wiley & Sons.

Birch, K. E., & Heffernan, K. J. (2014). Crowdsourcing for Clinical Research–An Evaluation of Maturity. *Proceedings of the Seventh Australasian Workshop on Health Informatics and Knowledge Management* (HIKM 2014).

Birch, D. (1987). *Job creation in America: How our smallest companies put the most people to work*. New York: The Free Press.

Bird, B. (1995). Toward a theory of entrepreneurial competency. In J. A. Katz & R. H. Brockhaus (Eds.), *Advances in entrepreneurship, firm emergence, and growth* (Vol. 2, pp. 51–72). Greenwich, CT: JAI press.

Black, S. (2009). Microloans and Micronarratives: Sentiment for a Small World. *Public Culture*, *21*(2), 269–292. doi:10.1215/08992363-2008-029

Boateng, R., Heeks, R., Molla, A., & Hinson, R. (2008). E-commerce and socio-economic development: Conceptualizing the link. *Internet Research, 18*(5), 562–594. doi:10.1108/10662240810912783

Bodea, C.N., Mogoş, R.I., Dascălu M.-I, Purnuş, A., & Ciobotar, N.G. (2015). Simulation-Based E-Learning Framework for Entrepreneurship Education and Training. *Amfiteatru Economic, 17*(38).

Bodea, C. N., Mogoş, R. I., & Dascălu, M.-I. (2012). How e-Learning Experiences Enhance the Social Presence in Community of Practices: an Empirical Analysis. In C. El Morr & P. Maret (Eds.), *Virtual Community Building and the information Society: Current and Future Directions* (pp. 75–120). Hershey, PA: Information Science Reference. doi:10.4018/978-1-60960-869-9.ch005

Bodea, C. N., & Purnus, A. (2012). Project Risk Simulation Methods – A Comparative Analysis. *Management & Marketing Challenges for the Knowledge Society, 7*(4), 565–580.

Booth, D. A., & Freeman, R. P. J. (2014). Mind-reading versus neuromarketing: How does a product make an impact on the consumer? *Journal of Consumer Marketing, 31*(3), 177–189. doi:10.1108/JCM-08-2013-0674

Borch, O. J., Førde, A., Rønning, L., Vestrum, I. K., & Alsos, G. A. (2008). Resource configuration and creative practices of community entrepreneurs. *Journal of Enterprising Communities, People and Places in the Global Economy, 2*(2), 100–123.

Borjas, G., & Bronars, S. (1989).Consumer Discrimination and Self-Employment. *Journal of Political Economy, 97*(3), 581–605. doi:10.1086/261617

Bose, R. (2008). Competitive intelligence process and tools for intelligence analysis. *Industrial Management & Data Systems, 108*(4), 510–528. doi:10.1108/02635570810868362

Bourdieu, P. (1977). *Outline of a Theory of Practice* (R. Nice, Trans.). doi:10.1017/CBO9780511812507

Bourdieu, P. (1984). *Distinction: A social critique of the judgement of taste.* Harvard University Press.

Bradford, N. J. (2004). *Creative cities: structured policy dialogue report.* Canadian Policy Research Network, Family Network.

Brenes, E. R. (2012). *The future of entrepreneurship in Latin America.* Palgrave Macmillan. doi:10.1057/9781137003324

Briscoe, G., & Marinos, A. (2009). Digital ecosystems in the clouds: Towards community cloud computing. *3rd IEEE International Conference Digital Ecosystems and Technologies, DEST '09.* doi:10.1109/DEST.2009.5276725

Brown, R., Mason, C., & Mawson, S. (2014). *Increasing the Vital 6%: Designing Effective Public Policy To Support High Growth Firms.* National Endowment for Science Technology & Arts (NESTA). Retrieved from: http://www.nesta.org.uk/sites/default/files/working_paper_increasing_the_vital_6_percent.pdf

Brynjolfsson, E., & Hee Oh, J. (2012). T*he Attention Economy: Measuring the Value of Free Digital Services on the Internet.Thirty Third International Conference on Information Systems*, Orlando, FL.

Brynjolfsson, E., & Kahin, B. (2000). *Understanding the digital economy: data, tools and research.* Massachusetts, USA: The MIT Press.

Bucherer, E., Eisert, U., & Gassmann, O. (2012). Towards systematic business model innovation: Lessons from product innovation management. *Creativity and Innovation Management, 21*(2), 183–198. doi:10.1111/j.1467-8691.2012.00637.x

Burger-Helmchen, T., & Pénin, J. (2010). *The limits of crowdsourcing inventive activities: What do transaction cost theory and the evolutionary theories of the firm teach us.* In *Workshop on Open Source Innovation*, Strasbourg, France.

Burgoa, T., Herrera, E., & Treviño, J. (2013). Study on Family Business Administration in Mexico: Main Problems and Challenges Confronted. *International Journal of Good Conscience*, *8*(2), 1–22.

Business, S. (2013). Govt targets SMEs contributing 40% to GDP by 2015. In The Staronline [Business news]. Kuala Lumpur, MY: Star Publications (M) Bhd.

Butler, J. K. Jr, & Cantrell, R. S. (1984). A behavioral decision theory approach to modeling dyadic trust in superiors and subordinates. *Psychological Reports*, *55*(1), 19–28. doi:10.2466/pr0.1984.55.1.19

Butler, M. J. R. (2008). Neuromarketing and the perception of knowledge. *Journal of Consumer Behaviour*, *7*(4-5), 415–419. doi:10.1002/cb.260

Buvat, J. & Bisht, A. (2015, May 28). *Going Digital: General Electric and its Digital Transformation.* Retrieved from https://www.capgemini-consulting.com/general-electric-and-its-digital-transformation

Caird, S. (1992). Problems with the identification of enterprise competencies and the implications for assessment and development. *Management Education and Development*, *23*(1), 6–17.

Calof, J., & Dishman, P. L. (2002). The intelligence process: Front-end to strategic planning. working paper. University of Ottawa.

Cameron, K. S., Bright, D., & Caza, A. (2004). Exploring the relationships between virtuousness and performance. *The American Behavioral Scientist*, *47*(6), 766–790. doi:10.1177/0002764203260209

Cantillon, R. (1755). *Essay on the Nature of Trade in General.* London: Macmillan.

Carayannis, E. G., & Von Zedtwitz, M. (2005). Architecting gloCal (global–local), real-virtual incubator networks (G-RVINs) as catalysts and accelerators of entrepreneurship in transitioning and developing economies: Lessons learned and best practices from current development and business incubation practices. *Technovation*, *25*(2), 95–110. doi:10.1016/S0166-4972(03)00072-5

Carmon, A., & Nachshon, I. (1973). Ear asymmetry in perception of emotional non-verbal stimuli. *Acta Psychologica*, *37*(6), 351–357. doi:10.1016/0001-6918(73)90002-4 PMID:4767606

Carr, J., Dickinson, E., McKinnon, S. L., & Chávez, K. R. (2016). Kiva's Flat, Flat World: Ten Years of Microcredit in Cyberspace. *Globalizations*, *13*(2), 143–157. doi:10.1080/14747731.2015.1062603

Casson, M. (1982). The Entrepreneur. Totowa, NJ: Barnes & Noble Books.

Cauveri, A. (2015). *Literature survey on distributed Computing.* Retrieved August 12, 2015, from http://www.academia.edu/8298875/Literature_survey_on_distributed_computing

Chae, C.-J., Shin, Y. J., Choi, K., Kim, K.-B., & Choi, K.-N. (2016). A Privacy Data Leakage Prevention Method in P2P Networks. *Peer-to-Peer Networking and Applications*, *9*(3), 508–519. doi:10.1007/s12083-015-0371-x

Cha, H.-J., Yang, H.-K., & Kim, Y.-H. (2016). A Node Management Scheme for Stable P2P Service in Mobile Ad-hoc Networks. *Peer-to-Peer Networking and Applications*, *9*(3), 558–565. doi:10.1007/s12083-015-0392-5

Chandler, G. N., & Jensen, E. (1992). The founder's self-assessed competence and venture performance. *Journal of Business Venturing*, *7*(3), 223–236.

Charney, A., & Libecap, G. (2000). *Impact of entrepreneurship education.* Kauffman Center for Entrepreneurial Leadership.

Chartrand, T. L., Huber, Shiv, & Tanner. (2008). Nonconscious Goals and Consumer Choice. *Journal of Consumer Research*, *35*.

Che-Ha, N., & Mohd-Said, S. (2012). Innovation practices by Malaysian SMEs. In *Malaysian SMEs in the new economy* (pp. 39–55). Cengage Learning.

Cheikhrouhou, N., Pouly, M., & Madinabeitia, G. (2013). Trust categories and their impacts on information exchange processes in vertical collaborative networked organizations. *International Journal of Computer Integrated Manufacturing*, *26*(1/2), 87–100. doi:10.1080/0951192X.2012.681913

Chen, O. (2012). Govts support SMEs against likely economic downturn. *China Daily Asia*. Retrieved from http://www.chinadailyasia.com/business/2012-03/10/content_72478.html

Cheong, Y.-W., & Lo, L.-J. (2011). Facial Asymmetry: Etiology, Evaluation, and Management. *Chang Gung Medical Journal*, *34*, 341–351. PMID:21880188

Chesbrough, H. (2009). *Open Innovation, The New Imperative for Creating and Profiting from Technology*. Harvard Business School Press.

Chesbrough, H. W. (2003). *Open Innovation. The New Imperative for Creating and Profiting from Technology*. Harvard Business School Press.

Chesbrough, H. W. (2003). *Open Innovation: The New Imperative for Creating and Profiting from Technology*. Boston: Harvard Business School Press Books.

Chia, M. (2008). The Aetiology, Diagnosis and Management of Mandibular Asymmetry. *Ortho Update*, *1*, 44–52.

Chia, R., & MacKay, B. (2007). Post-processual challenges for the emerging strategy-as-practice perspective: Discovering strategy in the logic of practice. *Human Relations*, *60*(1), 217–242. doi:10.1177/0018726707075291

Chicago City Treasurer. (n.d.). *Elevator Pitch Contest*. Retrieved from: http://www.chicagocitytreasurer.com/small-business-entrepreneurs/elevator-pitch-contest/

Chong, C. W., Chong, S. C., & Gan, G. C. (2011). The KM processes in Malaysian SMEs: An empirical validation. *Knowledge Management Research & Practice*, *9*(2), 185–196. doi:10.1057/kmrp.2011.8

Chua, J. H., Chrisman, J. J., & Steier, L. P. (2003). Extending the theoretical horizons of family business research. *Entrepreneurship Theory and Practice*, *27*(4), 331–338. doi:10.1111/1540-8520.00012

Chun, R. (2005). Ethical character and virtue of organizations: An empirical assessment and strategic implications. *Journal of Business Ethics*, *57*(3), 269–284. doi:10.1007/s10551-004-6591-2

Cilliers, F., & May, M. (2012). The directors' roles in containing the Robben Island Diversity Experience (RIDE). *SA Journal of Industrial Psychology*, *38*(2), 11–21. doi:10.4102/sajip.v38i2.986

CIPI. (2003). *Comisión Intersecretarial de Política Industrial. Observatorio PYME. Primer Reporte de Resultados*. Secretaría de Economía. Retrieved from: http://www.protlcuem.gob.mx/swb/work/models/economia/Resource/965/1/images/ESTUDIOPYMESCIPI.pdf

Coduras, A., Cuervo-Arango, C., & Saiz-Alvarez, J. M. (2016). Interregional Integration of Entrepreneurial Activity in Spain. *Strategic Change-Briefings in Entrepreneurial Finance*, *25*(2), 205–220.

Collins, D., Morduch, J., Rutherford, S., & Ruthven, O. (2009). *Portfolios of the Poor: How the world's poor live on $2 a day*. Princeton, NJ: Princeton University Press.

Collins, L., Swart, R., & Zhang, B. (2013). *The Rise of Future Finance: the UK's Alternative Finance Benchmarking Report*. London: NESTA.

Colombo, M., & Delmastro, M. (2001). Does InternetMake a Difference? *Small Business Economics*, *16*(3), 177–190. doi:10.1023/A:1011127205758

Columbus, L. (2016). *Roundup Of Cloud Computing Forecasts And Market Estimates*. Retrieved April 21, 2016, from http://www.forbes.com/sites/louiscolumbus/2016/03/13/roundup-of-cloud-computing-forecasts-and-market-estimates-2016/#4e794d7474b0

Constine, J. (2013, November 19). *MetroMile Launches Per Mile Car Insurance And Free Driving Analytics Device*. Retrieved from https://techcrunch.com/2014/07/16/per-mile-car-insurance/

Consultancy.uk. (2015). *80% of firms face lagging digital transformation maturity*. Retrieved from http://www.consultancy.uk/news/3040/80-percent-of-firms-face-lagging-digital-transformation-maturity

Cooke, P. (1996). The new wave of Regional Innovation Networks: Analysis, Characteristics and Strategy. *Small Business Economics*, *8*(2), 159–171. doi:10.1007/BF00394424

Cornell University. INSEAD, & World Intellectual Property Organization. (2014). Global innovation index 2014 rankings. In S. Dutta, B. Lanvin, & S. Wunsch-Vincent (Eds.), The global innovation index 2014 -The human factor in innovation. Geneva, Switzerland: Authors.

Cotton, J., & Gibb, A. A. (1992). *An Evaluation Study of Enterprise Education in the North of England, Enterprise and Industry Education Unit*. Durham City, UK: Durham University Business School.

Cowling, M., & Taylor, M. (2001). Two DifferentSpecies? *Small Business Economics*, *16*(3), 167–175. doi:10.1023/A:1011195516912

Crespi, G., & Zuñiga, P. (2012). Innovation and productivity: evidence from six Latin American countries. World Development, 40, 273–290.

Curran, D., Lynn, T., & O'Gorman, C. (2016). The Role of Personal Factors in the Location Decision of Software Services Start-up Firms. *European Planning Studies*, *24*(3), 551–567. doi:10.1080/09654313.2015.1046369

D'Aveni, R. A. (1994). *Hyper Competition. Managing the Dynamics of Strategic Maneuvering*. New York: The Free Press.

Damanpour, F. (1996). Organizational complexity and innovation: Developing and testing multiple contingency models. *Management Science*, *42*(5), 693–716. doi:10.1287/mnsc.42.5.693

Dapkevicius, A., & Melnikas, B. (2009). Influence of Price and Quality to Customer Satisfaction: Neuromarketing Approach. Science – Future of Lithuania, 1(3).

Daude, C. (2010). *Innovation, productivity and economic development in Latin America and the Caribbean*. OECD Publishing. doi:10.1787/5kmlcz254421-en

Davidson, E., & Vaast, E. (2010). *Digital Entrepreneurship and its Sociomaterial Enactment*. Paper presented at 43rd Hawaii International Conference on System Sciences (HICSS).

Davidson, E., & Vaast, E. (2010). Digital entrepreneurship and its sociomaterial enactment. In *Proceedings of the 43rd Hawaii International Conference on System Sciences*. IEEE Computing Society.

Davidsson, P. (2004). *Researching Entrepreneurship*. New York: Springer.

Davila, T., Epstein, M. y Shelton, R. (2008). *Making Innovation Work, how to Manage It, Measure It, and Profit from It*. Pearson Education.

De Jong, J., Bruins, A., Dolfsma, W., & Meijaard, J. (2003). *Innovation in services firms explored: what, how and why?* Zoetermeer: EIM, Business and Policy Research.

Delmar, F., & Davidsson, P. (2000). Where do they come from? Prevalence and characteristics of nascent entrepreneurs. *Entrepreneurship and Regional Development, 12*(1), 1–23.

Dennis, W. J. Jr. (2003). Raising response rates in mail surveys of small business owners: Results of an experiment. *Journal of Small Business Management, 41*(3), 278–295. doi:10.1111/1540-627X.00082

Denzin, N. K., & Lincoln, Y. S. (2005). *The Sage handbook of qualitative research.* London: Sage.

DeTienne, D. R., & Chandler, G. N. (2004). Opportunity identification and its role in the entrepreneurial classroom: A pedagogical approach and empirical test. *Academy of Management Learning & Education, 3*(3), 242–257. doi:10.5465/AMLE.2004.14242103

Djelic, M. L., & Quack, S. (Eds.). (2010). *Transnational communities: Shaping global economic governance.* Cambridge University Press. doi:10.1017/CBO9780511778100

Dobrzynski, J. H. (1993). Relationship Investing. *Business Week, 42*(2), 93-105.

Doganova, L., & Eyquem-Renault, M. (2009). What do business models do?: Innovation devices in technology entrepreneurship. *Research Policy, 38*(10), 1559–1570. doi:10.1016/j.respol.2009.08.002

Domen, B. (2011). Reimagining Charity: KIVA's Ideology of Entrepreneurial Charity. *Advances in Consumer Research. Association for Consumer Research (U. S.), 39*, 489–490.

Domínguez, M. C. (2016). Strategy Innovation, Collaboration and Competition in Mexico.*Global Conference and Finance Proceedings.*

Douglas, S. P., & Craig, C. S. (2007). Collaborative and iterative translation: An alternative approach to back translation. *Journal of International Marketing, 15*(1), 30–43. doi:10.1509/jimk.15.1.030

Drucker, P. F. (2002). The discipline of innovation. *Harvard Business Review, 80*, 95–104. PMID:12195923

Duke. (n.d.). Retrieved from: http://www.dukegen.com/angel-pitch-event

Durán-Encalada, J. A., San Martin-Reyna, J. M., & Montiel-Campos, H. (2012). Research Proposal to Examine Entrepreneurship in Family Business. *Journal of Entrepreneurship, Management and Innovation, 8*(3), 58–77.

Durán-Encalada, J.A., & San Martín-Reyna, J.M. (2013). Estudio comparativo de la empresa familiar mexicana en el contexto mundial. *Revista de Contaduría, Administración e Informática Año, 2*(5), 21–54.

Dutz, M. (2007). *Unleashing India's Innovation: Towards Sustainable and Inclusive Growth.* Washington, DC: World Bank. doi:10.1596/978-0-8213-7197-8

Dyer, W. G. Jr. (1986). *Cultural Change in Family Firms: Anticipating and Managing Business and Family Transitions.* San Francisco: Jossey-Bass.

Ebner, W., Leimeister, J. M., & Krcmar, H. (2009). Community engineering for innovations: The ideas competition as a method to nurture a virtual community for innovations. *R & D Management, 39*(4), 342–356. doi:10.1111/j.1467-9310.2009.00564.x

Economic Planning Unit. (2001). *The third outline perspective plan 2001–2010.* Putrajaya, Malaysia: Prime Minister's Department.

Econsultancy. (2014). *Digital Trends Briefing. Produced by Econsultancy in partnership with Adobe, highlights the key digital trends. challenges and opportunities which marketers need to be aware of.* Retrieved from https://econsultancy.com/reports/quarterly-digital-intelligence-briefing-2014-digital-trends

Ecorys. (2012). *EU SMEs in 2012: At the crossroads. Annual report on small and medium-sized enterprises in the EU, 2011/12.* Rotterdam, The Netherlands: Ecorys Nederland BV. Retrieved from http://ec.europa.eu/enterprise/policies/sme/facts-figures-analysis/performance-review/files/supporting-documents/2012/annual-report_en.pdf

Eddleston, K., Kellermanns, F. W., & Sarathy, R. (2008). Resource configuration in family firms: Linking resources, strategic planning and environmental dynamism to performance. *Journal of Management Studies*, *45*(1), 26–50.

Edelman, L. F., Brush, C. G., & Manolova, T. S. (2002). The impact of human and organizational resources on small firm strategy. *Journal of Small Business and Enterprise Development*, *9*(3), 236–244. doi:10.1108/14626000210438562

Edward, P. (2006). The Ethical Poverty Line: A Moral Quantification of Absolute Poverty. *Third World Quarterly*, *27*(2), 377–393. doi:10.1080/01436590500432739

Egrie & Bietsch. (2013). *Marketing Research Ethics: How Consumers feel about Neuromarketing.* Academic Press.

Elevator Pitch. (2012). Retrieved from: http://www.elevatorpitch2012.com/

Elsevier. (2015). *Next-Generation Personalized Learning Solution at NLN Education Summit.* Press Release. Retrieved from http://www.elsevier.com/about/press-releases/clinical-solutions/elsevier-to-introduce-next-generation-personalized-learning-solution-at-nln-education-summit-2015,-sept.-30-oct.-2

Engaming. (2012). *"Gamifying" Training* [With The Cheesecake Factory]. Retrieved from https://engaming.wordpress.com/2012/06/15/gamifying-training-with-the-cheesecake-factory/

Eriksson, P., & Kovalainen, A. (2008). Qualitative methods in business research. *Sage (Atlanta, Ga.).*

Ernst & Young. (2014). *Family Business Yearbook.* Retrieved on July 13th, 2015 from: https://familybusiness.ey-vx.com/insights/family-business-yearbook-2014.aspx

Etzkowitz, H., & Leydersdorff, L. (1995). The triple helix of university-industry-government relations: A laboratory for knowledge based economic development. *East Review*, *14*(1), 11–19.

EU. (2012). *Effects and impact of entrepreneurship programmes in higher education.* Entrepreneurship Unit, Directorate-General for Enterprise and Industry European Commission.

EU. (2013). *Entrepreneurship Education – A Guide for Educators.* Directorate-General for Enterprise and Industry European Commission.

Euchner, J. (2014). Innovation Culture and Open Space Meetings. *Research Technology Management, 57*(1), 9-10.

Expansion. (2015). *Las 500 empresas más importantes de México en 2015.* Retrieved on May 1st, 2016 from: http://expansion.mx/rankings/interactivo-las-500/2015

Facebook. (n.d.). *Kiva.* Retrieved from: https://www.facebook.com/KIVA

Facebook. (n.d.). Retrieved from: https://www.facebook.com/CEONationalElevatorPitch

Facebook. (n.d.). Retrieved from: https://www.facebook.com/pages/Hong-Kong-Young-Entrepreneur-Club/1

Facet, T. (2011). Lessons on business virtual incubation services. In INFODEV (Ed.), Information for Development Program. The Netherlands: infoDev.

Fai, F., & Von Tunzelmann, N. (2001). Industry-specific competencies and converging technological systems: Evidence from patents. *Structural Change and Economic Dynamics*, *12*(2), 141–170. doi:10.1016/S0954-349X(00)00035-7

Faul, F., Erdfelder, E., Buchner, A., & Lang, A.-G. (2009). Statistical power analyses using G*Power 3.1: Tests for correlation and regression analyses.[PubMed]. *Behavior Research Methods*, *41*(4), 1149–1160. doi:10.3758/BRM.41.4.1149

Fayolle, A. (1998). *Teaching of entrepreneurship: Outcomes from an innovative experience*. Internationalizing Entrepreneurship Education and Training Conference, Oestrich-Winkel, Alemania.

Fayolle, A., & Ulijn, J. (2001). *Comparing entrepreneurial and innovation cultures: The European perspective of French, German and Dutch engineers, some empirical evidence about their technology versus market orientation*. The Future of Innovation Studies Conference, Eindhoven, The Netherlands.

Fayolle, A. (2007). *Handbook of Research in Entrepreneurship Education, A General Perspective* (Vol. 1). Edward Elgar Publishing Limited. doi:10.4337/9781847205377

Feldhaus, F. (2015). *Cloud Interoperability*. Retrieved August 4, 2015, from https://www.ogf.org/OGF34/materials/2417/Cloud+Interoperability+BoF.pdf

Ferreira, F. H., Messina, J., Rigolini, J., López Calva, L. F., Lugo, M. A., & Vakis, R. (2013). *Economic Mobility and the Rise of the Latin American Middle Class*. Washington, DC: The World Bank.

Fettig, K., Robertson, L., & Priem, S. (2013). *Informatics in the Simulation Lab; Learn how faculty at Minnesota State College – Southeast Technical in Red Wing and Winona, MN have developed full integration of informatics into Sim Lab*. Minnesota Simulation Conference 2014. Available at http://www.riverland.edu/mnsim/Informatics%20in%20the%20Classroom%20with%20notes.pdf

Fichman, R. G. (1999). The Diffusion and Assimilation of Information Technology Innovations. In Framing the Domains of IT Management: Projecting the Future…Through the Past. Cincinnati, OH: Pinnaflex Educational Resources, Inc.

Finkle, T., & Deeds, D. (2001). Trends in the Market for Entrepreneurship Faculty. *Journal of Business Venturing*, (16): 613–663.

Flannery, M. (2007). KIVA and the Birth of Person-to-Person Microfinance. *Innovations: Technology, Governance, Globalization*, *2*(1-2), 31–56. doi:10.1162/itgg.2007.2.1-2.31

Flores, Baruca, & Saldivar. (2014). Is Neuromarketing Ethical? Consumers Say Yes. Consumers Say No. *Journal of Legal, Ethical and Regulatory Issues, 17*(2).

Fornell, C., & Larcker, D. (1987). A second generation of multivariate analysis: Classification of methods and implications for marketing research. *Revista de Marina (Valparaiso, Chile)*, *51*, 407–450.

Foster, C. & Heeks, R. (2013). Conceptualising Inclusive Innovation: Modifying Systems of Innovation Frameworks to Understand Diffusion of New Technology to Low-Income Consumers. European Journal of Development Research, 25, 333–355.

Free Personality Test. (2016). Retrieved on 20th may 2016 from https://www.16personalities.com/free-personality-test

Fugate, D. L. (2007). Neuromarketing: A layman's look at neuroscience and its potential application to marketing practice. *Journal of Consumer Marketing*, *24*(7), 385–394. doi:10.1108/07363760710834807

Fuld, L. M. (2006). *The secret language of competitive intelligence*. New York: Crown Business.

Fuld, L. M. (2010). *The secret language of competitive intelligence: How to see through and stay ahead of business disruptions distortions, rumors and smoke screens*. Indianapolis, IN: Dog Ear Publishing.

Gajjala, R., & Birzescu, A. (2011). Digital Imperialism through Online Social/Financial Networks. *Economic and Political Weekly*, *66*(13), 95–102.

Galloway, S. J. (2009). Simulation Techniques to Bridge the Gap Between Novice and Competent Healthcare Professionals. *Online Journal of Issues in Nursing*, *14*(2), 31.

Gámez-Gutiérrez, J., & Saiz-Alvarez, J. M. (2012). Microcredits for Women Entrepreneurship: Are They and Effective Tool to Avoid Family Impoverishment? In M. A. Galindo & D. Ribeiro (Eds.), *Women's Entrepreneurship and Economics: New Perspectives, Practices, and Policies* (pp. 167–177). Springer. doi:10.1007/978-1-4614-1293-9_11

García Fuentes, M. (2015). Negocios en familia. Herencia Multiplicada. *Entrepreneur, emprende, crece y triunfa*. Septiembre. Available at: https://mx.zinio.com/www/browse/product.jsp?rf=sch&productId=500662334#/

García-Tenorio, R. J. & Sánchez, Q. J. (2009). Amenazas y oportunidades: El reto de emprender en la empresa familiar. *Gestión de empresa*, *212*, 66-77.

Gartner, W. B. (1988, Spring). Who is an entrepreneur is the wrong question. *American Small Business Journal*.

Gartner, W. B. (1990). What are we talking about when we talk about entrepreneurship? *Journal of Business Venturing*, *5*(1), 15–28. doi:10.1016/0883-9026(90)90023-M

Gates, B. (1999). *Business@ the speed of thought*. London: Penguin Books.

Gem Consortium. (2014). *The level of early-stage entrepreneurship has staibilisied – and is one of the highest in the Central and Eastern Europe region*. Retrieved from: http://www.gemconsortium.org/country-profile/103

Gem Consortium. (n.d.). Retrieved from: http://www.gemconsortium.org/

Gibb, A. A. (1998). Entrepreneurial core capacities, competitiveness and management development in the 21st century. *Internationalizing Entrepreneurship Education and Training Conference*. European Business School.

Gibb, A. A. (2002). Creating conducive environments for learning and entrepreneurship – living with, dealing with, creating and enjoying uncertainty and complexity. *Industry and Higher Education*, *16*(3), 135–148. doi:10.5367/000000002101296234

Gibbs, S. (2012). Cloud Computing. *International Journal of Innovative Research in Engineering & Science*, *1*(1), 10–17. PMID:22397113

Giddens, A. (1979). *Central problems in social theory: action, structure and contradictions in social analysis* (Vol. 241). University of California Press. doi:10.1007/978-1-349-16161-4

Giddens, A. (1984). *The constitution of society: introduction of the theory of structuration*. University of California Press.

Gilad, B. (1996). *Business blindspots*. Chicago, IL: Probus Publishing Company.

Gilbert, C., & Bakan, P. (1973). Visual asymmetry in perception of faces. *Neuropsychologia*, *11*(3), 355–362. doi:10.1016/0028-3932(73)90049-3 PMID:4792186

Gill, R., & Larson, G. S. (2014). Making the ideal (local) entrepreneur: Place and the regional development of high-tech entrepreneurial identity. *Human Relations*, *67*(5), 519–542. doi:10.1177/0018726713496829

Glaser, J. E. (2015). *Inteligencia conversacional*. Spanish. Inteligencia Conversacional. Bibliomotion Inc.

Glavin, R. (2008). When simulation should and should not be in the curriculum. In R. Kyle & W. Murray (Eds.), *Clinical simulation: Operations, engineering and management*. New York: Elsevier, Inc. doi:10.1016/B978-012372531-8.50017-0

Global Web Index. (2013). Retrieved from https://www.globalwebindex.net/

Godelier, M. (1986). *The Mental and the Material.* New York: Verso.

Going, M., & Read, J. D. (1974). Effects of uniqueness, sex of subject, and sex of photograph on facial recognition. *Perceptual and Motor Skills, 39*(1), 109–110. doi:10.2466/pms.1974.39.1.109

Goldstein, H. (1986). Toward the integration of theory and practice: A humanistic approach. *Social Work, 31*(5), 352–357.

Gopal, R. D., Ramesh, R., & Whinston, A. B. (2003). Microproducts in a Digital Economy: Trading Small, Gaining Large. *International Journal of Electronic Commerce, 8*, 9–29.

Götz, O., Liehr-Gobbers, K., & Krafft, M. (2010). Evaluation of structural equation models using the partial least squares (PLS) approach. In V. E. Vinzi, W. W. Chin, J. Henseler, & H. Wang (Eds.), *Handbook of partial least squares* (pp. 691–711). Berlin: Springer; doi:10.1007/978-3-540-32827-8_30

Grammer, K., & Thornhill, R. (1994). Human facial attractiveness and sexual selection. *Journal of Comparative Psychology, 108*, 233–242. doi:10.1037/0735-7036.108.3.233 PMID:7924253

Grant, R. M. (2010). *Contemporary strategy analysis* (7th ed.). John Wiley & Sons Ltd.

Gredler, M. (1996). Educational games and simulations: a technology in search of a (research) paradigm. In Technology (Vol. 1, pp. 521–540). MacMillan. Retrieved from http://www.aect.org/edtech/ed1/17/index.html

Groen, A., Oakey, R. P., & Cook, G. (2008). *New Technology-based Firms in the New Millennium* (Vol. 6). Bingley, UK: Emerald Group Publishing.

Grossman, G. M., & Helpman, E. (1991). Quality Ladders in the Theory of Growth. *The Review of Economic Studies, 58*(1), 43–61. doi:10.2307/2298044

Grozev, N., & Buyya, R. (2012). Inter-Cloud architectures and application brokering: Taxonomy and survey. *Software, Practice & Experience*, 369–390.

Guglielmo, F., & Palsule, S. (2015). *El líder social.* Bibliomotion, Inc.

Guha, I. B. (2011). *Literature review of "Innovative financing for SMEs".* Retrieved from http://www.smeworld.org/story/seminary-special-supplement/innovative-financing-for-smes.php

Gupta, A. K., & Govindarajan, V. (1984). Business unit strategy, managerial characteristics, and business unit effectiveness at strategy implementation. *Academy of Management Journal, 27*(1), 25–41. doi:10.2307/255955

Habbershon, T. G., & Williams, M. (1999). A resource-based framework for assessing the strategic advantages of family firms. *Family Business Review, 9*(2), 157–170.

Hafeez, Shariff, & Lazim. (2013). Does innovation and relational learning influence SME performance? An empirical evidence from Pakistan. *Asian Social Science, 9*(15), 204.

Hair, J., Anderson, R., Tatham, R., & Black, W. (2007). Análisis Multivariante (5th ed.). Pearson Prentice Hall.

Hair, J. F., Black, W. C., Babin, B. J., & Anderson, R. E. (2010). *Multivariate data analysis.* Upper Saddle River, NJ: Pearson Education, Inc.

Hair, J. F., Hult, G., Tomas, M., Ringle, C. M., & Sarstedt, M. (2014). *A primer on partil least squares structural equation modeling (PLS-SEM).* Los Angeles, CA: SAGE Publications, Inc.

Hair, N., Wetsch, L. R., Hull, C. E., Perotti, V., & Hung, Y.-T. C. (2012). Market Orientation in Digital Entrepreneurship: Advantages and Challenges in A Web 2.0 Networked World. *International Journal of Innovation and Technology Management, 9*(6), 1250045. doi:10.1142/S0219877012500459

Hall, A., Melin, L., & Nordqvist, M. (2001). Entrepreneurship as radical change in the family business: Exploring the role of culture patterns. *Family Business Review*, *14*(3), 193–208. doi:10.1111/j.1741-6248.2001.00193.x

Hamel, G. (2006). The why, what, and how of management innovation. *Harvard Business Review*, (Feb): 72–84. PMID:16485806

Hamilton, L. C., & Asundi, R. (2008). Technology usage and innovation: Its effect on the profitability of SMEs. *Management Research News*, *31*(11), 830–845. doi:10.1108/01409170810913033

Hargadon, A., & Kenney, M. (2012). Misguided policy? Following venture capital into clean technology. *California Management Review*, *54*(2), 118–139. doi:10.1525/cmr.2012.54.2.118

Harper, D. A. (2003). *Foundation of Entrepreneurship and Economic Development*. New York: Routledge. doi:10.4324/9780203380215

Harris, H. (2001). Content analysis of secondary data: A study of courage in managerial decision making. *Journal of Business Ethics*, *34*(3/4), 191–208. doi:10.1023/A:1012534014727

Hartley, S. E. (2010). *Kiva.org. Crowd-Sourced Microfinance & Cooperation in Group Lending*. Harvard University's DASH Repository.

HBR. (2015). *The digital transformation of business*. Retrieved from https://hbr.org/resources/pdfs/comm/microsoft/the_digital_transformation_of_business.pdf

Helfat, C.E., Finkelstein, S., Mitchell, W., Peteraf, M., Singh, H., Teece, D. & Winter, S. G. (2009). *Dynamic capabilities: Understanding strategic change in organizations*. Wiley.com.

Hempel, J. (2015). *Facebook Launches M, Its Bold Answer to Siri and Cortana*. Retrieved from http://www.wired.com/2015/08/facebook-launches-m-new-kind-virtual-assistant/

Hermes, N., Lensink, R., & Habteab, T. M. (2006). Does the Group Leader Matter? The Impact of Monitoring Activities and Social Ties of Group Leaders on the Repayment Performance of Group-based Lending in Eritrea. *African Development Review*, *18*(1), 72–97. doi:10.1111/j.1467-8268.2006.00133.x

Herring, J. P. (1992). The role of intelligence in formulating strategy.[PubMed]. *The Journal of Business Strategy*, *13*(5), 54–60. doi:10.1108/eb039518

Hilbert, M. (2010). When is Cheap, Cheap Enough to Bridge the Digital Divide? Modeling Income Related Structural Challenges of Technology Diffusion in Latin America. *World Development*, *38*(5), 756–770. doi:10.1016/j.worlddev.2009.11.019

Hilmi, M. F., Ramayah, T., Mustapha, Y., & Pawanchik, S. (2011, 5-6 December). *Exploring intellectual capital of Malaysian small and medium entrepreneurs*. Paper presented at the IEEE Colloquium on Humanities, Science and Engineering Research (CHUSER), Penang, Malaysia. doi:10.1109/CHUSER.2011.6163808

Hindle, K. (2002, June). A grounded theory for teaching entrepreneurship using simulation games. *Simulation & Gaming*, *33*(2), 236–241. doi:10.1177/1046878102332012

Hindle, K., & Yencken, J. (2004). Public research commercialisation, entrepreneurship and new technology based firms: An integrated model. *Technovation*, *24*(10), 793–803. doi:10.1016/S0166-4972(03)00023-3

Hinson, R., Boateng, R., & Olav Jull, S. (2008). E-business financing: Preliminary insights from a developing economy context. *Journal of Information. Communication & Ethics in Society*, *6*(3), 196–215. doi:10.1108/14779960810916219

Hitt, L. M., & Brynjolfsson, E. (1996). Productivity, Business Profitability, and Consumer Surplus: Three Different Measures of Information Technology Value. *Management Information Systems Quarterly*, *20*(2), 121–142. doi:10.2307/249475

Höfer C.N., & Karagiannis G. (2011). Cloud computing services: taxonomy and comparison. *Journal of Internet Services and Applications*, 81-94.

Hofstede, G. (1984). *Culture's consequences: International differences in work-related values* (2nd ed.). Beverly Hills, CA: SAGE Publications.

Holt, D. (2004). *How Brands Become Icons*. Boston: Harvard Business School Press.

Holt, D., & Cameron, D. (2010). *Cultural Strategy: Using Innovative Ideologies to Build Breakthrough Brands*. New York: Oxford University Press.

Howe, J. (2006). The rise of crowdsourcing. *Wired Magazine, 14*(6), 1-4.

Howells, J., & Tether, B. (2004). *Innovation in Services: Issues at Stake and Trends, Final Report*. Brussels: Commission of the European Communities.

Hu, Y. (2004). *Estimating face pose by facial asymmetry and geometry*. Retrieved from http://www.ifp.illinois.edu/~hu3/FaceTracking_PoseEstimation/Documents/Head%20pose%20estimation-CameraReady.pdf

Hugos, M., & Hulitzky, D. (2011). *Business in the Cloud. What every business needs to know about cloud computing*. John Wiley & Sons.

Huizingh, E. K. R. E. (2010). Open innovation: State of the art and future perspectives. *Technovation*. doi:10.1016/j.technovation.2010.10.002

Hull, C. E. K., Hung, Y.-T. C., Hair, N., Perotti, V., & DeMartino, R. (2007). Taking advantage of digital opportunities: A typology of digital entrepreneurship. *International Journal of Networking and Virtual Organisations*, *4*(3), 290–303. doi:10.1504/IJNVO.2007.015166

Hung, D. W. L., & Chen, V. D. T. (2002). Understanding how thriving Internet quasi-communities work: Distinguishing between learning about and learning to be. *Educational Technology*, *42*(1), 23–27.

Hung, D., & Nichani, M. (2002). Differentiating between Communities of Practices (CoPs) and Quasi-Communities: Can CoPs exist Online?. *International Journal on E-Learning*, *1*(3), 23–29.

Hussein, R. D., Farzaneh, G., & Amiri, F. (2011). Analyzing the impact of competitive intelligence on innovation at scientific research centers In Isfahan science and technology town. *Interdisciplinary Journal of Contemporary Research in Business*, *3*(5), 940–948.

IAB. (2014). *Internet Advertising Revenue*. Retrieved from http://www.iab.net/media/file/IAB_Internet_Advertising_Revenue_FY_2014.pdf

Iancu, I., & Iancu, B. (2012). Future Internet: challenges and opportunities in advertising. In *PR Trend. New media challenges and perspectives International Conference*. Babeş-Bolyai University and Hanns Seidel Foundation.

Ilcan, S., & Phillips, L. (2010). Developmentalities and Calculative Practices: The Millenium Development Goals. *Antipode*, *42*(2), 844–874. doi:10.1111/j.1467-8330.2010.00778.x

INEGI. (2010). *Economic Census*. Mexican National Institute of Statistics and Geography. Retrieved from: http://www.inegi.org.mx/saladeprensa/boletines/2015/especiales/especiales2015_07_5.pdf

International Monetary Fund. (2014). *Economic growth rates of advanced economies*.

IPMA. (2015). *IPMA Individual Competence Baseline ICB4, International Project Management Association*. IPMA.

Ireland, R. D., & Webb, J. W. (2007). Strategic entrepreneurship: Creating competitive advantage through streams of innovation. *Business Horizons*, *50*(1), 49–59. doi:10.1016/j.bushor.2006.06.002

Isenberg, D. (2010). The big idea: How to start an entrepreneurial revolution. *Harvard Business Review*, *88*(6), 40–50.

Jabatan Perangkaan Malaysia. (2012). *Banci Ekonomi 2011 - Profil Perusahaan Kecil dan Sederhana*. Putrajaya, Malaysia: Jabatan Perangkaan Malaysia.

Janeway, W. H. (2012). *Doing capitalism in the innovation economy: markets, speculation and the state*. Cambridge, UK: Cambridge University Press; doi:10.1017/CBO9781139381550

Jangra, A., & Bala, R. (2011). Spectrum Of Cloud Computing Architecture: Adoption And Avoidance Issues. *International Journal of Computing and Business Research*. Retrieved August 2, 2015, from http://www.researchmanuscripts.com/PapersVol2N2/IJCBRVOL2N2P13.pdf

Jansen, W. & Grance, T. (2011). *Guidelines on security and privacy in public cloud computing*. NIST Special Publication, 800-144.

Jarzabkowski, P., Balogun, J., & Seidl, D. (2007). Strategizing: The challenges of a practice perspective. *Human Relations*, *60*(1), 5–27. doi:10.1177/0018726707075703

Javalgi, R. G., Todd, P. R., Johnston, W. J., & Granot, E. (2012). Entrepreneurship, muddling through, and Indian Internet-enabled SMEs. *Journal of Business Research*, *65*(6), 740–744. doi:10.1016/j.jbusres.2010.12.010

Jaworski, B. J., Macinnis, D. J., & Kohli, A. K. (2002). Generating competitive intelligence in organizations. *Journal of Market-Focused Management*, *5*(4), 279–307. doi:10.1023/B:JMFM.0000008071.19917.36

Johannessen, J.-A., Dolva, J. O., & Kolvereid, L. (1997). Perceived innovation success in the Russian market. *International Journal of Information Management*, *17*(1), 13–20. doi:10.1016/S0268-4012(96)00039-4

Johansson, B., Karlsson, C., & Stough, R. (Eds.). (2006). *The emerging digital economy: Entrepreneurship, clusters, and policy*. Springer-Verlag Berlin Heidelberg. doi:10.1007/3-540-34488-8

Jones, E. (2016). *The top five in-demand cloud skills for 2016*, Retrieved May 12, 2016 from http://www.cloudcomputing-news.net/news/2016/feb/16/top-five-demand-cloud-skills-2016/

Jones, B. C. (2001). Facial symmetry and judgments of apparent health. Support for a "good genes" explanation of the attractiveness–symmetry relationship. *Evolution and Human Behavior*, *22*, 417–429. doi:10.1016/S1090-5138(01)00083-6

Jose Silva, M., Sousa, G., Moreira, J., & Simões, J. (2011). Innovation Activities in the Service Sector: Empirical Evidence from Portuguese Firms. *Journal of Knowledge Management, Economics and Information Technology*, (6).

Jung, H. M., Hwang, I.-S., Moon, J.-K., & Park, H.-S. (2016). A Security Monitoring Method for Malicious P2P Event Detection. *Peer-to-Peer Networking and Applications*, *9*(3), 498–507. doi:10.1007/s12083-015-0369-4

Kahaner, L. (1996). *Competitive intelligence*. New York: Simon and Schuster.

Kahneman, D. (2011). *Think fast and slow*. New York: Farrar, Straus and Giroux.

Kamien, M. I., & Schwartz, N. (1982). *Market structure and innovation*. New York: Cambridge University Press.

Kamp, B. (2014). Overcoming Market and Network Failures by Means of Embedded Business Model Innovation. *Boletin de estudios economicos*, *69*(213), 423-442.

Kanter, R. M. (2012). Enriching the ecosystem. *Harvard Business Review*, *90*(3), 140–147. PMID:22299513

Kantis, H. D., & Federico, J. S. (2012). *Entrepreneurial Ecosystems in Latin America: The role of policies*. International Research and Policy Roundtable. Retrieved from http://www.innovacion.gob.cl/wp-content/uploads/2012/06/Entrepreneurial-Ecosystems-in-Latin-America_the-role-of-policies.pdf

Kantis, H., Ishida, M., & Komori, M. (2002). *Empresarialidad en economías emergentes: Creación y desarrollo de nuevas empresas en América Latina y el Este de Asia*. Banco Interamericano de Desarrollo, Departamento de Desarrollo Sostenible, División de Micro, Pequeñas y Medianas Empresas.

Kantis, H. (2005). The emergence of dynamic ventures in Latin America, Southern Europe and East Asia: An international comparison. *International Journal of Entrepreneurship and Small Business*, *2*(1), 34–56. doi:10.1504/IJESB.2005.006069

Kar, A. K. (2016). Measuring Competition in Microfinance Markets: A New Approach. *International Review of Applied Economics*, *30*(4), 423–440. doi:10.1080/02692171.2015.1106445

Karcher, B. (1998). *Does Gender Really Matter? The influences of gender and qualifications on self-employment and their implications on entrepreneurship education*. Internationalizing Entrepreneurship Education and Training Conference, Oestrich-Winkel, Alemania.

Kavis, M. (2014). *Architecting the Cloud: Design Decisions for Cloud Computing Service Models (SaaS, PaaS, and IaaS)*. John Wiley & Sons. doi:10.1002/9781118691779

Kayakutlu, G., & Büyüközkan, G. (2010). Effective supply value chain based on competence success. *Supply Chain Management: An International Journal*, *15*(2), 129–138. doi:10.1108/13598541011028732

Kellermanns, F. W., & Eddleston, K. A. (2006). Corporate entrepreneurship in family firms: A family perspective. *Entrepreneurship Theory and Practice*, *30*(6), 809–830. doi:10.1111/j.1540-6520.2006.00153.x

Kellermanns, F. W., Eddleston, K. A., Barnett, T., & Pearson, A. (2008). An exploratory study of family member characteristics and involvement: Effects on entrepreneurial behavior in the family firm. *Family Business Review*, *21*(1), 1–14. doi:10.1111/j.1741-6248.2007.00107.x

Kelley, D. J., Singer, S., & Herrington, M. (2015). *The Global Entrepreneurship Monitor 2015*. Global Report. Retrieved from: http://www.gemconsortium.org/

Kenny, C. (2003). The Internet and economic growth in less-developed countries: A case of managing expectations? *Oxford Development Studies*, *31*(1), 99–113. doi:10.1080/1360081032000047212

Kim Chan, W., & Mauborgne, R. (2005). *La Estrategia del Océano Azul: Cómo desarrollar un nuevo mercado donde la competencia no tiene importancia*. Harvard Business School Press y Grupo Editorial Norma.

Kimura, D. (1966). *Dual functional asymmetry of the brain in visual perception*. Retrieved from http://www.ling.upenn.edu/courses/Fall_2013/ling520/Kimura1.pdf

Kim, Y., & Ha, S. (2010). Innovation activities and innovation performances of SMEs: The Korean electronic parts industry 1990–19951). *Asian Journal of Technology Innovation*, *18*(1), 125–160. doi:10.1080/19761597.2010.9668685

Kirkman, A. (1999). The Americas. *Forbes Magazine*. Retrieved on May 13th, 2015 from: http://www.forbes.com/global/1999/0705/0213097a.html

Kiva Cup. (2009). Retrieved from: http://KIVAnews.blogspot.com/2009/07/paris-KIVA-cup-2009.html

Kiva. (n.d.). Retrieved from: http://www.Kiva.org/about

Knight, G. (1996). *Born global.* Wiley International Encyclopedia of Marketing.

Knight, P. T. (2011). e-Development in Brasil. *Journal of the Knowledge Economy*, *2*(1), 77–116. doi:10.1007/s13132-010-0024-5

Kohn, A. (1997). Students don't "work", they learn. *Education Week*, (Sept): 3.

Kollmann, T. (2006). What is e-entrepreneurship?–fundamentals of company founding in the net economy. *International Journal of Technology Management*, *33*(4), 322–340. doi:10.1504/IJTM.2006.009247

Kolvereid, L., & Moen, O. (1997). Entrepreneurship among business graduates: Does a major in entrepreneurship make a difference? *Journal of European Industrial Training*, *21*(4-5), 154–157.

Kotter, J. P. (2015).*Acelerar*. Harvard Business School Publishing Corporation.

Kourilsky, M. L. (1996). *Mini-Society*. Kauffman Center for Entrepreneurial Leadership, E.M. Kauffman Foundation.

Kozinets, R. V. (2008). Technology/Ideology: How Ideological Fields Influence Consumers' Technology Narratives. *The Journal of Consumer Research*, *34*(6), 865–881. doi:10.1086/523289

Kritikos, A. (2014). Entrepreneurs and their impact on jobs and economic growth. *IZA World of Labor*, *8*. doi:10.15185/izawol.8

Krücken-Pereira, L., Debiasi, F., & Abreu, A. F. (2001). Technological innovation and competitive intelligence: An interactive process. *READ - Electronic Journal of Directors, 7.*

La Porta, R., Lopez-de-Silanes, F., & Shleifer, A. (1999). Corporate ownership around the world. *The Journal of Finance*, *54*(2), 471–517. doi:10.1111/0022-1082.00115

Lafuente, A., & Salas, V. (1989). Types of entrepreneurs and firms: The case of newspanish firms. *Strategic Management Journal*, *10*(1), 17–30. doi:10.1002/smj.4250100103

Laine, P. M. (2010). *Toimijuus strategiakäytännöissä: Diskurssi-ja käytäntöteoreettisia avauksia*. Väitöskirja, Sarja A-1.

Laine, P. M., & Vaara, E. (2007). Struggling over subjectivity: A discursive analysis of strategic development in an engineering group. *Human Relations*, *60*(1), 29–58. doi:10.1177/0018726707075279

Lampel, J., Jha, P. P., & Bhalla, A. (2012). Test-driving the future: How design competitions are changing innovation. *The Academy of Management Perspectives*, *26*(2), 71–85. doi:10.5465/amp.2010.0068

Largacha-Martinez, C., von Kimakowitz, E., Spitzeck, H., Pirson, M., Kierksmeier, C., & Amann, W. (2014).*Tendencias Gerenciales, Humanistic Management Network*. Palgrave Macmillan.

Larivet, S. (2009). *Economic intelligence in small and medium business in France: A survey* Paper presented at the 3rd European Competitve Intelligence Symposium, Stockholm, Sweden.

Lastres, H. M. M., & Cassiolato, J. E. (2003). *Systems of innovation and development from a South American perspective: a contribution to Globelics*. Paper presented at the First Globelics Workshop on Innovation Systems. Aalborg University, Aalborg, Denmark.

Laursen, K., & Salter, A. (2006). Open for innovation: The role of openness in explaining innovation performance among UK manufacturing firms. *Strategic Management Journal*, *27*(2), 131–150. doi:10.1002/smj.507

Le Breton-Miller, I., & Miller, D. (2006). Why Do Some Family Businesses Out-Compete? Governance, Long-Term Orientations, and Sustainable Capability. *Entrepreneurship: Theory and Practice*, *30*(6), 731–746.

Lederman, D., Messina, J., Pienknagua, S., & Rigolini, J. (2014). *Latin American Entrepreneurs. Many firms but little Innovation*. Washington, DC: International Bank of Reconstruction and Development/The World Bank.

Lee, N., Broderick, A. J., & Chamberlain, L. (2007). What is 'neuromarketing'? A discussion and agenda for future research. *International Journal of Psychophysiology*, *63*(2), 199–204. doi:10.1016/j.ijpsycho.2006.03.007 PMID:16769143

Lee, S.-H., Dewester, D., & Park, S. R. (2008). Web 2.0 and opportunities for small businesses. *Service Business*, *2*(4), 335–345. doi:10.1007/s11628-008-0043-5

Lenartowicz, T., & Johnson, J. (2003). A Cross-National Assessment of the Values of Latin America Managers. *Journal of International Business Studies*, *34*(3), 266–281. doi:10.1057/palgrave.jibs.8400032

Leong, B. D., & Clark, H. (2003). Culture-based knowledge towards new design thinking and practice—A dialogue. *Design Issues*, *19*(3), 48–58. doi:10.1162/074793603768290838

Leontief, W. (1983). Technological Advance, Economic Growth, and the Distribution of Income. *Population and Development Review*, *9*(3), 403. doi:10.2307/1973315

Levie, J., Brown, W., & Steele, L. (2001). How entrepreneurial are Strathclyde Alumni? In International Entrepreneurship: Researching New Frontiers Conference. Universidad de Strathclyde.

Li, G., Park, E.-M., & Park, S.-T. (2016). Effects of Social Media Usage on Country Image and Purchase Intention from Social P2P Network Perspective. *Peer-to-Peer Networking and Applications*, *9*(3), 487–497. doi:10.1007/s12083-015-0379-2

Limpanitgul, T., & Robson, M. (2009). *Methodological considerations in a quantitative study examining the relationship between job attitudes and citizenship behaviors*. Paper presented at the 18th EDAMBA Summer Academy, Soreze, France.

Linden, G., Kraemer, K. L., & Dedrick, J. (2009). Who captures value in a global innovation network? The case of Apple's iPod. *Communications of the ACM*, *52*(3), 140–144. doi:10.1145/1467247.1467280

LinkedIn. (n.d.). *Kiva*. Retrieved from: https://www.linkedin.com/company/KIVA-org

Litvak, I., & Maule, C. (1976). Some perspectives. *Journal of International Business Studies*, *7*(1), 31–38. doi:10.1057/palgrave.jibs.8490692

Li, X. (2009). *Entrepreneurial competencies as an entrepreneurial distinctive: An examination of the competency approach in defining entrepreneurs (Master of Science)*. Singapore: Singapore Management University.

Lockett, N. J., & Brown, D. H. (2000). eClusters: The potential for the emergence of digital enterprise communities enabled by one or more intermediaries in SMEs. *Knowledge and Process Management*, *7*(3), 196–206. doi:10.1002/1099-1441(200007/09)7:3<196::AID-KPM85>3.0.CO;2-7

Lora, E. & Castellani, F. (2014). *Entrepreneurship in Latin America: A Step Up the Social Ladder?*. World Bank Publications.

Lorah, K. P. (2013). *Disruptive by design: the promise of the tech-enabled design competition* (Doctoral dissertation). Massachusetts Institute of Technology.

Lounsbury, M., & Crumley, E. (2007). New practice creation: An institutional perspective on innovation. *Organization Studies*, *28*(7), 993–1012. doi:10.1177/0170840607078111

Low, M. B., & McMillan, I. C. (1988). Entrepreneurship: Past research and future challenges. *Journal of Management*, 14.

Lu, K., Wang, J., & Li, M. (2016). An Eigentrust Dynamic Evolutionary Model in P2P File-sharing Systems. *Peer-to-Peer Networking and Applications*, *9*(3), 599–612. doi:10.1007/s12083-015-0416-1

Lumpkin, G. T., & Dess, G. G. (1996). Clarifying the entrepreneurial orientation construct and linking it to performance. *Academy of Management Review*, *21*(1), 135–172.

Lumpkin, G. T., & Dess, G. G. (1996). Clarifying the Entrepreneurial Orientation Construct and Linking it to Performance. *Academy of Management Review*, *21*(1), 135–172.

Lusch, R., & Nambisan, S. (2015). Service Innovation: A Service-Dominant Logic Perspective. *Management Information Systems Quarterly*, *39*(1), 155–175.

Lüthje, C., & Franke, N. (2002). *Fostering entrepreneurship through university education and training: Lessons from Massachusetts Institute of Technology.Second Annual Conference of European Academy of Management, Suecia.*

Macal, C., & North, M. J. (2010). Tutorial on agent-based modeling and simulation. *Journal of Simulation*, *4*, 151–162.

Macal, C., & North, M. J. (2009). Agent-based modeling and simulation. In *Proceedings of the 2009 Winter Simulation Conference*. IEEE. doi:10.1109/WSC.2009.5429318

Maertens, A., & Barrett, C. B. (2013). Measuring Social Networks' Effects on Agricultural Technology Adoption. *American Journal of Agricultural Economics*, *95*(2), 353–359. doi:10.1093/ajae/aas049

Maloney, W. F., & Sarrias, M. (2012). *What Makes Latin American Managers So Bad?* World Bank.

Maloney, W. F., & Valencia-Caicedo, F. (2012). *Part II: Engineers, Innovative Capacity and Development in the Americas. Background paper for this report.* World Bank.

Manafi, M., & Subamanian, I. D. (2015). The role of perceived justice in the relationship between human resource management practices and knowledge sharing: A study of Malaysian Universities Lecturers. *Asian Social Science*, *11*(12), 131–143. doi:10.5539/ass.v11n12p131

Manintveld, B., & Schalekamp, J. (2014, November 3). *Make money by buying and selling company data.* Retrieved from http://www2.deloitte.com/nl/nl/pages/data-analytics/articles/make-money-buying-selling-company-data.html

Man, T. W. Y., Lau, T., & Chan, K. F. (2002). The competitiveness of small and medium enterprises: A conceptualization with focus on entrepreneurial competencies. *Journal of Business Venturing*, *17*(2), 123–142. doi:10.1016/S0883-9026(00)00058-6

Man, T. W. Y., Lau, T., & Snape, E. (2008). Entrepreneurial competencies and the performance of small and medium enterprises: An investigation through a framework of competitiveness. *Journal of Small Business and Entrepreneurship*, *21*(3), 257–276. doi:10.1080/08276331.2008.10593424

Mărginean, I. (2010). Calitatea vieții în România: Present și perspective. *Revista Calitatea Vie*, *2*.

Markovitch, S., & Willmott, P. (2014*). Accelerating the digitization of business processes.* Mckinsey.com. Retrieved from http://www.mckinsey.com/insights/business_technology/accelerating_the_digitization_of_business_processes

Marom, D., & Sade, O. (2013). *Are the life and death of a young start-up indeed in the power of the tongue?* Lessons from Online Crowdfunding Pitches.

Marshall, A. (1890). *Principles of Economics*. London: MacMillan.

Martha, G. M. (2004). Institutions and the Adoption of Electronic Commerce in Mexico. *Electronic Commerce Research*, *4*(3), 201–219. doi:10.1023/B:ELEC.0000027980.16492.af

Mason, C., & Brown, R. (2013). Creating good public policy to support high growth firms. *Small Business Economics*, *40*(2), 211–225. doi:10.1007/s11187-011-9369-9

Mastor, K. A., Jin, P., & Cooper, M. (2000). Malay culture and personality: A big five perspective. *The American Behavioral Scientist*, *44*(1), 95–111. doi:10.1177/00027640021956116

Mather, T., Kumaraswamy, S., & Latiif, S. (2009). *Cloud Security and Privacy: An Enterprise Perspective on Risks and Compliance (Theory in Practice)*. O'Reilly Media.

Matiz, F. (2006). Emprendimiento Como Un Pilar para La Competitividad de las Naciones. *Revista EAN*, (59), 119-129.

Matiz, F. (2009). Investigaci6n en Emprendimiento, un reto para la Construcci6n de Conocimiento. *Revista EAN*, (66), 169-182.

Matlay, H. (2004). E-entrepreneurship and small e-business development: Towards a comparative research agenda. *Journal of Small Business and Enterprise Development*, *11*(3), 408–414. doi:10.1108/14626000410551663

McClure, S. M., Li, J., Tomlin, D., Cypert, K. S., Montague, L. M., & Read Montague, P. (2004). Neural correlates of behavioral preference for culturally familiar drinks. *Neuron*, *44*(2), 379–387. doi:10.1016/j.neuron.2004.09.019 PMID:15473974

McCrae, R. R., & Terracciano, A. (2005). Personality profiles of cultures: Aggregate personality traits.[PubMed]. *Journal of Personality and Social Psychology*, *89*(3), 407–425. doi:10.1037/0022-3514.89.3.407

McCrindle, M. (2014). The ABC of XYZ: Understanding the Global Generations (3rd ed.). McPherson's Printing Group.

Meet Up. (n.d.). *PitchForce*. Retrieved from: http://www.meetup.com/IGNITE-Fundraisers-monthly-startup-pitch-events/

MeetUp. (n.d.). Retrieved from: http://www.meetup.com/New-York-Startup-Pitch-Night

Mell, P., & Grance, T. (2009). *NIST Definition of Cloud Computing Version 15*. NIST.

Middleton, K. L., & Byus, K. (2011). Information and communications technology adoption and use in small and medium businesses. *Management Research Review*, *34*(1), 98–110. doi:10.1108/01409171111096496

Miller, D. (1983). The correlates of entrepreneurship in three types of firms. *Management Science*, *29*(7), 770–791. doi:10.1287/mnsc.29.7.770

Miller, D., & Garnsey, E. (2000). Entrepreneurs and technology diffusion: How diffusion research can benefit from a greater understanding of entrepreneurship. *Technology in Society*, *22*(4), 445–465. doi:10.1016/S0160-791X(00)00021-X

Miller, S. (2001). *Competitive intelligence: An overview*. Alexandria, VA: Society of Competitive Intelligence Professionals.

Mills, C., & Pawson, K. (2012). Integrating motivation, risk-taking and self-identity: A typology of ICT enterprise development narratives. *International Small Business Journal*, *30*(5), 584606. doi:10.1177/0266242610390594

Millwarbrown. (2014). *Adreaction*. Retrieved from http://www.millwardbrown.com/adreaction/2014/report/Millward-Brown_AdReaction-2014_Global.pdf

MIT Global Start Up. (2013). *Elevator Pitch Contest*. Retrieved from: http://mitgsw.org/2013/workshop/elevator-pitch-contest-page/

MIT Global. (2013). *MIT Global Startup Workshop*. Reviewed March 4, 2014 from http://mitgsw.org/2013/

MIT. (n.d.). Retrieved from: http://www.mit100k.org/

MITI. (2006). *IMP3 Third industrial master plan 2006-2020 Malaysia -Towards global competitiveness*. Kuala Lumpur: Ministry of International Trade and Industry Malaysia.

Mohd-Said. (2012). SMEs in the Malaysian economy. In *Malaysian SMEs in the new economy* (pp. 1–17). Cengage Learning.

Moisander, J., & Valtonen, A. (2006). Qualitative marketing research: A cultural approach. *Sage (Atlanta, Ga.)*.

Moody, M. (2008). Building a Culture: The Construction and Evolution of Venture Philanthropy as a New Organizational Field. *Nonprofit and Voluntary Sector Quarterly*, *37*(2), 324–352. doi:10.1177/0899764007310419

Morin, C. (2011). Neuromarketing: The New Science of Consumer Behavior. *Soc*, *48*(2), 131–135. doi:10.1007/s12115-010-9408-1

Morse, J. M., & Richards, L. (2002). *Read me first for a user's guide to qualitative research.* Sage.

Moss, T.W., & Meyskens, M. (2015). The Effect of Virtuous and Entrepreneurial Orientations on Microfinance Lending and Repayment: A Signaling Theory Perspective. *Entrepreneurship Theory and Practice*, 27-52.

Moss, J. P. (1991). Methods of three dimensional analysis of patients with asymmetry of the face. *Proceedings of the Finnish Dental Society. Suomen Hammaslaakariseuran Toimituksia*, *87*(1), 139–149. PMID:2057480

Motoring. (2014). *BMW rolls out ConnectedDrive Services and Apps.* Retrieved from http://www.motoring.com.au/bmw-rolls-out-connecteddrive-services-and-apps-43096/

Muller. (2002b). *Creating intelligence.* Johannesburg, South Africa: Knowres Publishing.

Muller, M. L. (2002). *Managing competitive intelligence.* Johannesburg, South Africa: Knowres Publishing.

Murphy, E. R., Illes, J., & Reiner, P. B. (2008). Neuroethics of neuromarketing. *Journal of Consumer Behaviour*, *7*(4-5), 293–302. doi:10.1002/cb.252

Myers, K. K., & Sadaghiani, K. (2010). Millennials in the workplace: A communication perspective on millennials' organizational relationships and performance. *Journal of Business and Psychology*, *25*(2), 225–238. doi:10.1007/s10869-010-9172-7 PMID:20502509

Nair, M., Kuppusamy, M., & Davison, R. (2005). A Longitudinal Study on the Global Digital Divide Problem: Strategies to Close Cross-Country Digital Gap. *The Business Review, Cambridge*, *4*, 315–326.

Naldi, L., Nordqvist, M., Sjoberg, K., & Wiklund, J. (2007). Entrepreneurial orientation, risk taking, and performance in family firms. *Family Business Review*, *20*(1), 33–47. doi:10.1111/j.1741-6248.2007.00082.x

National SME Development Council. (2014). *SME annual report 2013/2014 - Transitioning for productivity-led and innivation-driven growth.* Kuala Lumpur: SMECorp.

NBIA. (2001). *Best Practice in Action: Guidelines for Implementing First Class Business Incubation Programs.* NBIA.

Nelson, R. R., & Winter, S. G. (2002). Evolutionary theorizing in economics. *The Journal of Economic Perspectives*, *16*(2), 23–46. doi:10.1257/0895330027247

Neumann, I. B. (2002). Returning practice to the linguistic turn: The case of diplomacy. *Millennium*, *31*(3), 627–651. doi:10.1177/03058298020310031201

NextBerlin. (2014). *This is the New Normal.* Review March 4, 2014 from http://nextberlin.eu/

Ngoasong, M. Z. (2015). *Digital Entrepreneurship in Emerging Economies: The role of ICTs and local context.42nd AIB-UKI Conference*, Manchester, UK.

Ngwenyama, O., & Morawczynski, O. (2009). Factors affecting ICT expansion in emerging economies: An analysis of ICT infrastructure expansion in five Latin American countries. *Information Technology for Development, 15*(4), 237–258. doi:10.1002/itdj.20128

NMSBA. (n.d.a). *The Neuromarketing Science & Business Association*. Retrieved June 2015, from http://www.nmsba.com/

NMSBA. (n.d.b). *The Neuromarketing Science & Business Association*. Retrieved June 2015, from http://www.nmsba.com/Ethics

NMSBA. (n.d.c). *The Neuromarketing Science & Business Association*. Retrieved April 2016, from http://www.nmsba.com/neuromarketing-companies

Nobel, C. (2014). HBS Working Knowledge, Neuromarketing: Tapping Into the 'Pleasure Center' of Consumers. In Consumer Neuroscience and Neuromarketing (2nd ed.). Academic Press.

Nordqvist, M., & Merlin, L. (2010). Entrepreneurial families and family firms. *Entrepreneurship and Regional Development, 22*(3–4), 211–239. doi:10.1080/08985621003726119

Npr.org. (2016). *NPR*. Retrieved from http://www.npr.org/sections/alltechconsidered/2016/04/09/473422686/the-newest-public-transportation-in-town-uber

Nuortila, E. (2013). *Crowdfunding as the forms of financing in independent movie*. Helsinki: Metropolia University of Applied Sciences.

O'Neil, H. F., Wainess, R., & Baker, E. L. (2005). Classification of learning outcomes: Evidence from the computer game literature. *Curriculum Journal, 16*(4), 455–474. doi:10.1080/09585170500384529

O'Reilly, T. (2007). What is web 2.0: Design patterns and business models for the next generation of software. *Communications & Strategies*, 17.

OECD. (1997). *Proposed Guidelines for Collecting and Interpreting Technological Innovation Data*. Paris: Oslo Manual.

OECD. (1999). *Fostering Entrepreneurship*. Paris: OECD.

OECD. (2000). *A new economy? The changing role of innovation and information technology in growth*. Paris: OECD.

OECD. (2001). *Entrepreneurship, Growth and Policy*. OECD DSTI/IND.

OECD. (2013). *An international benchmarking analysis of public programmes for high-growth firms*. Paris: OECD LEED Programme.

OECD. (2014). *Latin American Economic Outlook 2015*. OECD iLibrary. Available at: http://www.oecd-ilibrary.org/development/latin-american-economic-outlook-2015_leo-2015-en

Oestreicher-Singer, G., & Zalmanson, L. (2013). Content or Community? A Digital Business Strategy for Content Providers in the Social Age. *Management Information Systems Quarterly, 37*(2), 591–616.

Ohmae, K. (1990). *The borderless world*. New York.

Ohme & Matukin. (2012). *Small Frog That Makes a Big Difference*. DOI: .10.1109/MPUL.2012.2189169

Ojala, A., & Tyrvainen, P. (2011). Developing Cloud Business Models: A Case Study on Cloud Gaming. *Software, IEEE, 28*(4), 42–47. doi:10.1109/MS.2011.51

Olick, J. K., & Robbins, J. (1998). Social memory studies: From "collective memory" to the historical sociology of mnemonic practices. *Annual Review of Sociology, 24*(1), 105–140. doi:10.1146/annurev.soc.24.1.105

Oliner, S. D., & Sichel, D. E. (2000). The Resurgence of Growth in the Late 1990s: Is Information Technology the Story? *The Journal of Economic Perspectives*, *14*(4), 3–22. doi:10.1257/jep.14.4.3

Olteanu, B. (2015, January). Neuroethics and Responsibility in Conducting Neuromarketing Research. *Neuroethics*. doi:10.1007/s12152-014-9227-y

Onetti, A., Zucchella, A., Jones, M., & McDougall-Covin, P. (2012). Internationalization, innovation and entrepreneurship: Business models for new technology-based firms. *Journal of Management & Governance.*, *16*(3), 337–368. doi:10.1007/s10997-010-9154-1

Orehn, T. (2009). Modeling and simulation: a comprehensive and integrative view. In Agent-based Simulation and Systems Engineering. Wiley-VCH, GmbH&Co, KgaA.

Osterwalder, A., Lagha, S. B., & Pigneur, Y. (2002). *An ontology for developing e-business models*. IFIP DsiAge.

Osterwalder, A., & Pigneur, Y. (2002). *An e-business model ontology for modeling e-business.15th Bled Electronic Commerce Conference*, Bled, Slovenia.

Owen, S., Anil, R., Dunning, T., & Friedman, E. (2011). *Mahout in Action*. Manning Publications.

PageFair. (2013). *The Rise of AdBlocking*. Retrieved from http://downloads.pagefair.com/reports/the_rise_of_adblocking.pdf

Pagés-Serra, C. (2010). *The Age of Productivity: Transforming Economies from the Bottom Up*. Inter-American Development Bank. doi:10.1057/9780230107618

Paine, L. S. (1991). Ethics as Character Development: Reflections on the Objective of Ethics Education. In R. E. Freeman (Ed.), *Ethics and Governance* (pp. 67–88). Oxford University Press.

Pamula N. B., Jairam K., & Rajesh B. (2014). Cache-Aside Approach For Cloud Design Pattern. *International Journal of Computer Science and Information Technologies*, 1423-1426.

Passman, R. (2001). Experiences with student-centered teaching and learning in high-stakes assessment environments. *Education*, *122*(1), 189–199.

Pateli, A., & Giaglis, G. (2003). *A framework for understanding and analysing e-business models.Bled Electronic Commerce Conference*.

Patiño, O. (2016). Key Aspects to Develop Long-Term Microfinance and Financial Inclusion for Social Development. In Handbook of Research on Social Entrepreneurship and Solidarity Economics (pp. 260–281). Hershey, PA: IGI Global.

Payne, G. T., Brigham, K. H., Broberg, J. C., Moss, T. W., & Short, J. C. (2011). Organizational virtue orientation and family firms. *Business Ethics Quarterly*, *21*(2), 257–285. doi:10.5840/beq201121216

PEMANDU. (2010). *Economic transformation programme. A roadmap for Malaysia*. Putrajaya, Malaysia: Performance Management and Delivery Unit, Prime Minister's Department.

Pérez-Uribe R. (2012). *El ambiente laboral y su incidencia en el desempeño de las organizaciones: estudio de las mejores empresas para trabajar en Colombia*. (Tesis doctoral). Universidad Nebrija. Retrieved from http://edicionesean.ean.edu.co/index.php/es/productos-de-investigacion1/tesis-doctorales/23-publicaciones/192-el-ambiente-laboral-y-su-incidencia-en-el-desempeno-de-las-organizaciones-estudio-en-las-mejores-empresas-para-trabajar-en-colombia

Pérez-Uribe, R., & Vargas, H. (2015). El uso del método MicMac, para la definición de procesos de intervención en las organizaciones. Revista Ciencia y Poder Aéreo. *Escuela de Postgrados de la Fuerza Aérea Colombiana, 10*(1). Retrieved from http://www.publicacionesfac.com/index.php/cienciaypoderaereo/article/view/156

Pérez-Uribe, R., Nieto, M., Vargas, H., & Díaz, H. (2015). *MIC MAC, aplicado a los componentes del PENTEX, una aproximación a la validación de las matrices*. Vicerrectoria de Investigación. Universidad EAN. Retrieved from https://www.researchgate.net/publication/280881459_MIC_MAC_APLICADO_A_LOS_COMPONENTES_DEL_PENTEX_UNA_APROXIMACIN_A_LA_VALIDACIN_DE_LAS_MATRICES

Pérez-Uribe, R. (2009). *Modelo de Modernización para la Gestión de Organizaciones MMGO*. Bogotá: Universidad EAN.

Petrişor, I., & Străin, N. A. (2013). Approaches on the competitive intelligence. *USV Annals of Economics & Public Administration, 13*(1), 100–109.

Phan, V. (2010). *Neuromarketing. Who Decides to Buy?* The Triple Helix.

Pietrobelli, C., & Rabellotti, R. (2010). *The global dimension of innovations systems and enterprise upgrading–Linking innovation systems and global value chains* (No. 025). SLPTMD Working Paper Series.

Pistrui, D., Welsch, H. P., & Roberts, J. S. (1997). The re-emergence of family businesses in the Transforming former Soviet-Bloc: Family Contributions to entrepreneurship development in Romania. *Family Business Review, 10*(3), 221–237. doi:10.1111/j.1741-6248.1997.00221.x

Pitch Event. (n.d.). Retrieved from: http://www.pitchevent.com/

Pitch Me Green. (2014). *Pitch Me Green. Mentoring program for early stage green and clean technology ventures*. Reviewed March 4, 2014 from http://www.pitchmegreen.com/

Plassmann, H., Ramsøy, T. Z., & Milosavljevic, M. (2012). Branding the brain: A critical review and outlook. *Journal of Consumer Psychology, 22*(1), 18–36. doi:10.1016/j.jcps.2011.11.010

Podsakoff, P. M., MacKenzie, S. B., Lee, J.-Y., & Podsakoff, N. P. (2003). Common method biases in behavioral research: A critical review of the literature and recommended remedies.[PubMed]. *The Journal of Applied Psychology, 88*(5), 879–903. doi:10.1037/0021-9010.88.5.879

Podsakoff, P. M., MacKenzie, S. B., & Podsakoff, N. P. (2012). Sources of method bias in social science research and recommendations on how to control it.[PubMed]. *Annual Review of Psychology, 63*(1), 539–569. doi:10.1146/annurev-psych-120710-100452

Pop, Dabija, & Iorga. (2014). Ethical Responsibility Of Neuromarketing Companies In Harnessing The Market Research – A Global Exploratory Approach. *Ethical Responsibility of Neuromarketing Companies in Harnessing the Market Research – a Global Exploratory Approach. Organizations' responsibility to Consumers, 16*(35).

Pop, N. A. (2012). Ana Maria Iorga. A New Challenge For Contemporary Marketing – Neuromarketing. *Management & Marketing Challenges for the Knowledge Society, 7*(4), 631–644.

Porter, M. E. (1990a). The Competitive Advantage of Nations. *Harvard Business Review, 68*(2), 73–93.

Porter, M. E. (1990b). *The Competitive Advantage of Nations*. Free Press. doi:10.1007/978-1-349-11336-1

Prahalad, C. K., & Hart, S. (2002). The Fortune at the Bottom of the Pyramid. *Strategy & Business, 26*, 54–67.

Prahalad, C. K., & Ramaswamy, V. (2004a). Co-Creating unique value with customers. *Strategy and Leadership, 32*(3), 4–9. doi:10.1108/10878570410699249

Press, G. (2015). *6 Predictions About The Future Of Digital Transformation*. Retrieved from http://www.forbes.com/sites/gilpress/2015/12/06/6-predictions-about-the-future-of-digital-transformation/2/#e64562d750c

Priporas, C. V., Gatsoris, L., & Zacharis, V. (2005). Competitive intelligence activity: Evidence from Greece. *Marketing Intelligence & Planning*, *23*(7), 659–669. doi:10.1108/02634500510630195

Puchol, L., Martín, M., & Núñez, A. (2006). *El libro de las habilidades directivas*. España: Díaz de Santos.

Purdue University. (n.d.). *Elevator Pitch Contest*. Retrieved from: http://www.purdue.edu/discoverypark/entr/competitions/elevator_pitch_contest.php

PwC. (2015). *The Sharing Economy. Consumer Intelligence Series*. Retrieved from https://www.pwc.com/us/en/industry/entertainment-media/publications/consumer-intelligence-series/assets/pwc-cis-sharing-economy.pdf

Quek, E. E. (2003). A survey of TQM practices in the Malaysian electrical and electronic industry. *Total Quality Management and Business Excellence*, *14*(1), 63–78. doi:10.1080/14783360309708

Quinones, G., Nicholson, B., & Heeks, R. (2013). Positioning research on e-entrepreneurship in developing economies: a study of latin-american digital ventures, Paper for ICIER 2013, Rio de Janeiro.

Quiroga, R. et al. (2014). *Panorama del ecosistema del emprendimiento de Bogotá*. IMPACT-A.

Ragalie, S., Neagu, C., Muscalu, M. S., Russu, C., Badileanu, M., Bulearca, M., & Neacsu Dalu, I. et al. (2007). Modalitati de imbunatatire a productivitatii la nivel de organizatie, sectorial si national, *Journal of Industrial Eonomics. Centre for Industrial Economics and Services*, *5*(4), 7–14.

Rahman, M., Hasan, M. R., & Floyd, D. (2013). Brand orientation as a strategy that influences the adoption of innovation in the bottom of the pyramid market. *Strategic Change*, *22*(3-4), 225–239. doi:10.1002/jsc.1935

Ramírez, M. P. (2016). Modelo de Innovación Abierta Colaborativa. Universidad EAN.

Ramírez, J., & Anderson, S. (2007). *Kiva Case Study*. UC Berkeley.

Ramsøy, T. (2014). Consumer Neuroscience and Neuromarketing (2nd ed.). Academic Press.

Ramsøy, T. Z., & Skov, M. (2014). Brand preference affects the threshold for perceptual awareness. *Journal of Consumer Behaviour*, *13*(1), 1–8. doi:10.1002/cb.1451

Ranjan, J. (2009). Business intelligence: Concepts, components, techniques and benefits. *Journal of Theoretical and Applied Information Technology*, *9*(1), 60–70.

Rappa, M. (2003). *Business models on the web*. Available at http://digitalenterprise.org/models/models.html

Rapp, J., Hill, R. P., Gaines, J., & Mark Wilson, R. (2009, Winter). Advertising and Consumer Privacy. Old Practices and New Challenges. *Journal of Advertising*, *38*(4), 51–61. doi:10.2753/JOA0091-3367380404

Ratten, V. (2013). Cloud computing: A social cognitive perspective of ethics, entrepreneurship, technology marketing, computer self-efficacy and outcome expectancy on behavioral intentions. *Australasian Marketing Journal*, *21*(3), 137–146. doi:10.1016/j.ausmj.2013.02.008

Reckwitz, A. (2002). Toward a Theory of Social Practices A development in culturalist theorizing. *European Journal of Social Theory*, *5*(2), 243–263. doi:10.1177/13684310222225432

Reich, B. H., Sauer, C., & Wee, S. Y. (2008). Innovative practices for IT projects. *Information Systems Management*, *25*(3), 266–272. doi:10.1080/10580530802151210

Rensselaer. (n.d.). Retrieved from: http://www.eship.rpi.edu/elevator.php

Renvoisé, P., & Morin, C. (2007). *Neuromarketing: Understanding the "Buy Button" in Your Customer's Brain*. Thomas Nelson.

Reynolds, P., Camp, S., Bygrave, W., Autio, E., & Hay, M. (1999). *Global Entrepreneurship Monitor: Executive Report*. Kauffman Center for Entrepreneurial Leadership/Babson College/London Business School.

Rhodes, G., Proffitt, F., Grady, J. M., & Sumich, A. (1998). Facial symmetry and the perception of beauty. *Psychonomic Bulletin & Review*, *5*(4), 659–669. doi:10.3758/BF03208842

Ries, E., & San Julián, J. (2013). *El método lean startup: cómo crear empresas de éxito utilizando la innovación continua*. Barcelona: Deusto.

Robinson, P., & Sexton, E. (1994). The effect of education and experience on self-employment success. *Journal of Business Venturing*, *9*(2), 141–157. doi:10.1016/0883-9026(94)90006-X

Rogers, E. (2008). *Diffusion of innovations*. New York: Free Press.

Roitner, A. (2008). Competitive intelligence in Austria: An empirical study. *(Masters of Arts Unpublished)*. Wien University.

Roper, S., & Hart, M. (2013). *Supporting Sustained Growth Among SMEs – Policy Models and Guidelines*. University of Warwick, Enterprise Research Centre, White Paper No. 7.

Røpke, I. (2009). Theories of practice — New inspiration for ecological economic studies on consumption. *Ecological Economics*, *68*(10), 2490–2497. doi:10.1016/j.ecolecon.2009.05.015

Rossberger, R. J. (2014). National Personality Profiles and Innovation: The Role of Cultural Practices. *Creativity and Innovation Management*, *23*(3), 331–348. doi:10.1111/caim.12075

Rowe, C. (1996). Evaluating management training and development: Revisiting the basic issues. *Industrial and commercial. Training (New York, N.Y.)*, *28*(4), 17–23.

Roy, A. (2010). *Poverty Capital: Microfinance and the Making of Development*. New York: Routledge.

Roy, A. (2012). Subjects of Risk: Technologies of Gender in the Making of Millennial Modernity. *Public Culture*, *24*(1), 131–155. doi:10.1215/08992363-1498001

Rusko, R. (2013). The Redefined Role of Consumer as a Prosumer: Value Co-Creation, Coopetition, and Crowdsourcing of Information Goods. In P. Renna (Ed.), *Production and Manufacturing System Management: Coordination Approaches and Multi-Site Planning*. IGI Global. doi:10.4018/978-1-4666-2098-8.ch009

Rusko, R. (2014). Crowdsourcing for Transcultural Marketing and Innovation. In B. Christiansen, S. Yıldız, & E. Yıldız (Eds.), *Transcultural Marketing for Incremental and Radical Innovation* (pp. 382–394). Hershey, PA: Business Science Reference; doi:10.4018/978-1-4666-4749-7.ch018

Rutherford, S. (2008). Micropensions: Old Age Security for the Poor? In *New Partnerships for Innovation in Microfinance* (pp. 241–264). Springer.

Saaty, T. L. (1994). How to make a decision: The analytic hierarchy process. *Interfaces*, *24*(6), 19–43. doi:10.1287/inte.24.6.19

Saayman, A., Pienaar, J., de Pelsmacker, P., Viviers, W., Cuyvers, L., Muller, M.-L., & Jegers, M. (2008). Competitive intelligence: Construct exploration, validation and equivalence. *Aslib Proceedings*, *60*(4), 383–411. doi:10.1108/00012530810888006

Sábato, J., & Botana, N. (1968). La ciencia y la tecnología en el desarrollo futuro de América Latina. *Revista de la Integración*, *1*(3), 15–36.

Saiz-Álvarez, J. M. (2009). La organización de Cooperación de Shangai (ocs) claves para la creación de un futuro líder mundial. *Revista De Economía Mundial*, (23), 307-326. (in Spanish)

Saiz-Alvarez, J. M. (Ed.). (2016). Handbook of Research on Social Entrepreneurship and Solidarity Economics. Hershey, PA: Information Science Reference (IGI Global).

Saiz-Alvarez, J. M., & Colvin-Diez, J. (2015). Purple Oceans and the M&A Virtuous Circle: New Concepts for the Theory of Mergers and Acquisitions. In A. Tavidze (Ed.), *Progress in Economics Research* (pp. 139–154). Nova Press.

Saiz-Álvarez, J. M., & Colvin, J. (2015). *Purple Oceans and the Mergers & Acquisitions Virtuous Circle. New Concepts for the Theory of Mergers & Acquisitions. In Progress in Economic Research* (vol. 29, pp. 45–58). Nova Press.

Sako, M. (2012). Business Models for Strategy and Innovation. *Communications of the ACM*, *55*(7), 22–24. doi:10.1145/2209249.2209259

Saleh, A. S., & Ndubisi, N. O. (2006). An evaluation of SME development in Malaysia. *International Review of Business Research Papers*, *2*(1), 1–14.

Samad, N. A. (2007). *Positioning Malaysian SMEs in the global.* Paper presented at the Proceedings of Persidangan Kebangsaan IKS 2007, Kota Kinabalu, Malaysia.

Samal, A., & Iyengar, P. A. (1992). Automatic recognition and analysis of human faces and facial expressions. *Pattern Recognition*, *25*(1), 65–77. doi:10.1016/0031-3203(92)90007-6

Sarasvathy, D. K., Simon, H. A., & Lave, L. (1998). Perceiving and managing business risks: Differences between entrepreneurs and bankers. *Journal of Economic Behavior & Organization*, *33*(2), 207–225. doi:10.1016/S0167-2681(97)00092-9

Saunders, M., Lewis, P., & Thornhill, A. (2007). *Research methods for business students* (4th ed.). London: Prentice Hall.

Sawyer, F. H. (1975). A conceptual analysis of empathy. *The Annual of Psychoanalysis*, *3*, 37–47.

Schenk, E., & Guittard, C. (2009, December). *Crowdsourcing: What can be Outsourced to the Crowd, and Why*. In *Workshop on Open Source Innovation*, Strasbourg, France.

Schindehutte, M. (2006). *Entrepreneurship Simulations*. Accessed September 2, 2014, http://old.ied.econ.msu.ru/cmt2/lib/c/198/File/simulations.pdf

Schumpeter, J. (1942). *Creative destruction. Capitalism, socialism and democracy*. Harper Perennial Modern Classics.

Schumpeter, J. A. (1934). *The Theory of Economic Development; an Inquiry into Profits, Capital, Credit, Interst, and the Business Cycle*. Cambridge, MA: Harvard University Press.

Schumpeter, J. A. (1939). *Business cycles* (Vol. 1). New York: McGraw-Hill.

Schumpeter, J. A. (1939). *Business Cycles: A Theoretical, Historical, and Statistical Analysis of the Capitalist Process*. New York: McGraw-Hill.

Schumpeter, J. A. (1943). *Capitalism, socialism and democracy* (6th ed.). Routledge.

Schumpeter, J. A. (1994). *Teoría del desenvolvimiento económico*. México: FCE.

Schwittay, A. F. (2013). Making Poverty into a Financial Problem: From Global Poverty lines to Kiva.org. *Journal of International Development*, *26*(4), 508–519. doi:10.1002/jid.2966

Secure-retail.com. (2015). *Secure Retail*. Retrieved from http://www.secure-retail.com/blog/reverse_showrooming_why_the_web_is_actually_driving_traffic_into_the_store

Senge, P., Smith, B., Kruschwitz, N., Laur, J., & Schley, S. (2011). The Necessary Revolution. *Necessary Revolution - Business Book Summaries, 1*.

Sewdass, N., & Du Toit, A. (2014). Current state of competitive intelligence in South Africa. *International Journal of Information Management, 34*(2), 185–190. doi:10.1016/j.ijinfomgt.2013.10.006

Shaker, S., & Gembicki, M. (1999). *The war room guide to competitive intelligence*. New York: McGraw-Hill.

Shane, S. A. (2003). *A general theory of entrepreneurship: the individual-opportunity nexus*. Cheltenham, UK: Edward Elgar. doi:10.4337/9781781007990

Shane, S. A., & Venkataraman, S. (2000). The Promise of Entrepreneurship as a Field of Research. *Academy of Management Review, 25*(1).

Sharma, P., Chrisman, J. J., & Chua, J. H. (1997). Strategic Management of the Family Business: Past Research and Future Challenges. *Family Business Review, 10*(1), 1–34. doi:10.1111/j.1741-6248.1997.00001.x

Shastri, R., Waddiparthi, K., Nagarajan, S., Narayanan, A., & Ahamed, A. (2014, December 1). *Building a Code Halo Economy for Insurance*. Retrieved from https://www.cognizant.com/InsightsWhitepapers/building-a-code-halo-economy-for-insurance-codex1072.pdf

Shiv, B., & Fedorikhin, A. (1999, December). Heart and Mind in Conflict: The Interplay of Affect and Cognition in Consumer Decision Making. *The Journal of Consumer Research, 26*(3), 278–292. doi:10.1086/209563

Shukran, M. K., Sultana, M. R., & Rahman, F. (2011). *Moving towards the knowledge-based economy: A comparative study between Malaysia and South Korea*. Paper presented at the Research and Innovation in Information Systems (ICRIIS), 2011 International Conference.

Simonson, I. (2005). In Defense of Consciousness: The Role of Conscious and Unconscious Inputs in Consumer Choice. *Journal of Consumer Psychology, 15*(3), 211–217. doi:10.1207/s15327663jcp1503_5

Simpson, B. (2009). Pragmatism, Mead and the practice turn. *Organization Studies, 30*(12), 1329–1347. doi:10.1177/0170840609349861

Singapore Government. (2013). *SPRING Singapore. Performance indicators*. Retrieved from http://www.spring.gov.sg/aboutus/pi/pages/performance-indicators.aspx#.Ud4_nkHI0-J

Slush. (2014). *SLUSH*. Reviewed March 4, 2014 from http://www.slush.org/

SMECorp Malaysia. (2013). *New definition of small and medium enterprises (SMEs)*.

Smidts, A., Hsu, M., Sanfey, A. G., Boksem, M. A. S., Ebstein, R. B., Huettel, S. A., & Yoon, C. et al. (2014). Advancing consumer neuroscience. *Marketing Letters, 25*(3), 257–267. doi:10.1007/s11002-014-9306-1

Sobel, D. (2004). Longitude: The True Story of a Lone Genius who Solved the Greatest Scientific Problem of His Time. *Academy of Management Learning & Education, 3*(2), 220–220. doi:10.5465/AMLE.2004.13500465

Souitaris, V., & Cohen, M. (2003). Internet-Business or Just Business?: Impact of 'Internet-Specific' Strategies on Venture Performance. *European Management Journal, 21*(4), 421–437. doi:10.1016/S0263-2373(03)00080-X

Spiegel, M. R., & Stephens, L. (2009). *Estadística. 4ta edición*. D.F, México: Mc Graw-Hill.

Srinivasan, S. (2014). *Cloud Computing Basics*. SpringerBriefs in Electrical and Computer Engineering. DOI 10.1007/978-1-4614-7699-3_1

Starta-Etan Expet Group. (2003). *Higher education and research for the ERA: Current trends and challenges for the near future*. Bruxelles, Belgium: European Commission, DG Research.

StartUp Boot Camp. (n.d.a). *Istanbul*. Retrieved from: http://www.startupbootcamp.org/events/istanbul-pitch-day-3.html

StartUp Boot Camp. (n.d.b) Retrieved from: http://www.startupbootcamp.org/events/startupbootcamp-pitch-day-athens.html

StartUp Pitch Night. (n.d.). Retrieved from: http://www.startuppitchnight.com/startuppitchnight/

Sterbenz, C. (2015). Here's who comes after Generation Z — and they'll be the most transformative age group ever. *Business Insider*. Retrieved from http://www.businessinsider.com/generation-alpha-2014-7-2

Strauss, A. C., & du Toit, A. S. A. (2010). Competitive intelligence skills needed to enhance South Africa's competitiveness. *Aslib Proceedings*, *62*(3), 302–320. doi:10.1108/00012531011046925

Suomala, J., Palokangas, L., Leminen, S., Westerlund, M., Heinonen, J., & Numminen, J. (2012, December). Technology Innovation. *Management Review*.

Sveiby, K. (1997). *The new organizational wealth: Managing and measuring knowledge-based assets*. San Francisco, CA: Berrett-Koehler Publishers.

Swan, M. (2012). Crowdsourced Health Research Studies: An Important Emerging Complement to Clinical Trials in the Public Health Research Ecosystem. *Journal of Medical Internet Research*, *14*(2), e46. doi:10.2196/jmir.1988 PMID:22397809

SwitchPitch. (n.d.). *SwitchPitch is coming to New York*. Retrieved from: http://switchpitch.com/switchpitch-nyc/

Tackey, N., Perryman, S., & Connor, H. (1999). *Graduated business star-ups*. The Institute for Employment Studies, Reino Unido. UK Business Incubation (UKBI). Retrieved from: www.ukbi.co.uk

Tanev, S., & Bailetti, T. (2008). Competitive intelligence information and innovation in small Canadian firms. *European Journal of Marketing*, *42*(7-8), 786–803. doi:10.1108/03090560810877150

Tapscott, D. (1996). *The digital economy: Promise and peril in the age of networked intelligence*. McGraw-Hill New York.

Taylor, R. (2015). Creativity at Work. Kogan Page Limited.

Taylor, M., & Murphy, A. (2004). SMEs and e-business. *Journal of Small Business and Enterprise Development*, *11*(3), 280–289. doi:10.1108/14626000410551546

Taylor, S., & Novelli, L. Jr. (1991). Telling a Story About Innovation. *Issues & Observations*, *11*(1), 6–9.

TCS Digital Enterprise. (2014). *Digital Reimagination ™ - The Art of Reimagining Business*. Retrieved from http://www.tcs.com/SiteCollectionDocuments/White%20Papers/Art-reimagining-business-0714-1.pdf

Teece, D. J. (1992). Foreign Investment and Technological Development in Silicon Valley. *California Management Review*, *34*(2), 88–106. doi:10.2307/41166695

Teece, D. J. (2010). Business models, business strategy and innovation. *Long Range Planning*, *43*(2), 172–194. doi:10.1016/j.lrp.2009.07.003

Telefonica. (2014). *Telefonica survey of Millennials*. Retrieved from http://es.slideshare.net/TelefonicaEurope/telefonica-2014-gms-global-master-deck-final-100614?related=1

Tervo, T. (2013). *The Content is King! How to Sell the Documentary Idea for the Production Company?*. Helsinki: Metropolia University of Applied Sciences.

Tether, B. (2005). Do Services Innovate (Differently)? Insights from the European Innobarometer Survey. *Industry and Innovation*, *12*(2), 153–184. doi:10.1080/13662710500087891

The Entrpreneur Challenge. (2016). *Elevator Pitch Competition*. Retrieved from: http://challenge.ucsd.edu/elevator-pitch

Thewarapperuma, A. (2013). *East African virtual incubation pilot to launch in Nairobi, Kenya*. infoDev. Available: http://www.infodev.org/en/Article.968.html

Thomas, R., Kass, A., & Davarzani, L. (2015). *How digital technologies are changing the way we work | Accenture Outlook*. Accenture.com. Retrieved from https://www.accenture.com/us-en/insight-outlook-how-digital-technologies-are-changing-the-way-we-work.aspx

Tienari, J., & Meriläinen, S. (2009). *Johtaminen ja organisointi globaalissa taloudessa*. Helsinki: WSOYpro Oy.

Timmers, P. (1998). Business models for electronic markets. *Electronic Markets*, *8*(2), 3–8. doi:10.1080/10196789800000016

Tom, G., Nelson, C., Srzentic, T., & King, R. (2007). Mere Exposure and the Endowment Effect on Consumer Decision Making. *The Journal of Psychology*, *141*(2), 117–125. doi:10.3200/JRLP.141.2.117-126 PMID:17479582

Tornatzky, L., & Fleischer, M. (1990). *The Process of Technological Innovation*. Lexington.

Tsai, W. T., Sun, X., & Balasooriya, J. (2010). Service-Oriented Cloud Computing Architecture. In *Proceedings of the 2010 Seventh International Conference on Information Technology: New Generations.*

Ultra Light Start Ups. (n.d.). Retrieved from: http://ultralightstartups.com/

University of Idaho. (n.d.). Retrieved from: http://www.uidaho.edu/cbe/experientiallearning/view/elevatorpitch

Upton, N., Sexton, D., & Moore, C. (1995). *Have we made a difference? An examination of career activity of entrepreneurship majors since 1981. In Trabajo presentado en la Entrepreneurship Research Conference*. Babson College.

Ur-Rahman. (2013). SMEs can be engine of economic growth. *Pakistan's Premier Financial Daily*. Retrieved from http://www.brecorder.com/articles-a-letters/187/1181556/

Utterback, J. M. (1971). The Process of Technological Innovation within the Firm. *Academy of Management Journal*, *14*(1), 75–88. doi:10.2307/254712

Vaara, E., Kleymann, B., & Seristö, H. (2004). Strategies as discursive constructions: The case of airline alliances. *Journal of Management Studies*, *41*(1), 1–35. doi:10.1111/j.1467-6486.2004.00419.x

Valerio, A., Parton, B., & Robb, A. (2014). *Entrepreneurship Education and Training Programs around the World: Dimensions for Success*. Washington, DC: World Bank. doi:10.1596/978-1-4648-0202-7

Valles, Y, (2014). De baby boomers a millennials, la necesaria evolución de las estructuras sociales. *Revista Tecnológico de Monterrey "Generaciones HOY"*.

van der Kuip, I., & Verheul, I. (2003). *Early Development of Entrepreneurial Qualities: the Role of Initial Education. SCALES-paper N200311*. EIM Business and Policy Research.

Van Stel, A., Carree, M., & Thurik, R. (2005). The effect of entrepreneurial activity on national economic growth. *Small Business Economics*, *24*(3), 311–321. doi:10.1007/s11187-005-1996-6

VanGundy, A. (2007). *Getting to Innovation: How Asking the Right Questions Generates Great Ideas Your Company Needs*. Amacom.

Varela, R., Moreno, J., & Bedoya, M. (2015). *Global Entrepreneurship Monitor 2006 – 2013*. Cali: Universidad ICESI.

Vasilchenko, E., & Morrish, S. (2011). The Role of Entrepreneurial Networks in the Exploration and Exploitation of Internationalization Opportunities by Information and Communication Technology Firms. *Journal of International Marketing*, *19*(4), 88–105. doi:10.1509/jim.10.0134

Veciana, J. M. (1999). Creación de empresas como programa de investigación científica. *Revista Europea de Dirección y Economía de la Empresa*, *8*(3), 11–36.

Veciana, J. M. (2002). *Función Directiva*. Alfaomega, Universidad Autónoma de Barcelona.

Vedder, R. G., Vanecek, M. T., Guynes, C. S., & Cappel, J. J. (1999). CEO and CIO perspectives on competitive intelligence. *Communications of the ACM*, *42*(8), 108–116. doi:10.1145/310930.310982

Venkataraman, S. (1997). *The distinctive domain of entrepreneurship research*: An 23 editor's perspective. In J. Katz & J. Brockhaus (Eds.), *Advances in Entrepreneurship, Firm Emergence, and Growth*. Greenwich, CT: JAI Press.

Verner, R. (2001). Functional asymmetry in the human face: Perception of health in the left and right sides of the face. *Laterality*, *6*(3), 225–231. doi:10.1080/713754415 PMID:15513171

Vesper, K. H. (1996). *New Venture Experience*. Seattle, WA: Vector Books.

Vicens, L., & Grullón, S. (2011). *Innovation and Entrepreneurship: A Model Based on Entrepreneur Development*. Inter-American Development Bank. Retrieved from http://www.iadb.org

Vouk, M. (2008). Cloud Computing – Issues, Research and Implementations. *Journal of Computing and Information Technology*, 235-246.

Wadie, N. (2011). Competitive intelligence in Tunisian companies. *Journal of Enterprise Information Management*, *24*(1), 53–67. doi:10.1108/17410391111097429

Wagner, L. (2002). The 'new' donor: Creation or evolution? *International Journal of Nonprofit and Voluntary Sector Marketing*, *7*(4), 343–352. doi:10.1002/nvsm.191

Wall, B., Jagdev, H., & Browne, J. (2007). A review of eBusiness and digital business—applications, models and trends. *Production Planning and Control*, *18*(3), 239–260. doi:10.1080/09537280601127245

Wang, C. L. (2008). Entrepreneurial orientation, learning orientation, and firm performance. *Entrepreneurship Theory and Practice*, *32*(4), 635–657. doi:10.1111/j.1540-6520.2008.00246.x

Ward, J. L., & Aronoff, C. E. (1991). The Power of Patient Capital. *Nation's Business*, *79*(9), 48–49.

Webb, J. W., Ketchen, D. J. Jr, & Ireland, R. D. (2010). Strategic entrepreneurship within family-controlled firms: Opportunities and challenges. *Journal of Family Business Strategy*, *1*(2), 67–77. doi:10.1016/j.jfbs.2010.04.002

Weill, P., & Woerner, S. L. (2013). Optimizing Your Digital Business Model. *MIT Sloan Management Review*, *54*, 71–78.

Wenger, E. (1998). *Communities of Practice: Learning, Meaning and Identity*. Cambridge, MA: Cambridge University Press. doi:10.1017/CBO9780511803932

Whittington, R. (2006). Completing the practice turn in strategy research. *Organization Studies*, *27*(5), 613–634. doi:10.1177/0170840606064101

Wiklund, J., & Shepherd, D. (2003). Knowledge-Based Resources, Entrepreneurial Orientation, and the Performance of Small and Medium-Sized Businesses. *Strategic Management Journal*, *24*(13), 1307–1314. doi:10.1002/smj.360

Wiklund, J., & Shepherd, D. (2005). Entrepreneurial orientation and small business performance: A configurational approach. *Journal of Business Venturing*, *20*(1), 71–91. doi:10.1016/j.jbusvent.2004.01.001

Willems, R. M., Van der Haegen, L., Fisher, S. E., & Francks, C. (2014). On the other hand: Including left-handers in cognitive neuroscience and neurogenetics. *Nature Reviews. Neuroscience*, *15*(3), 193–201. doi:10.1038/nrn3679 PMID:24518415

Wilson, R. M., Gaines, J., & Hill, R. P. (2008). Neuromarketing and Consumer Free Will. *The Journal of Consumer Affairs*, *42*(3), 389–410. doi:10.1111/j.1745-6606.2008.00114.x

Wind, J., & Rangaswamy, A. (2001). Customerization: The next revolution in mass customization. *Journal of Interactive Marketing*, *15*(1), 13–32. doi:10.1002/1520-6653(200124)15:1<13::AID-DIR1001>3.0.CO;2-#

Wittrock, M. C. (1990). Generating processes of comprehension. *Educational Psychologist*, *24*(4), 345–376. doi:10.1207/s15326985ep2404_2

World Bank. (2016). *World Development Indicators.* World Bank. Available at: http://data.worldbank.org/data- catalog/world-development-indicators

World Bank. (n.d.). Retrieved from: http://www.worldbank.org

World Economic Forum. (2014). *The global competitiveness report 2014 - 2015.* Geneva, Switzerland: World Economic Forum.

Wright, S., Pickton, D. W., & Callow, J. (2002). Competitive intelligence in UK firms: A typology. *Marketing Intelligence & Planning*, *20*(6), 349–360. doi:10.1108/02634500210445400

Wright, T. A., & Cropanzano, R. (2004). The Role of Psychological Well-being in Job Performance: A Fresh Look at an Age-Old Quest. *Organizational Dynamics*, *33*(4), 338–351. doi:10.1016/j.orgdyn.2004.09.002

Xenikou, A., & Simosi, A. (2006). Organizational culture and transformational leadership as predictors of business unit performance. *Journal of Managerial Psychology*, *21*(6), 566–579. doi:10.1108/02683940610684409

Yartey, F. N. A. (2014). Microfinance, Digital Media and Social Change: A visual analysis of Kiva.org. *Communication & Social Change*, *1*(1), 63–89.

Yin, R. K. (1983). *Case Study Research: Design and Methods* (3rd ed.). Sage Publications.

Yoon, C., Gonzalez, R., Bechara, A., Berns, G. S., Dagher, A. A., Dubé, L., & Spence, C. et al. (2012). Decision neuroscience and consumer decision making. *Marketing Letters*, *23*(2), 473–485. doi:10.1007/s11002-012-9188-z

Young, E. C., & Welsch, H. P. (1993). Major elements in entrepreneurial development in central Mexico. *Journal of Small Business Management*, *31*(4), 80.

Youtube, R. S. (2015). Retrieved from https://www.youtube.com/watch?v=R8WkIC7bMnE

Youtube. (n.d.). Retrieved September 2015, from https://www.youtube.com/watch?v=IGQmdoK_ZfY

Yunus, M. (1999). *Banker to the Poor: Micro-Lending and the Battle against World Poverty.* New York: Public Affairs.

Zahra, S. A., & Nambisan, S. (2011). Entrepreneurship in global innovation ecosystems. *AMS Review, 1*(1), 4-17.

Zahra, S. A. (2005). Entrepreneurial risk-taking in family firms. *Family Business Review*, *18*(1), 23–40. doi:10.1111/j.1741-6248.2005.00028.x

Zahra, S. A., Hayton, J. C., & Salvato, C. (2004). Entrepreneurial in Family vs. Non-Family Firms: A Resource-Based Analysis of the Effect of Organizational Culture. *Entrepreneurship: Theory and Practice*, *28*(4), 363–382.

Zahra, S. A., Neubaum, D. O., & Huse, M. (2000). Entrepreneurship in medium-size companies: Exploring the effects of ownership and governance systems. *Journal of Management*, *26*(5), 947–976. doi:10.1177/014920630002600509

Zahra, S., & Dess, G. G. (2001). Entrepreneurship as a field of research: Encouraging dialogue and debate. *Academy of Management Review*, *26*(1), 8–10. doi:10.5465/AMR.2001.4011916

Zahra, S., Newey, L., & Li, Y. (2014). On the frontiers: The implications of social entrepreneurship for international entrepreneurship. *Entrepreneurship Theory and Practice*, *38*(1), 137–158. doi:10.1111/etap.12061

Zarate, R. (2015). *Liderazgo como factor clave en la sostenibilidad empresarial. In Gestión de la Sostenibilidad en el marco de las organizaciones*. Bogotá, Colombia: Editorial EAN.

Zaugg, R., & Thom, N. (2003). Excellence through implicit competencies: Human resource management, organisational development, knowledge creation. *Journal of Change Management*, *3*(3), 199–212. doi:10.1080/714042541

Zhu, K., Kraemer, K. L., & Xu, S. (2006). The Process of Innovation Assimilation by Firms in Different Countries: A Technology Diffusion Perspective on E-Business. *Management Science*, *52*(10), 1557–1576. doi:10.1287/mnsc.1050.0487

Zucchermaglio, C., & Talamo, A. (2003). The development of a virtual community of practices using electronic mail and communicative genres. *Journal of Business and Technical Communication*, *17*(3), 259–284. doi:10.1177/1050651903017003001

Zylkifli, Kamal, & Char, Rushdan, Yasoa, & Hassan. (2010). Small and medium enterprises (SMES) competing in the global business environment: A case study of Malaysia. *International Business Research*, *3*(1), 66–75.

About the Contributors

Ioana Iancu graduated from Political Science. She has MA studies in Management of Political Organizations and in Advertising at Babes-Bolyai University, Cluj-Napoca and MA studies in Political Science at Central European University, Budapest. She has a PhD in Sociology. She currently holds a Lecturer position at the Faculty of Political, Administrative and Communication Sciences, Babes-Bolyai University, Cluj-Napoca, Romania.

* * *

Ainul Mohsein Abdul-Mohsin is a senior lecturer at School of Management, Universiti Sains Malaysia. She is attached to the Department of International Business. She has 15 years of corporate experience from various industries. Her work experience varies from the financial services, international non-profit organisation, manufacturing, shared services and consultancy. She has work in Malaysia and abroad. Her interest in research are in competitive intelligence, change management, entrepreneurial competencies, and innovative performance.

Catalina Lucía Ruiz Arias is a Master of Science in Strategic Product Design from Delft University of Technology (Netherlands), a Specialist in Design and Product Management for Export, and an Industrial Designer. She has obtained her business experience in product design, development, and consultancy at national and international companies from various sectors (including companies located in Brazil and the Netherlands). Ms. Ruiz has also worked as a professor at various universities of the country. At present, she works for Universidad EAN as an Associate Professor at the Institute for Entrepreneurship, Leadership and Innovation - IDELI (from the Spanish Instituto de Emprendimiento, Liderazgo e Innovación), where she leads projects associated with consulting and training on innovation and design thinking for entrepreneurs and enterprises, and she coordinates the entrepreneurship incubator program (EAN IMPACTA).

Verónica Baños-Monroy is a Professor of Entrepreneurship at Tecnológico de Monterrey (Mexico). She received her M.Sc. A. from Tecnológico de Monterrey (ITESM), a Ph.D. in Administrative Science from the Universidad Nacional Autónoma de México (UNAM) and a second Ph. D. in International Business from the Universidad Autónoma de Madrid (UAM) in Spain. She studies the succession process and innovation in family firms. She has served as a Visiting Professor at the University of San Diego. She has also published articles in different international academic journals. Currently serves as a Full Professor and as a Family Business Consultant in the Entrepreneurship Division at Tecnológico de Monterrey, campus Guadalajara; she teaches courses in Family Business and Service Management at the undergraduate and masters levels.

Constanta-Nicoleta Bodea is a Professor of artificial intelligence and project management at the Academy of Economic Studies (AES), Bucharest, Romania. She directs a Masters Program in Project Management, and is the head of the Economic Research Department at AES. She managed more than 20 R&D and IT projects in the last ten years, contributor and author of 11 books and more than 50 papers on Project Management, Information Systems, and Artificial Intelligence, being Honored by IPMA with the Outstanding Research Contributions (2007).

Francisco Javier Matiz Bulla is a PhD (c) in New Economy Management from Universidad de Las Palmas de Gran Canaria - ULPGC (Spain), a Master in Organizational Management from Université du Québec à Chicoutimi - UQAC (Canada) and Universidad EAN, a Specialist in Consumer Psychology, and a Professional in International Trade. He coordinates the Young Export Entrepreneurs Program of the Ministry of Foreign Trade and has worked as a professor at various national universities. At present, he works for Universidad EAN as an Associate Professor, and as the Director of the Entrepreneurship Research Group and the Entrepreneur Program belonging to the Institute for Entrepreneurship, Leadership and Innovation - IDELI (from the Spanish Instituto de Emprendimiento, Liderazgo e Innovación) at the said University.

Omar Alonso Patiño Castro is a Doctor of Business Administration from the Nebrija University, Business Administrator at Pontificia Universidad Javeriana, a Specialist in Quality Management and Process Innovation from the Center for Postgraduate Studies in Business Administration of the Polytechnic University of Madrid, has a Master in Management from the University of Quebec Chicoutimi, with experience in financial and educational sectors. He is a Professor and the Director of the Research Group at the University EAN wherein he has also been the Dean and the Director of the Program of Business Administration.

Jorge Colvin-Díez has a PhD in Business Administration, Nebrija University, Spain. Diploma in Management Development Program, IESE-University of Navarre, Spain. Master in Finance at IEB (Instituto de Estudios Bursátiles), Spain. He is Finance and Accounting Professor at the American Schiller International University, Madrid Campus, and Associate Professor, Nebrija Business School, Spain. As researcher, he has developed either academic collaborations with IESE and have publications in scientific indexed journals. He has over 15 years of professional experience as auditor for PricewaterhouseCoopers and KPMG. Also he was responsible for mergers and acquisitions in Spain and member of the Chairman's Office for Codere. He currently works as an advisor in mergers and acquisitions for Spanish multinationals, and for advisory firms, investment banks, and private equity firms.

Alexandru Aurel Costan is a Ph.D. candidate at the Tehnical University of Cluj-Napoca, Computer Science department. His research focus is on cloud computing and inter-cloud technologies.

Vasile Teodor Dadarlat holds a Professor position, since 1998, at the Computer Science Department, Technical University of Cluj-Napoca. He is an invited professor at ICAM Lille, France, internal and external examiner for MSc and PhD thesis at University of Limerick, Ireland. He also serves as expert for national/international bodies. His main research interests are in communication networks protocols, telecommunication systems, integrated digital circuits, next generation networks and security in computer networks.

Spoaller F. Dorin holds a Bachelor Degree in Advertising and PR at Babeş-Bolyai University. Masters Degree in Advertising at Babeş-Bolyai University.

Katja Härkönen has long experience as entrepreneur. She has familiarized with innovation projects and events. Her education is focused on management and entrepreneurship, especially in the University of Lapland. Research interests are in the branches of entrepreneurship, innovations and management.

Bogdan Iancu graduated from the Technical University of Cluj-Napoca, Romania in 2006, with a bachelor degree in Computer Science. In 2008 he graduated a master course in Software Engineering at the same University and in 2011 he received his Ph.D. from the Technical University of Cluj-Napoca, with the thesis "Contributions to Quality of Service Assurance and Performance Enhancement in Computer Networks". He currently holds an Senior Lecturer position at the Computer Science Department, TUCN. His main research interests are in computer networks, network management, Quality of Service in homogeneous and heterogeneous networks, NGNs, sensor networks, and their applications to VANETs and smart grids. Also, a special interest is in the field of applied technology to innovative business.

José Emilio Jiménez Ibáñez is a PhD (c) in Industrial Project Design, Manufacturing, and Management; a Master in Design Engineering from Universitat Politècnica de València (Spain); a Master in Industrial Design Management from Universidad de Oviedo (Spain); and a Graphic Designer. He has worked as a professor both at national and international universities. The Spanish government granted him with two patents. In 2012 and 2014, Mr. Jiménez was selected to represent Colombia in the category of digital design at the Ibero-American Biennial on Design, and was granted an honorable distinction at the Motiva 2012 awards recognizing the design work in the Principality of Asturias (Spain). He is an independent strategic designer and participated in the Integral Design Accompanying Project promoted by the Ministry of Commerce, Industry and Tourism of Colombia, and Confecámaras. At present, he works as an Associate Professor for Universidad EAN at the Institute for Entrepreneurship, Leadership and Innovation - IDELI (from the Spanish Instituto de Emprendimiento, Liderazgo e Innovación).

Radu-Ioan Mogos is assistant professor at Bucharest University of Economic Studies. He is teaching applied artificial Intelligence, computer programming and genetic programming. His research is focused on multi-agent systems.

Laura Elly Naghi is associate professor at Bucharest University of Economic Studies, Faculty of Cibernetycs, Statistics anf Economic Informatics. Subjects taught: Insurance and Reinsurance, Insurance and Social protection, Risk Management in Insurance, International financial relations, Financial audit, Managerial finance. Interest in: Operational risk modeling, Actuarial models implemented in financial sector.

Alina Neamtu is Student PhD at Bucharest University of Economic Studies, Faculty of Cibernetisc, Statistics and Economic Informatics. Interest in: Microeconomics, Macroeconomics, International financial relations, Managerial finance, Artificial Neuronal Networks.

Adrian Peculea graduated from the Technical University of Cluj-Napoca, Romania in 1996, with a bachelor degree in Applied Electronics. In 2010 he received his Ph.D. in Computer Science from the Technical University of Cluj-Napoca. He currently holds a Senior Lecturer position at the Computer Science Department, TUCN. His main research interests are in computer networks, Quality of Service, Security and mobility in Computer Networks.

Rafael Ignacio Pérez-Uribe is a Doctor in Managerial Sciences, Universidad Antonio de Nebrija. Diploma of Advanced Studies in Applied Economics, Universidad Antonio de Nebrija. Master in Management of Organizations, Universidad EAN. Maître es Sciences, University of Quebec a Chicoutimi. Specialist in Evaluation and Creation of Managerial Indicators for Tertiary Education, Escuela de Administración de Negocios. Graduate studies as Kenkyusei (Assigned researcher) in Total Quality Control and Quality Circles, University of Fukushima, Japan. Business Administrator, Universidad Jorge Tadeo Lozano. Director of the research group in Management of large, medium and small companies (G3pymes) and Professor assigned at the Research and Knowledge Management Division Universidad EAN.

Sofia Petäjäniemi is specialized in the themes of financing. In addition, she has experiences as arranger of elevator pitch events. She has also experiences as an entrepreneur. Her research interests are based on the themes of management, innovations and entrepreneurship.

Oana Mădălina Popescu is assistant professor at Bucharest University of Economic Studies, Faculty of Cibernetics, Statistics and Economic Informatics. Subjects taught: Microeconomics, Macroeconomics, Data Analysis. Interest in: Artificial Neuronal Networks.

Alexandru Radu graduated from the Tehnical University of Cluj Napoca, Romania in 2015 with a bachelor degree in Computer Science. Currently he is pursuing a M.Sc. degree in Computer Networks and Distributed Systems at the Tehnical University of Cluj Napoca. His research interest are in cloud computing, distributed systems and applications to IoT.

Maria del Pilar Ramirez-Salazar has a PhD. in Managerial Sciences, Universidad Antonio de Nebrija. PhD. in Management, Universidad EAN. Master in Management of Organizations, Universidad EAN. Maître es Sciences, University of Quebec a Chicoutimi. Specialist in Evaluation and Creation of Managerial Indicators for Tertiary Education, Escuela de Administración de Negocios. Business Administration, Escuela de Administración de Negocios (EAN). Profesional in Pedagogy, Universidad Pedagógica Nacional. Researcher of the research group in Management of big, medium and small companies (G3pymes) and Professor at the School of Online Education School at Universidad EAN.

Edgar Ramírez-Solís is a Professor of Management at Tecnológico de Monterrey (Mexico). He received his Master in Planning and Communication from ITESO University, a Ph.D. in Administrative Science from the Universidad Nacional Autónoma de México (UNAM) and a second Ph. D. in International Business from the Universidad Autónoma de Madrid (UAM) in Spain. His research interest includes family firms with special interest in strategy and corporate governance practices. He has served as a Visiting Professor at the University of San Diego. He has also published articles in different international academic journals. Currently serves as a Full Professor and as a Family Business Consultant in the Marketing and Business Department at Tecnológico de Monterrey, campus Guadalajara; he teaches courses in Family Business and Strategic Management at the undergraduate and masters levels.

Petru Cosmin Rasa graduated from the Tehnical University of Cluj Napoca, Romania in 2015 with a bachelor degree in Computer Science. Currently he is pursuing a M.Sc. degree in Computer Networks and Distributed Systems at the Tehnical University of Cluj Napoca. His research interest are in IoT, cloud computing and mobile applications.

Lucia Rodríguez-Aceves is a Professor of Entrepreneurship at Tecnológico de Monterrey (Mexico). She received her Ph.D. in Business Administration from EGADE Business School. Her research interests includes Knowledge management and strategy in family firms, with a special focus in using social network analysis. She participates with the Global Knowledge Research Network and the Iberoamerican Knowledge Systems Community. Currently, she serves as a full time professor in the Entrepreneurship Division at Tecnológico de Monterrey, campus Guadalajara; she teaches courses in entrepreneurship and innovation at the undergraduate level.

Rauno Rusko is a lecturer of management at the University of Lapland. His research interests are in the cooperative, competitive and coopetitive features of business and management, especially in the branches of ICT, forest industry and tourism. His works have been published e.g. in the Forest Policy and Economics, the International Journal of Business Environment, International Journal of Technoentrepreneurship, the International Journal of Tourism Research, Industrial Marketing Management, Managing Leisure, European Management Journal, International Journal of Innovation in the Digital Economy and Global Business and Management Research: An International Journal.

José Manuel Saiz-Alvarez has a Ph.D. in Economics & Business Administration, Autonomous University of Madrid (Spain). Ph.D. in Sociology, Pontifical University of Salamanca (Spain). He has also studied at the Royal Complutense College at Harvard, University of South Florida, University Lumière-Lyon II (France), Tallinn University (Estonia), and University of Amsterdam (The Netherlands). He has been the Academic Director for Business Administration Doctoral Studies, Nebrija University (Spain). Dr. Saiz-Alvarez is Visiting Professor at TEC de Monterrey (Mexico), EAN University (Colombia), and the Catholic University of Santiago de Guayaquil (Ecuador). Honorary Professor, Autonomous University of Madrid (Spain). President of FIEC Institute. Member of the Latin-American Family Business Net. Co-founder of the International Center of Research on Solidarity Economy. Certificate of Recognition, Capitol House of Representatives of Puerto Rico. Diploma of Honor, Valahia University of Targoviste (Romania). Member of the Scientific Research Council, Kielce University (Poland). Member of the Editorial Board and referee of numerous national and international journals. Who's Who in the World, from the 2011 Edition. Polyglot in five languages, he has given more than one hundred lectures in Europe and America, and is author of more than a two hundred publications in his specialty. To date he has supervised 70 doctoral thesis from students from Argentina, Colombia, Ecuador, Spain, USA, Guatemala, Italy, Mexico, and Venezuela.

Farhan Shaikh works as a Strategic Research Consultant in Mumbai office of an international consulting firm, headquartered in France. He has completed his Executive Program in Management (Marketing) from SJM School of Management at IIT Bombay. His areas of interest in research are marketing strategy, Islamic marketing, and everything digital. He has extensively worked on projects which involve understanding the impact of digital on industries, such as finance, transportation and manufacturing, and functions such as marketing, HR, and entrepreneurship.

Stelian Stancu is a Professor of Microeconomics, Macroeconomics, Artificial Neuronal Networks at the Academy of Economic Studies (AES), Bucharest, Romania. He managed more than 20 R&D projects in the last ten years, contributor and author of 37 books and more than 100 papers on Microeconomics, Macroeconomics, Artificial Neuronal Network.

Index

U

V

Y

www.ingramcontent.com/pod-product-compliance
Lightning Source LLC
LaVergne TN
LVHW060425300926
849162LV00006B/1530
9781522509530